SALINE IRRIGATION
FOR
AGRICULTURE AND FORESTRY

WORLD ACADEMY OF ART AND SCIENCE

4

SALINE IRRIGATION FOR AGRICULTURE AND FORESTRY

edited by
HUGO BOYKO

SPRINGER-SCIENCE+BUSINESS MEDIA, B.V
1968

© SPRINGER SCIENCE+BUSINESS MEDIA DORDRECHT 1968
ORIGINALLY PUBLISHED BY WORLD ACADEMY OF ART AND SCIENCE IN 1968
SOFTCOVER REPRINT OF THE HARDCOVER 1ST EDITION 1968

ISBN 978-94-017-5696-9   ISBN 978-94-017-6016-4 (eBook)
DOI 10.1007/978-94-017-6016-4

Proceedings

of the
International Symposium on Plantgrowing with Highly Saline or Sea-water,
with and without Desalination
held in Rome
5–9 September 1965
organized by
The World Academy of Art and Science
in cooperation with the
Accademia Nazionale di Agricoltura
and the
Consiglio Nazionale delle Ricerche
and
Co-sponsored by UNESCO

Contents

PART III: SALINE SOILS AND BIOTOPES

PART IV: DESALINATION

APPENDIX

Editor's Preface

Two important Symposia were held in connection with the IIIrd Plenary Meeting of the World Academy of Art and Science.[1] One was the symposium on "Causes of Conflicts and World Education," with which the 3rd volume of our publication series is dealing. The other symposium had as its subject the utilization of highly saline water or sea-water by either its direct use for irrigation or by desalting it. The lectures of this latter symposium are being presented in the present volume.

Science and technology in their powerful drive forward have already shown two different ways to utilize the great amounts of saline waters in the deserts, as well as the practically unlimited resources of sea-water from the ocean. Both, the methods of direct use of salty waters as well as the sophisticated techniques of desalting them, open new and hopeful vistas for mankind and shed new rays of hope into the sad picture presented by our population statisticians for the next two or three decades.

To productivize the deserts with these saline waters for food production would mean the conquest of new regions of the total size of a large continent. According to the official U.S.D.A. Yearbook of 1948, sand deserts alone occupy an area seven times as large as the total agricultural area of the United States.

If farming can be introduced into these vast wastelands of our globe, and if, on the other hand, adequate birth control can be brought about by world education, then we shall have won a major victory for human welfare and for a lasting peace.

These words are so frequently repeated that they begin to sound like empty phrases. However, we have to consider that hundreds of millions of people will – without such a victory – out of necessity be forced to rise against hundreds of millions of other people in the foreseeable future, in order to save themselves from starvation.

1. See Appendix: Informatory Notes on "The World Academy of Art and Science" (page 296).

Furthermore, if our thoughts project towards the means at our disposal for such a fight for survival, then we have every reason to further any research work, which may devise a solution without such a desperate war.

Even today the masses are aware of the significance of sending hydrogen bombs and death rays by space ship to any place on earth, of chemical and bacterial war, and perhaps still worse methods of life extermination in the future.

In order to attain that major victory for humanity, all our economic and political powers, whether they be national or international, have to be mobilized for the support of this research.

This is particularly necessary for the research of converting the vast stretches of sand deserts by the direct use of saline waters into highly productive lands.

Up-to-date, with the help of UNESCO, most successful results could be achieved in the laboratory as well as in small scale field experiments in various countries along a climatic profile from the Tropics of India and the hot desert regions of South West Asia and North Africa to the cool temperate, humid regions of the North Sea and Baltic Sea coasts.

The new principles proved to be correct and instead of the much feared salt accumulation, a reduction of salt content in the sand could be observed.

This scientific breakthrough stands now before its last step, which almost all scientific breakthroughs have to encounter: it must overcome preconceived ideas, professional jealousies and administrative red herrings in order to bring its enormous and predictable benefits to waiting mankind.

The international symposium on saline and sea-water irrigation was a great stride forward in breaking former prejudices. Thanks for these achievements must be extended to the Special Agencies of UNO, UNESCO and FAO, the Accademia Nazionale di Agricoltura and the Consiglio Nazionale delle Ricerche of Italy, and other organizations, which together with the World Academy of Art and Science sponsored this Symposium.

One hundred and four experts from twenty four countries unanimously expressed these thoughts in a resolution at this symposium, saying that:

"They firmly believe that results already achieved by irrigation with highly saline water indicate clearly that those arid areas at least where such a water supply is available, can be rendered capable of crop production, and they therefore strongly recommend to international and national organizations concerned with human welfare that financial provision continue to be made for the necessary expansion of research and field trials."

Let us hope that this fourth volume will contribute in broadening the scope of this new line of research and will help to achieve its aims.

HUGO BOYKO
President,
World Academy of Art and Science

Opening Addresses

Ladies and Gentlemen,

It is indeed a great pleasure and privilege for me, in opening this important international Symposium on Problems of Irrigation by Salt Water and Desalination of Sea Water, to greet the participants on behalf of the National Research Council and on my own behalf and to thank them, wholeheartedly, for having accepted our invitation to come here and communicate the results of their studies and experience. My thanks are extended, also, to the Institutions that they represent.

Special thanks are due to the World Academy of Art and Science and to the National Academy of Agriculture for having taken the initiative of organizing an International Symposium on a theme of such particular actuality and great importance.

Before continuing with the few introductory remarks that I have to make, I wish to recall, briefly, to my colleagues the memory of Dr. SCHWEITZER. As you know, yesterday the world lost this great apostle of peace, a man who so truly belonged to the world of ideals from which the Academy draws its best elements. I request all those present here to dedicate a minute of silence to his memory. [A minute's silence is observed.]

The problem of the availability of water of good quality, in sufficient quantity and at the right time and place is by now a precondition of the economic and social development of many of the nations of the world. Desalting, which already, in many areas with a particular lack of hydric resources, represents an economic source of supply, will be able in the future to supply water for drinking purposes. But it is clear that, at least for the present, it cannot constitute a complete solution of the problem.

Agriculture certainly represents a major sector of water consumption, either because the greater part of soil water absorbed by plants is lost through evaporation or because it is lost through spilling. Expansion of agriculture is closely connected with the fertilization of high precipitation areas on the one hand and with irrigation of soils, which are arid but rich in nutritive elements, on the other hand.

In irrigational agriculture the volume of available water and the tolerance of plants to salt water are two equally important and complementary factors. If tolerance to salt could be augmented by means of genetic studies in those cultures which today have an economic value for mankind, or if it would be possible to utilize, economically, the vegetable species which tolerate salt, then agricultural development would receive a decisive impetus. Even a slight increase in the salt tolerance of existing cultures would improve present and future hydric availability as it would make possible the agricultural use of resources which are today constantly discarded because of their saline content and the channelling of a greater percentage of water to those sectors which cannot do without it.

These considerations sufficiently explain the importance, for the future of humanity, of the studies and the research which constitute the theme of this Symposium. May I be permitted to emphasize the fact that such studies and researches are liable to produce results of great scientific and technological importance in the field of desalination as they concern a sector of biology which presents an infinity of small models of small natural desalinators, biological organisms having the capacity to separate salts from water. The methods of desalination through membranes, and particularly the so-called inverted osmosis, indeed belong to that category of methods which reproduce natural phenomena; therefore it is not to be excluded that from observation and study of biological organisms and from observation and understanding of their mechanism, there can be obtained indications which might lead to this type of procedure being the most progressive and economic in the least possible time.

Italy, too, is interested in the water problem, which is assuming ever-growing proportions with regard to the increase in population, the improvement in the standard of living and the growing industrialization of irrigated cultures.

The National Research Council, in its desire to contribute to the research for suitable solutions of this problem, has initiated a nation-wide program in this direction. The aim of this program is the ascertainment of existing hydric resources, and experiments with methods and techniques which may increase the availability of usable water, with particular stress on seawater and the desalination of salt water. As a matter of fact, pioneering studies were also undertaken, in days gone by, here at the High Institution of Health, and other studies have been carried out by the National Committee for Nuclear Energy, as you will hear from Engineer DI MENZA.

Also industrial research institutes, such as Breda's, have started technological research into most advanced distillation processes. In developing this kind of activity we rely on the willing collaboration, at a scientific level, of the part of other countries interested in the same problem. Useful contacts have already been established with the Salinity Office of the U.S. Department of the Interior and with the Ministry of Development of the State of Israel. In the near future there will be held, in Washington, the First International Symposium on Water Desalination with the participation of delegates representing 58 nations.

Today, our Symposium will provide a useful opportunity for an exchange of views between representatives of various countries which, I am sure, will contribute to the establishment of a most desirable collaboration on an international level. Such meetings really constitute a valid instrument for the development of science and technology and I therefore express the wish that they may become ever more frequent in the future.

Such frequent meetings would, however, not be necessary if we could but know the secret of Moses' tree and, here, allow me to read the beautiful verses depicting the miraculous action of this tree. I am referring to the Book of Exodus in the Bible, first part, second section where it is told of the departure of the People of Israel from Egypt, their journey through the desert to the Red Sea and from there to Mount Sinai. I refer to Chapter 15, Verses 22–27:

"And Moses led Israel onward from the Red Sea, and they went out into the Wilderness of Shur; and they went three days in the wilderness, and found no water. And when they came to Marah, they could not drink of the waters of Marah for they were bitter: therefore the name of it was called Marah. And the people murmured against Moses, saying, What shall we drink? And he cried unto the Lord; and the Lord shewed him a tree, and he cast it into the waters, and the waters were made sweet. There he made for them a statute and an ordinance, and there he proved them."

There is no doubt that if we had knowledge of the miracle of the mechanism of that tree there would be no need for frequent meetings on water desalination.

Let me express the wish that this Meeting may bring forth contributions to our knowledge of the functioning of membranes and with that purpose in mind I repeat my heartiest greetings to all the participants and express my very best wishes for the success of the Symposium.

Professor VINCENZO CAGLIOTI
President, National Research Council of Italy

Your Excellencies, Ladies and Gentlemen, Dear Friends,
In recent years the international development has been rather worrying. The large amounts of money which are being used for armaments have a tendency to increase, and, as a result, the possibilities to give technical and financial aid to underdeveloped countries, where people are starving and suffering from severe diseases, may be reduced. Much more ought to be

done in order to raise the social, cultural and economic standard in the underdeveloped countries. A close cooperation in order to promote a peaceful development in the world is more urgent in our times than it was at any time previously.

Consequently, the aims of the World Academy of Art and Science as an Agency "for Human Welfare" it also more urgent than before. The two groups of problems, to which most of the time during these meetings will be devoted, represent different branches of research, but they are as a matter of fact, from our point of view, closely related to each other.

A successful application of the results obtained from the research concerning "Salinity problems" will be extremely valuable for raising quite considerably the food production in arid zones. Thereby one of the main causes of conflicts [1] can be removed. Starvation or lack of food has since immemorial times been one of the main causes of conflicts. May I remind you of an old Chinese saying: "It is difficult to see the difference between right and wrong when the stomach is empty."

Men have never fought about the oxygen of the air, because there is an abundance of it, enough for all of us, but they have been fighting about fertile soil, water, energy material, coal, oil, etc., and valuable ores of different kinds. However, if all these natural resources be utilized in an efficient way, there wouldn't be any shortage and, consequently, no cause to fight about the resources.

The different branches of research which the WAAS is trying to promote, for instance by creating a World University, all have the ultimate goal to explore the resources of our planet in the most efficient way, thereby removing several of the main causes of conflicts.

For a peaceful development in the world it is not enough to give the peoples in the underdeveloped countries political freedom, they must also get an opportunity to develop their natural resources in order to raise the material and cultural standard of their communities. If we can assist them in this respect, then we are in fact removing an important group of causes of conflicts between states and peoples.

The main purpose of the so called peace research, which in recent years has been discussed in several countries, is to make clear the reasons of conflicts, thereby promoting a peaceful development to *"human welfare."* A successful research of this kind is as a matter of fact a very important condition for creating an international state of justice.

We have met again in order to listen to and discuss a number of lectures on some most important subjects.

May I, finally express the hope, or rather conviction, that all of you will

1. See Volume III in this series: Conflict Resolution and World Education, Dr. W. Junk Publishers, The Hague 1966. American (offset) edition by Indiana University Press, Bloomington, Indiana, 1967.

get a valuable profit from participating in this Symposium and in the following Plenary meeting of the World Academy.

Professor HUGO OSVALD
Vice President of the World Academy of Art and Science

Ladies and Gentlemen,

I would like to apologize on behalf of the Secretary General of UNESCO that he was not able to come personally because he is, at present, away from Headquarters on another important mission; as his Representative and on behalf of UNESCO in general, I wish to extend a very hearty welcome to the scientists of many countries who are gathered today here in Rome in order to study and contribute to the further development of the complex problem of irrigation with saline water. In co-sponsoring this Symposium, the Department of Advancement of Science entertains a foremost interest in all matters relating to both the theoretical and practical progress in this highly important field.

Within the framework of UNESCO's former major project on Arid Lands, careful attention was devoted to the complex relationship of soil, plant and water in arid and semi-arid regions, with special reference to crop-yields.

Four large Meetings were devoted to an exhaustive examination of the theoretical aspects of the salinity problem in arid areas as well as of the practical effects of man's effort for improvement in agricultural production:

In 1958 salinity problems in the arid zone were analysed in detail at Teheran; In 1959, at Madrid, salinity was seen as one of the chief processes which determine the plant-water relationship in dry regions; In 1962 at Tashkent and in 1964 at Budapest, the dynamic interdependence of salinity, irrigation and alkaline soils was brought to light. Preliminary reviews of research were discussed in these Symposia, and the proceedings of these were published in the UNESCO's Arid Zones Research publications. As regards UNESCO's role in sponsoring research and exchange of information in the salinity and irrigation field, I would like to stress that the great interest and competence of one of our sister Agencies, namely the United Nations Food and Agricultural Organisation, has been a continuous source of support for our own efforts in this scientific field.

This UNESCO-FAO collaboration is best examplified by the joint project on the International Sourcebook on Irrigation and Drainage.

I presume, you already know that since 1962 UNESCO has been executing a United Nations Special Fund Project in Tunisia, on Research and Training on Irrigation with Saline Water. I am happy to report to you that the manager of this project is here with us, representing the Natural Resources Research Division of UNESCO. I have no doubt that with the wealth of his practical experience in this field he will be able to contribute significantly to the ensuing discussions of this learned gathering.

In conclusion, I wish to state on behalf of the Director General of UNES-CO that it is a great pleasure for me to collaborate with you in co-sponsoring this Symposium. May this Symposium, held under the auspices of the World Academy of Art and Science, proceed well and achieve the greatest success.

Dr. THA HLA
Director, Department of Advancement of Science, UNESCO

Mister Chairman, Ladies and Gentlemen,

Dr. DE MERIDIEU, who was originally appointed by the Director General of the United Nations Food and Agricultural Organizations to present his greetings, has suddenly been prevented to be here at the Opening Session of this important Symposium. I have therefore been asked to present the greetings of the Director General on his behalf.

I would only like to say that I am very glad, indeed, to have been invited to participate in this Symposium. There are quite a number of units in FAO, which are particularly interested in the forthcoming lectures and discussions on saline irrigation problems. It is for this reason that our Production Section (Plant Production Division), our Land and Water Development Division and our Forest and Forest Product Division, are represented here by experts.

FAO, as you know, is not a research organization. We do not carry out basic research but we are very interested in any research which can serve as a background for the practical side we are concerned with, and particularly as regards the increase of production.

It is quite exciting that in spite of the fact that the continents of the world are surrounded by so much water, water is the limiting and most critical factor in plant growing. We hope to gain from this Symposium much information which will prove useful for our various programs around the world. So I am glad to have the opportunity to repeat that we thank very much the organizers of the Meeting for this occasion.

Dr. M. ANDERSEN
Forestry Division, Food and Agricultural Organization (FAO) of UNO

On the Use of our Water Sources and the Problem Complex of Brackish and Sea-water for Irrigation

by

Professor GIUSEPPE MEDICI

President, Academy of Agricultural Sciences of Italy and former (at the time of organizing this Symposium) Minister of Commerce and Industry of Italy

Mister Chairman, Ladies and Gentlemen,

As President of the National Academy of Agriculture of Italy, which took part in the organization of this conference, I feel greatly honoured for the privilege of delivering a brief opening address on the problems confronting the modern world in connection with the ever growing water requirement. The more so as the problems dealt with in the reports, being of a prevalent technical and scientific nature, will make my attempt of taking a short and synthetic view of the fundamental data, concerning the problem of water in the modern world, not altogether useless or worthless. One may believe that, at least from a general point of view, there was a time, when water supplies available to men were so plentiful that most of these problems, specifically problems of economic nature, could not possibly be raised. It is true that water, unlike to air, was in no time available to man all over the world in practically unlimited quantities, but it is also true that there were periods in the history of mankind, when the quantity of water provided by springs and raised ground water was largely sufficient for the food, agricultural and industrial requirements of man.

If one considers that, in Napoleon's time, the world population is estimated to have been only 900 million inhabitants, that such a figure increased to 1600 millions on the eve of our century, and to 3 billions, i.e. 3000 millions in 1963, and now more or less, as everybody knows, to three billions and a half, it is manifest that the increasing water shortage of the modern world is mainly due to the substantial increase in population.

After World War II, the industrialized countries have become gravely aware that their economic development was seriously hampered by the insufficient supply of water at disposal for this purpose. In the United States of America, for instance, the aggregate water consumption is estimated at 1.420 million cubic meters per day, of which only 9% is utilized for city and home purposes. If one considers that 1.420 million cubic meters correspond to a canal with a flow of 16.000 cubic meters per second, a broad idea may be had of the huge amount of water required today in the whole world.

One should also not forget that, although water is one of the most widespread raw materials on our planet, the amount of fresh water utilized directly by man for his food, agriculture and industrial requirements represents but a small fraction of the whole, and that it is often located quite apart from the consumption area. The problem of finding first quality water fit for human consumption is as old as our world itself, and I would certainly refrain from dwelling on this subject, were it not for the fact that in recent times two separate water lines were required to be installed in some important cities, like New York, Paris, London, in order that the water designed to human consumption and that for sanitary requirements might be kept apart.

This shows that the problem of water for direct human consumption does not present the same dramatic features that characterizes the water for food production. The quantity of potable water is so plentiful that some countries exist where only first quality water is drunk. But, however high the increase in the consumption of drinking water may be, it can easily be coped with, due to the fact that drinking water for direct consumption may at any time be paid at least $ 0.16 per cubic meter. I beg your pardon if I dare mention such data which will be the object of large discussions for you, but I thought that my opening address ought to be a little bit of a compromise. Drinkable water can be produced if necessary even by extracting it from the sea. A point of this opening address is the following evidence, namely that we are already now able to produce drinking water from sea-water at a competitive cost of production. A lot of people here will not be in agreement with me, but this exchange of opinions is one of the purposes of this meeting.

Quite as important is the problem of irrigation water. The great urge of early civilization sprang up and developed along the water courses that crossed and still cross the desert regions. We just heard President CAGLIOTI speaking about the desert of Egypt and of Israel, and it is hardly necessary to recall that the old Egyptian, Babylonian and Chinese civilizations depended on the irrigation of deserts or of semi-arid regions: the Nile, the Tigris, the Euphrates, the Indus, the Ganges, and many others, all point to lands where the life of the local population depended on the life of the river.

The amount of fresh water used for irrigation is important, and is imposing. The available estimates, although, like all world estimates, not to one hundred per cent accurate, may induce to believe that the quantity of water used for irrigation reaches an aggregate volume of 170 000 cubic meter per second, utilized in a number of ways during the farming season. It can be said that most of the available water is utilized today for irrigation purposes. This second point to which I dare to draw your attention, is the more significant if one considers that besides the irrigation carried on by means of canals taking the water from great rivers, there is also much irrigation carried on from wells and especially from a large number of small and medium sized storage tanks that collect the rainfall during the wet season for use in periods of draught. Well-known and of significant importance is such a water reserve for the irrigation of a substantial part of Asia, for instance, in India. The area irrigated in India alone is certainly between 25 million and 32 million hectares. A large part of these lands are irrigated with water gathered during the rainy season in such tanks.

Although such estimates are always to be considered cautiously, it has to be borne in mind that the global world consumption of fresh water is roughly about 210 000 cubic meter per second. Therefore almost 80% or 75% of the latter are utilized for irrigation purposes. That means that more or less three quarters of the whole water utilized by men in the world have an irrigation utilization.

The remarkable increase of population occurs especially in the large agricultural countries such as China, India, Indonesia, and it is this which brings the food problem of today to a serious point. The fight against hunger, which FAO is presently organizing with great dedication, is only one of the many facets of the dramatic situation that is being experienced by millions of human beings who are faced with the problem of the daily bread.

It is beyond any doubt, that particularly in some regions in Asia, Africa and Latin America, the increase of agricultural production is conditioned by a more abundant supply of fresh water. Although water shortage does not represent in this instant an immediate obstacle that hinders the solution of the food problem, it will do so in the near future, and it is a frightening raise in population figures that have been foreseen by the experts in demography. It is expected that the world population will reach the mark of 7 billion inhabitants in only 35 years. Even if a lot of people are not in agreement with these forecasts of the scientists in demography, certainly a very great amount of population will be added to the number of population living now in the world.

To this great problem, drinkable water for men and water for irrigation, another ever increasing and unforeseen request for fresh water will now be added. The several hundred thousand cubic meters of fresh water representing the daily consumption of American industry are a clear indication of what the industrial consumption will be when most of the other countries too will be in possession of industrial plants in line with the requirements of modern civilization. On the whole, however, water consumption by industry is very low compared with the world's total consumption of fresh water. It will, however, be appropriate to recall that in some European regions, around the Mediterranean basin, one limiting factor in the development of the steel and chemical industry is constituted by a shortage of fresh water. Thus, a large steel mill built in Taranto, Southern Italy, had to be achieved with a plant for desalting of sea-water. This steel mill needs for its operation 4,000 cubic meters of fresh water daily. The development of the great chemical industry has always required the construction of imposing water systems to extract from rivers the water necessary for running their plants.

This factor is new in the life of most countries and it deserves to be mentioned here, because in the next decade, we will try and take an ever increasing share of fresh water from irrigation; this is another point, which is, in my opinion, extremely important, particularly in relation to the fact that the development of both fields and of the economy everywhere in the world will ask increasing quantities of fresh water, and the competition between agricultural purposes and industrial purposes will be very acute. The rapid depletion of the water supply of the Po (the Po river is the most important river in Italy) has made competition keen among farmers who want to use it for irrigation purposes, manufacturers who want it for their industrial plants, and public authorities whose concern is to ensure a sound minimum for keeping this main inland waterway open. This point means that one must include into the considerations also the need of water for waterways. And this is just one of the problems which have to be tackled now as well as in the future.

We have now entered a period in which fresh water is increasingly considered a raw material and it should be treated as such, especially from the economic point of view. Gone are the days when water was, and still is in part, considered as one of the resources to be had practically free of charge like air. It may be assumed today that the problems connected with water economy should, all over the world, be dealt with in the same way as those related to iron or oil economy; we must also be increasingly conscious of the fact that we are lucky enough to possess a limited reserve in the form of Sea-water. I say "limited" because, today, the problem of its use is essentially of an economic nature. This Congress has devoted some important reports also to economic aspects, but evidently, it is mainly the technical and scientific point of view which will be stressed in the discussion. But it is important to underline that the problem now is of an essentially economic nature. And such it remains even though it is dependent on scientific and technological developments that will lower the production costs of fresh water. Economic problems are essentially problems of cost. Since the supply of brackish and sea-water is practically unlimited, the cost of production prevails over the fresh water problem. The reduction occurring in the cost of desalting of sea-water has allowed our horizon to grow larger. However, serious problems of transportation are still in existence as it is also the case for fresh water which is often located hundreds of kilometers away from the utilization center. The problems to be tackled are therefore essentially the following:

(1) Cost of fresh water transportation. Very often we have great quantities of fresh water, but they are so far away that the cost of transportation is not competitive

(2) Cost of purification of undrinkable fresh water. Cost for desalting brackish or sea-water. The impressive advancement accomplished in the production of pipe-lines and pumps have remarkably lowered water transportation costs by reducing water losses on the way; this enables us today to carry fresh water to places over one thousand kilometers away; that is another point, on which I want to draw your attention. In Europe at least, in Italy for instance, if I would say in our Parliament: "It is possible to transport the fresh water of the Po to Puglia," where mills are working with sea-water, then everybody will think that I have got mad. On the other hand, in California, we have fresh water transportation over a distance that is near to one thousand kilometers. This point may not be directly included in the problems of irrigation with brackish or salt water, but it is in close relation with the whole complex for which we are gathered here. There is an impressive advancement accomplished in the production of pipe-lines and pumps, as I have mentioned before; although this is a point which is not included in the program of our Congress, it has to be kept in mind. We believe that the transportation problem is of utmost importance, not the least because it has above all a permanent character; even when sea-water and brackish water may be desalted at low costs, the problem of transportation will still exist for the reason that sea-water is not always available at the utilization point. It is quite possible that in the years to come, water shortages may be relieved by means of exchanges, carried through a policy of international cooperation, but the events that are taking place in the Near East with regard the water of the Jordan river and the problems raised by the populations of the great river regions, the regulation of these great river flows in Asia, America and Africa, do remind that solutions of these problems are extremely difficult to achieve.

On this account the possibility of coping with the requirement of fresh water by means of exchanges, which would eliminate long and useless transportations, should not be neglected; because very often it is neglected for, evidently, the human being does not like to tackle the most difficult problems that really are or are not of a technical but often of a political nature.

This possibility of eliminating long and useless transportation of fresh water simply by exchange should not be neglected also in this meeting. It will, however, take many decades before such a policy may give results, and the importance of the problem of water transportation, and the need of appropriating adequate financial means for research work in this direction cannot be stressed strongly enough. The fact that some large consumption areas, such as London, New York, and Tokyo, are in the proximity of the sea, do not influence the point as such; the most difficult water transport problem lies with regard to irrigation water, which is usually used at considerable distances from the sea. It may be added that the sources of fresh water still available in important amounts such as the flow of the Amazonas river, or the ice of the poles, are located far apart from optimum utilization centers.

These broad outlines of the main features of the problem allow us to place also the problem of desalination in the proper framework. It is one of the fundamental problems of the whole complex of desalting of brackish and sea-water and the construction of industrial plants that are already on the mark.

The probable advancement which will take place during the next years as a consequence of the reduction of the cost of the kilowatt per hour obtained by nuclear energy, will ease the supply of fresh water from traditional sources; and it will also bring the supply of fresh water extracted from the ocean to a competitive level. Partly this has already been done; for example in countries of the Mediterranean Basin, where the traditional available water sources are almost exhausted. It should also be added that the construction of double acting, high power nuclear plants for the production of electric power and for the distillation of fresh water, have first contributed to the decrease of desalting cost and thus allowed the implementation of important projects.

On the other hand, thousands of small islands throughout the world are apparently permanently supplied with fresh water at a very high cost by means of tankers; it is to be assumed, therefore, that in many cases the construction of small-size desalting plants may conveniently replace more expensive transportation. Important research will be necessary with the aims of placing them, thus replacing old fashioned if not archaic methods by less expensive ones.

While studies and research work aim at the development of less expensive methods for desalting and transporting water, it is imperative that the utilization of the existing supplies of fresh water be carefully planned. The resolute pledge of the United States of America to develop economic methods of desalting of sea-water and the influence that such prospects exercise on our spirit, must not make us forget that an equally urgent task has to be accomplished, pending the solution of other technical problems, namely an international planning for the utilization of fresh water. Such an international planning will test the good will of all peace loving peoples.

The problem of water is a threefold one:

(1) It is a problem of foreign policy; in that it calls for broad international co-operation.

(2) It is a problem of internal policy: in that it calls for planning in the use of all available sources.

(3) It is a problem of research, and in that it calls also for adequate funds in order that the cost of production of fresh water may be lowered.

Thus, it is quite understandable why the White House advisor for scientific methods, Dr. DONALD HERNING, while commenting the successful Gemini and other similar projects, expressed his doubts and perplexity on the usefulness of new and costly space projects, when he said: "There are so many and more important things still to be done here on the earth." A point of view, expressed also by leading scientists of the Soviet Union. It is perhaps not altogether out of place to quote the witty remark credited to a well-known scholar in brackish water Chemistry, who recently said: "Billions are spent to send the men to the moon, and yet we are no longer in a condition to water our own garden tomatoes." In some way, evidently, this remark was related to the fact that the people living now in New-York are not always in the position to use fresh water for normal common purposes.

The public opinion all over the world is demanding insistingly that an ever increasing amount of money should be allocated to research work for the desalting of sea-water, and for the use of brackish water for the irrigation of agricultural land. It is a fortunate occurrence that the National Council for Nuclear Energy in Italy, thanks to the substantial funds made available for carrying out all the necessary research work and studies, did succeed in reducing the cost of production of kilowatt per hour to a considerable extent. Although this opens new and wide horizons for the desalting of sea-water, it is not yet enough. Because in the desalination processes, the cost of electric power presents one half of the entire cost while the other half goes for the amortization of the plant and for operation expenses. Every advancement that may be accomplished in the technology of desalination will therefore be accompanied by remarkable economic results.

Most important is also agronomic research in the field of irrigation by means of fresh water. It was found out, for instance, that the waste of fresh water in the present irrigation process is so great, that it can be assumed that in the course of the next decade the acreage presently irrigated could be doubled without increase of the quantity of water used now. From the same point of view we should also be concerned that no desalted water should be wasted in the irrigation process.

I have deliberately avoided to dwell on this specific economic-agronomic technique since irrigation problems will be discussed at length in the present Symposium. It is clear, however, that further studies based on the research carried out so far would considerably help to implement the program which we have seen and which aims at increasing the irrigated area without correspondingly increasing non-saline water consumption. It is not sufficient to know the best methods of irrigation, it is also indispensable that these thousands of millions of poor peasants everywhere in the world, that carry on irrigation in Asia, in Africa, in Latin America, and also in Europe be enabled to make use of the technical assistance. This is a problem that perhaps may be regarded as a fundamental problem in view of the fact that so many human creatures are condemned to suffer from hunger. And it is for this reason also that complementary agricultural research should be pushed forward in order to obtain the best utilization of the existing sea-water resources, while, on the other hand, also those technical assistance programs should be implemented, that FAO has developed with so great success and that competent authorities in particular are stimulating with greatest zeal and dedication. Agricultural experimentation should also be enabled to face systematically the problem of direct irrigation with brackish water, which is largely available in many countries of the world. And this experimental work should be extended to the field of genetics in order to allow suitable varieties to be developed which prove resistant to the high salinity of soil and water.

Much has already been done by highly capable researchers and a good number of them are fortunately gathered at this Symposium. We wish to express the general feeling of gratitude to them. May their work be of valuable help in the struggle against hunger!

PART I: PRINCIPLES, PROBLEMS AND LABORATORY EXPERIMENTS

Principles of Saline and Marine Agriculture

by

HUGO BOYKO

It is one of the significant signs of our time that gradually the political leaders are listening to the warning calls of scientists: that one of the most basic commodities for human life and, at the same time, one of the most abundant natural resources on our earth – water – will be wanting for our daily use within a foreseeable future if we do not find the methods to prevent this in time.

This danger is, of course, particularly acute in the drier regions of our globe, but during the last decades it is becoming a threat also in countries with a humid climate where urbanization and industrialization is accompanied by an accelerating pollution and poisoning of our natural water resources. This process starts already with the raindrops in the air and continues on their way back to the oceans.

In order to counteract these dangers, science has searched for and has found several ways towards solutions which can be grouped into two main categories:

One way is the demineralization of sea-water and of other highly saline waters: successful trials by freezing or distillation can be traced back into prehistoric times but it is only now that science and technology opened the possibilities for large scale demineralization by various methods. At present, we are looking for the last but not less important break-through in this field, namely the *economic* solution, and for potable water and for certain industrial purposes this economical side seems to be near its solution. It is, however, the agriculturist who, by far, is the main consumer of fresh water in arid and humid regions alike, and for whom the current costs for artificial desalination seem to be prohibitive with the exception of a few insignificant branches of horticulture.

On the other hand, plant ecological research is since a few years offering a completely new way of solving the problem: it found a scientific solution for the direct use of highly saline and even of sea-water for irrigation without involving agriculture in serious economic problems. Only the usual *technological* problems such as finding the most suitable economic crops and their selection according to the respective local climate, protection of irrigation installation against corrosion, etc. have to be overcome.

But this new branch of plant ecology offers more than only the use of natural saline waters for agriculture. Parallel to one of the new principles, the "Biological Desalination," it has been applied successfully also for "Biological Purification" of artificially polluted waters, foremost industrial waste waters (SEIDEL, 1966) thus complementing the benefits to agriculture (in arid and humid regions alike) with the benefits to their industrial and urbanized centres.

We shall restrict ourselves in this article to one field only: the principles enabling the direct use of highly saline water up to sea-water concentration in agriculture (including horticulture and forestry), the details of which were extensively dealt with in the book "Salinity and Aridity – New Approaches to Old Problems" in the chapter on "basic ecological principles" (H. BOYKO, 1966).

The great vistas for future development opened by this new line of research are of particular importance for arid regions. There, the possibilities to expand agriculture were up to now very limited by several natural phenomena dangerous to crop plants. The most significant difficulties existing are the dry climate, the salinity of most of their water sources, the salinization of normal agricultural soils even by irrigation with so-called fresh water, and last but not least the shifting dunes which cover, on our globe, a total area twice as large as the area of the United States. Since 1929 the author had the opportunity to study the ecology of steppes and deserts, dune lands and salt swamps in many countries north and south of the equator.

Always when glykophytic (i.e. non-salt tolerant) vegetation was found in the midst of halophytic vegetation areas, it was either near a spring of fresh water or another fresh water source, or it was on an island having soil with a particularly good permeability offered by sand and gravel. Thus, I came to the conclusion that a combination of the two geophysical dangers for plant life, saline water on the one hand and dune sand on the other hand, may lead to the solution of agricultural problems in large desert areas.

Gradually the validity of the basic principles could be proved by exact and verifiable experiments carried out by myself together with my wife and co-worker, Dr. Elisabeth BOYKO (H. BOYKO & E. BOYKO, 1966, and 1968). For this purpose we intentionally used extreme conditions of soil permeability as well as of salt concentrations, taking dune sand as soil and natural sea-water of four different types of concentration for irrigation.

These four types are: 1) East-mediterranean sea-water, 2) Oceanic concentration, 3) North Sea concentration and 4) that of the Caspian Sea; all with parallel control experiments for each species separately with fresh water (see Table on page 87 of this volume (BOYKO & BOYKO, 1968).

A few well known facts and a number of new principles worked together to achieve the results of these experiments. Since centuries it is known that sand allows *quick percolation and good aeration*, on the one hand and on the other, we know of the easy solubility of sodium chloride (NaCl) and magnesium chloride ($MgCl_2$) which are the only actually dangerous components to plants; but sodium chloride dissolves already at an air humidity of 80% as known to every housewife and magnesium chloride at even a still lower one. All other components constitute an excellent fertilizer which supplies the rootsystem with a very well *balanced ionic environment*. (For details of this new principle see HEIMANN, 1966).

One of the main causes of soil salinization by irrigation is also known since long, namely the high capacity of clay to adsorb sodium. This adsorption results in the swelling of the clay particles whereby the soil becomes impermeable; thus even small amounts of salt in the irrigation water will accumulate continuously and soon lead to a degree of soil salinization beyond the limit of salt tolerance in plants.

The *lack of clay particles* in any significant percentage in sand is one of the main causes of the higher salt tolerance of all species grown on sand.

In addition to the principles mentioned before (marked by italics) a number of new principles were instrumental to the success of our own experiments with sea-water and other highly saline waters as well as of those made since then by others in various countries. It is possible and even probable that there may also be other factors of influence of a physical, chemical and/or biological nature of which we know nothing yet or little only, such as for instance *microbiological influences*. In any case, on the basis of our present knowledge *"General Rules"* could already be presented to the UNESCO-WAAS-ITALY Symposium and accepted there as a valid basis for further experimental work as well as for practical application (see BOYKO, 1966).

The scientific details of the new principles mentioned in the following were extensively dealt with in Vol. XVI of the Monographiae Biologicae (H. BOYKO, 1966) which complements this volume. The respective pages are given in brackets. (See also BOYKO, 1967).

1. *Partial Root Contact* (Mon. Biol. XVI, p. 147–152): In sand microscopically small feeder roots having a diameter of only one-tenth to one-hundredth of the space between the sand particles are, in time as well as in space, only partly in contact with the brine. After quick percolation of the salty water they are mainly surrounded by moist air and frequently, particularly in the cool morning hours, by "subterranean dew" i.e. by fresh water (see No. 3).

2. *Viscosity* (Mon. Biol. XVI, p. 152–158): Contrary to the prevailing opinion that the feeder roots would soon be suffocated by a surrounding film of brine or salt after percolation and evaporation, the author was able to prove that this film has a thickness of only 10^{-5} to 10^{-6} mm i.e. of a few Angström only (1 Angström (Å) = 10^{-7} mm). In view of the very low viscosity of

sea-water, the film cannot remain as a continuous layer of brine, but may leave only tiny patches or separate infinitesimally small accumulations of salt crystals, which are no hindrance at all for the roots of a living and active plant.

3. *Subterranean Dew* (Mon. Biol. XVI, p. 158–165): This principle discovered by the author during his studies in the Negev, particularly in the years 1944–1949 was proved by an excellent laboratory experiment carried out by A. ABRAHAM (1954) at the Hebrew University. Observations in sandy deserts and also in parts of the Negev showed that a layer beneath the surface was moist in the cool morning hours in spite of the dry, hot rainless season. This was caused by the difference between day and night temperatures in the soil layer three to eight centimetres beneath the surface, leading to a temporary condensation of air moisture. Many plants reveal this layer by developing their root systems in accordance with it (BOYKO, BUGOSLAV-HILLEL & TADMOR, 1957).

4. *Biological Desalination* (Mon. Biol. XVI, 165–168): The sodium and chloride content of very sandy soil is significantly diminished by harvesting salt-accumulating crop plants – in spite of continuous irrigation with saline water.

Soon after publication of this principle by the author, entirely independent studies on another natural process were described by Professor Aaron KATZIR, Head of the Weizmann Institute's Department of Polymer Research and his co-workers, Drs. Benzion and Margaret GINZBURG. Both discoveries are of very great importance.

The biological desalination I have referred to specifically in this article concerns the soil and the ground-water in it, and the reduction of the sodium and chlorine content in both by cultivation of salt-accumulating plants. The second – KATZIR & B. and M. GINZBURG – process is effected by microscopic organisms living in water. To distinguish them – they are two entirely different processes – I suggest that we call the first "biological soil desalination" (although ground-water in contact with root systems is also involved) and the second "biological water desalination."

5. *Adaptability of plants to fluctuating osmotic pressure* (Mon. Biol. XVI, p. 168–173): This principle was discovered by a number of scientists – including the author – in different countries. Such an adaptability is specific for each species but is much higher and more frequent than assumed up to now. In our experiments with sea-water the plants had to overcome a particularly high osmotic pressure, sea-water of oceanic concentration, for instance, having an osmotic value of twenty atmospheres. This was too much for some of the species, but other species withstood it surprisingly well. Under more fortunate conditions, for example at Orinon, on the northern coast of Spain, where rainfall, albeit in small quantities, is common in the summer also, dozens of varieties of vegetables and flowers are today being grown successfully in the sandy soil, irrigated directly with water from the

ocean (about 35,000 ppm total salt content, with 75% of it sodium chloride).

6. *Raised Vigour* (Mon. Biol. XVI, p. 175–178): After four years of irrigation with sea-water we started a "dying experiment," the first of its kind. No irrigation at all was given and two particularly long, rainless seasons, one of nine months and the other of eight, made the results even more significant. It appeared that after the complete cessation of irrigation, a higher percentage of plants which had previously been given sea-water survived – and with greater vigour – than those of the same species in the control series previously irrigated with fresh water.

7. The geophysical theory of *Global Salt Circulation* (Mon. Biol. XVI, p. 180–195): Since this principle is of more general interest, the relevant graph (see Mon. Biol. XVI, page 191, fig. 15) may help in explaining it. Global salt circulation and global water circulation are phenomena of the same category and are partly simultaneous. If the former did not take place, no life would be possible on earth, at least not on terra firma. The salt brought with every raindrop onto the earth's surface would accumulate and would cover all continents with a thick layer in a geologically short period. Global salt circulation is a very complicated phenomenon but the whole complex is only one of the innumerable proofs of the final and all-embracing principle:

8. The *Basic Law of Universal Balance* (Mon. Biol. XVI, p. 195–196): This natural law, formulated by the author in 1960, reads: Wherever and whenever the natural equilibrium is disturbed, readjustment takes place until a state of equilibrium is restored.

Proof of this law is furnished in every macrocosmic event as well as in the microcosmos of every atom; in all geophysical phenomena, as well as in the chemo-physiological equilibrium of every living being, and in the equilibrium of every biocoenosis as a whole.

The time factor, so important for each individual being, plays a very insignificant role. An earthquake, for example, may restore a disturbed geological equilibrium in a matter of a few seconds and may have the same effect as gradual erosion over the course of millions of years. Both time spans, however, are of similar significance in relation to eternity – which is the working time of this basic law.

SUMMARY

Extensive ecological studies and observations of desert and steppe areas as well as of many saline habitats through several decades and in all continents induced the author to carry out several series of experiments with various plant species and to combine in them dune sand as soil with natural sea-water in various concentrations (10,000–50,000 mg/l Total Salt Concentration with 75% NaCl and 10% $MgCl_2$) for irrigation.

A few well known facts and a number of new principles worked together to make these experiments successful. The facts known since long are the quick percolation and good aeration in sand on the one hand, and the easy solubility of the two dangerous components of sea-water: sodium chloride (NaCl) and magnesium chloride ($MgCl_2$), on the other hand. Further also the lack of sodium adsorption for the lack of clay particles.

The new principles discussed here are:

The Principles of Partial Root Contact;

The Viscosity Principle;

The phenomenon of "Subterranean Dew";

The Adaptability to the Factors of Erratics (in this case mainly to fluctuating osmotic pressure);

The Principle of "Raised Vitality";

Biological Desalination,

The Global Salt Circulation, and

The Basic Law of Universal Balance.

The new principles may be applicable to desert areas of a total size similar to that of a large continent. Many food and fodder crops as well as industrial raw materials can possibly be grown on the basis of these principles, as experiments have already shown.

In view of this potential global impact on Human Welfare, the World Academy of Art and Science has established a Working Group on this subject in order to organize close international cooperation for this new line of research and it is to be hoped that the successful small scale experiments carried out up to now will soon be enlarged on a global scale with the help of the specialized Agencies of UNO, ICSU and the interested scientific Unions as well as by the newly founded World University of WAAS.

RÉSUMÉ

Des vastes études écologiques et des observations sur des zones désertiques ou steppiques de même que sur des "habitats" salins pendant beaucoup de decennies et sur tous les continents ont amené l'Auteur à entreprendre beaucoup d'épreuves du sable de dune comme terrain avec de l'eau de mer naturelle à de différentes concentrations (10,000–50,000 mg/l concentration totale de sel avec 75% de NaCl et 10% $MgCl_2$) pour l'irrigation.

Certains faits bien connus et un nombre de nouveaux principes ont rendu ces épreuves complètement satisfaisantes. Les faits connus depuis longtemps sont la rapide infiltration et la bonne aération du sable d'une part et d'autre part la haute solubilité du chlorure de sodium (NaCl) et du chlorure de magnesium ($MgCl_2$). En outre le manque d'adsorption de sodium à cause du manque de particules d'argiles.

Voici les nouveaux principes en discussion:

Le Principe du Contact Partiel des Racines;
Le Principe de la Viscosité;
Le Phénomène de la Rosée Souterraine;
L'Adaptation au Facteur Erratique;
Le Principe de la "Vitalité Augmentée";
Le Dessalement Biologique,
La Circulation Globale du Sel, et
La Loi fondamentale de la Balance Universelle.

Ces nouveaux principes peuvent être appliqués sur des zones désertiques de dimensions pareilles à celles d'un vaste continent. Beaucoup de cultures destinées à l'alimentation et de cultures fourragères ainsi que des matières premières industrielles peuvent être cultivés d'après ces principes.

En vue de cette potentielle poussée mondiale sur le Bien-être Humain, la "World Academy of Art and Science" a etabli un Groupe de Travail à ce sujet pour etablir une organisation internationale stricte sur la base de ce nouveau type de recherche et il faut espérer que les succès obtenus avec les épreuves faites sur petite échelle jusqu'à présent seront bientôt obtenus sur une échelle mondiale grâce aux Agences spécialisées de l'UNO (United Nations Organisation), et de l'ICSU et toutes les Organisations scientifiques intéressées, comme aussi l'Université Mondiale de la WAAS récemment fondée.

RIASSUNTO

Ampi studi ecologici ed osservazioni compiute per parecchi decenni in zone desertiche e steppiche, così come in vari ambienti salini e in tutti i continenti, hanno indotto l'autore ad intraprendere una serie di esperimenti con varie specie di piante usando congiuntamente come terreno le dune sabbiose e per l'irrigazione acqua di mare naturale in varie concentrazioni (10.000–50.000 mg/litro di Total Salt Concentration con il 75% di NaCl.)

Alcuni fatti ben conosciuti e una serie di nuovi principii hanno contribuito al successo di questi esperimenti. I fatti conosciuti da parecchio tempo sono, da un lato, la rapida percolazione e la buona aereazione della sabbia, dall'altro la facile solubilità dei due pericolosi componenti dell'acqua di mare: il cloruro di sodio (NaCl) e il cloruro di magnesio ($MgCl_2$). Inoltre il mancato assorbimento di sodio dovuto all'assenza di particelle di argilla.

I nuovi principii che saranno illustrati nel corso delle proiezioni, sono: il principio del contatto radicale parziale, il principio della viscosità, il fenomeno della rugiada sotterranea, la adattabilità al fattore dei terreni anomali, il principio della "vitalità insorgente," la desalinazione biologica e la circolazione globale del sale.

Questi nuovi principii sono applicabili in zone desertiche di dimensioni simili a quelle di un vasto continente. Molte coltivazioni di prodotti destinati all'alimentazione umana e del bestiame, così come coltivazioni di interesse industriale, possono essere effettuate seguendo questi principii, come i hanno già dimostrato gli esperimenti compiuti.

In vista di questo potenziale sviluppo del benessere umano, la World Academy of Art and Science ha istituito un Gruppo di Lavoro con il compito di organizzare una stretta collaborazione internazionale in questa nuova linea di ricerche ed è auspicabile che i successi ottenuti con esperimenti su piccola scala saranno presto portati su scala mondiale con l'aiuto delle agenzie specializzate dell'ONU, dell'ICSU e delle organizzazioni scientifiche interessate.

REFERENCES

ABRAHAM, A. 1954. Examples of Root Adaptability to Aridity. *VII. Congr. Int. Bot Rapports. Communications 11 et 12*, Paris. p. 237–239.

BOYKO, ELISABETH. 1952. The Building of a Desert Garden; *J.Roy. Hortic. Soc. 76:* 1–8.

BOYKO, ELISABETH & HUGO BOYKO. 1968. The Desert Garden of Eilat 15 years after its creation. p. 133–160. *In* H. BOYKO [Ed.] Saline Irrigation in Agriculture and Forestry. Dr W. Junk N.V., The Hague.

BOYKO, HUGO (ed.). 1966. Salinity and Aridity – New Approaches to Old Problems. Monographiae Biologicae XVI, pp. 408. Dr. W. Junk N.V., The Hague. (H. BOYKO: Principles, p. 131–200; H. & E. BOYKO: Experiments, p. 214–282).

BOYKO, HUGO. 1967. Salt-water Agriculture. *Scient. Amer.* March 1967. New York. vol. 216 (3): 89–96.

BOYKO, HUGO. 1967. Irrigation with Brackish and Sea-water. *Ariel, Review of the Arts and Sciences in Israel*, Jerusalem. No. 18: 45–58.

BOYKO, HUGO & ELISABETH BOYKO. 1968. Experiments with Seawater and other saline waters in Israel and other countries. *In* H. BOYKO [Ed.] Saline Irrigation in Agriculture and Forestry. Dr. W. Junk N.V., The Hague.

BOYKO, H., D. BUGOSLAV (HILLEL) & N. TADMOR. 1957. Pasture Research (in the Negev). Final Report to Ford Foundation, sub-project C-1d, pp. 295–356, Jerusalem.

HEIMANN, H. 1966. Plant Growth under Saline Conditions and the Balance of the Ionic Environment. p. 201–213. *In* H. BOYKO [Ed.] Salinity and Aridity – New Approaches to Old Problems, Dr. W. Junk N.V., The Hague.

SEIDEL, K. 1966. Reinigung von Gewässern durch höhere Pflanzen. *Naturwissenschaften 53:* 289–298.

Author's address: Dr. Hugo Boyko, World Academy of Art and Science, 1 Ruppin Street, Rehovoth, Israel.

Germination Capacity of Seeds in Saline Soil

by

GIACOMO LOPEZ

The ever-growing food requirements make necessary the utilization of new lands, previously abandoned or little utilized by man because of their unfavourable conditions or because use was made, in those areas of scarce atmospheric precipitation or lack of perennial water, of more or less saline water.

To those cases belong those salty soils which are saline due to natural causes or have become saline as a result of unsuitable irrigational practices with saline waters or, furthermore, by absorption from the depths, through capillary force, of possible saline deposits.

In these soils the growth of the plants is characterized by uneven and stunted development and by the presence of sometimes very extensive totally sterile areas. These phenomena are especially evident in the last stages of vegetative development but if we are following the plant cycle from the time of sowing it will be noted that actually the above phenomena may have started with germination and emergence, and therefore, according to AYERS (1952), it is just in the first phases of life that we must seek the *critical period* of plant growth in soils of anomalously high salinity.

Moreover, establishing a scale of salt tolerance for cultivated plants involves some specific problems because certain plants which are sensitive to salt in some periods of their development become resistant in other periods. Maize, for example, is much more resistant to salinity during the vegetative phase than in the stages of the reproductive phase while sugar beet, a plant very tolerant to salt during the last periods of its growth, turns out to be very sensitive during germination (RICHARDS et al. 1954).

According to AYERS (1952), the excess of salts may influence seed germination either by diminishing the absorption rate of the water or by facilitating the influx of ions in quantities large enough to make them toxic. Such phenomena result in a lower rate of seed emergence and a drop in germination percentage.

Another negative effect is the augmented osmotic pressure of the medium. The negative effect on growth may be due, in fact, not so much to the nature of the individual saline components in the solution, as to their total concentration which, because of the increased osmotic pressure, seems to hamper the entry of water itself into the vegetal cells. As a matter of fact, the reduced growth may be caused by a process which tends to increase the internal osmotic pressure of the plants in order to compensate and balance the osmotic pressure of the medium in which the roots are living and so maintain unchanged their turgor.

However, according to ALLISON (1964) resistance of a plant to salt may be defined as the degree at which this osmotic equilibrium can be reached without sacrificing growth.

Another theory is that formulated by REPP, ALLISTER, DE VERE & WIEBE (1959), according to whom the adaptation and the resistance of plants in saline soils is due to the capacity of resistance of the protoplasm to salt, i.e. to the characteristic of plants of being able to accumulate the salts in the cellular fluid without causing the death of the cells themselves. The increase of saline concentration in the cellular fluid results in an increase of the internal osmotic pressure of the plants and hence in a greater capacity of absorption of water from the saline soil.

Moreover, we should not forget that in the system plant-water-soil, the physical, chemical and mineralogical characteristics of the soil may influence in a more or less positive manner both the germination and the growth of the plants even in presence of rather concentrated circulating saline solutions. Without entering into details it may be sufficient to mention the action of calcium, whether exchangeable or in the form of carbonate, which can neutralize or eliminate the negative effects of sodium.

In Puglia, the custom of irrigation with saline waters is widespread along the Adriatic coastal strip south of the Ofanto's mouth and along the Ionian coasts of Salento. It is not impossible, therefore, that salinity might develop under such conditions, even though temporarily, owing to the nature of the soils undergoing irrigation and owing to their geological substrata. Moreover, due to different pedogenesis, along the coastal zone which extends from the above-mentioned mouth of the Ofanto to Manfredonia, there exists a vast zone (about 10,000 hectares) of salty soils of marine origin (BOTTINI & LISANTI, 1951).

The aim of the present research has been therefore to study the capacity of germination of the seeds of some plants more commonly cultivated in such zones in a typical soil imbued with salt of various degrees of salinity.

MATERIALS AND EXPERIMENTAL METHODOLOGY

For the researches involved in the present study, we have chosen the type of soil which in view of its diffusion in the irrigated areas of the region was

sufficiently representative. The choice fell though on a "red transfer earth"
on Pleistocene calcareous tuff the physico-chemical characteristics of which
are as follows:

Coarse sand	5.3%	Nitrogen	0.84%
Fine sand	18.9%	P_2O_5, total	1.54%
Silt	24.7%	K_2O, total	11.38%
Clay	33.6%	P_2O_5 assimilable	70 mg/kg
Limestone	15.2%	K_2O exchangeable	288 mg/kg
Organic substances	2.24%	H_2O at ½ atm	24.56%
pH	7.46	H_2O at 15 atm	13.15%

The above data show that the soil is "silty-clayey," moderately calcareous,
and that its physical, chemical and hydrological characteristics are not
much different from the average prevailing for this type of soil.

The water with which the soil was imbued with salt, was prepared in the
laboratory, taking into account in this case too the general characteristics
of the saline waters of the region's wells.

The highest concentration of the water from which, as it will be seen
further below, the less concentrated waters were prepared, turned out to
be 32 g/l; in it the single saline components are present, roughly, in the
same quantities and in the same proportions in which they exist in the
natural waters of the area, influenced, due to the carstic phenomenon widespread in the region, by sea-waters.

The composition of the water prepared in the laboratory as compared
with a highly saline water sampled from a well located at Macchie, near
Bari, (B), is given below:

	A		B	
	mg/l	Meq/l	mg/l	Meq/l
Fixed residue at 105°C	32,000	—	10,910	—
Chlorine Ion	17,000	493.2	5,470	154.22
Sulphuric Ion	2,238	46.6	768	15.99
Bicarbonic Ion	234	3.8	504	8.26
Carbonic Ion	—	—	—	—
Calcium Ion	411	20.5	239	11.92
Magnesium Ion	590	48.6	151	12.19
Sodium Ion	10,740	466.9	3,450	150.00
Potassium Ion	298	7.6	140	3.58

Minor elements, chiefly boron, were not taken into account in the preparation of the water in order to avoid their influence in the germination
tests.

From the maximum concentration the desired dilutions were prepared
so as to obtain the following concentrations: 2, 4, 8, 16 and 32 g/l. Moreover,
this was supplemented by a control with 0 g/l.

The seeds taken into consideration, belong to plants commonly cultivated in the area. These were: durum wheat (*Triticum durum*, var. *Cappelli*),

common wheat (*Triticum vulgare*, var. *Mara*), barley (*Hordeum sativum*, cv. *Motto-la*), Alfalfa (*Medicágo sativa*, cv. *Ascolana*), Alexandrine clover (*Trifolium alexan-drinum*), vetch (*Vicia sativa*), tomato (*Solanum lycopersicum*) with its two varieties: *S. Marzano* and *P. 19*, broccoli (*Brassica oleracea*, var. *capitata*), lettuce (*Lactuca sativa*), endive (*Cichorium endivia*) and beet (*Beta vulgaris* var. *Cycla*).

The technique followed was that of AYERS & HAYWARD (1948), modified by DEWEY (1960). The following is a brief description of the method:

200 g of dry soil are introduced into Petri dishes having a size of 26 × 170 mm. The soil is then humidified, until its field capacity is reached, with the saline solutions according to the following scheme:

(1) Sol. A,   0 ppm
(2)   ,,   B,   2,000   ,,
(3)   ,,   C,   4,000   ,,
(4)   ,,   D,   8,000   ,,
(5)   ,,   E, 16,000   ,,
(6)   ,,   F, 32,000   ,,

In practice, 50 ml of saline solution were added to the soil of each dish, reaching an average humidity of 24–25% or a figure equal to that found experimentally in the laboratory during determination of water content at $\frac{1}{3}$ atm which, as it is known, is very near to the real field hydric capacity for this type of soil.

Thus the actual amounts of salt added for each treatment were 0, 0.1, 0.2, 0.4, 0.8 and 1.6 g, respectively.

Each treatment was repeated three times.

Emergence was recorded daily until the records failed to show any further variation. The test ended 10 to 12 days after seeding. Emergence was calculated as percentage of the number of seeds emerging and the control. It will be useful to mention here that the control even without the addition of any salt always showed a low content in soluble salts due to that of the original soil.

Upon completion of each germination test, the electrical conductivity of the soil's saturation extract was established. The equivalent in total soluble salts present in the soil was calculated according to the following formula (JACKSON, 1958):

$$S = \frac{L \times 0.64 \times a}{1,000}$$

in which S = soluble salts present in 1,000 g of soil;

  L = electrical conductivity of the saturation extract in mmhos/cm;

  a = water content at saturation of 1,000 g soil.

Considering that the soil is a "red soil" of transfer, silty-clayey, which at saturation shows a content of 57.5% of water of its dry weight, we get: a = 575.

The saturation extract was taken into account instead of the 1 to 5 extract of the soil because the former, though showing values of soluble salts slightly inferior to those obtained with the latter, seems to be in closer relation with field hydric capacity and with the solutions actually circulating in the soil.

RESULTS AND DISCUSSION

The results obtained with each individual plant taken into consideration are given below. With a view to facilitating interpretation, resistance to salt has been related to the electrical conductivity of the saturation extract of the soil as it resulted at the end of each test, and not to the saline concentration of the water supplied to the germination capsules.

The maximum limit of toleration in seed germination turned out to be, with the exception of barley, 16 g/l saline concentration; in fact, beyond that limit the concentration becomes toxic, the seeds fail to absorb water and do not germinate at all. Only the local variety of barley showed an 8% germination capacity at a maximum saline concentration of 32 g/l.

On the other hand, the electrical conductivity of the saturation extract of the soil turns out to be 12.1–16.4 mmhos/cm equaling a saline content of about 4.5–6.3 g of soluble salts per kg of soil.

Graminaceae

Graminaceae in general turned out to be resistant or moderately resistant in germination in saline environment.

Table I and Fig. 1 show the data obtained concerning the capacity of germination of the two wheats and the barley.

The two wheats (*T. durum* and *T. vulgare*) show moderate resistance up to values of electrical conductivity of the saturation extract of the soil of 3.31 and 5.65 mmhos/cm, respectively. Beyond these values their capacity of germination falls suddenly, declining to about 50% at the level of 12.1

TABLE I

Germination of wheat and barley in saline soil

Solution added to the soil	Electrical conductivity of the saturation extract of the soil in mmhos/cm	Equivalent in soluble salts in g/kg of soil	Germination, %		
			Durum wheat (*Cappelli*)	Common wheat (*Mara*)	Barley
A	0.72	0.27	100	100	100
B	1.81	0.67	73.7	93.9	86.2
C	3.31	1.22	71.3	88.8	91.0
D	5.65	2.08	59.0	87.1	93.1
E	12.10	4.45	45.0	55.2	93.1
F	23.72	8.73	0.0	0.0	8.3

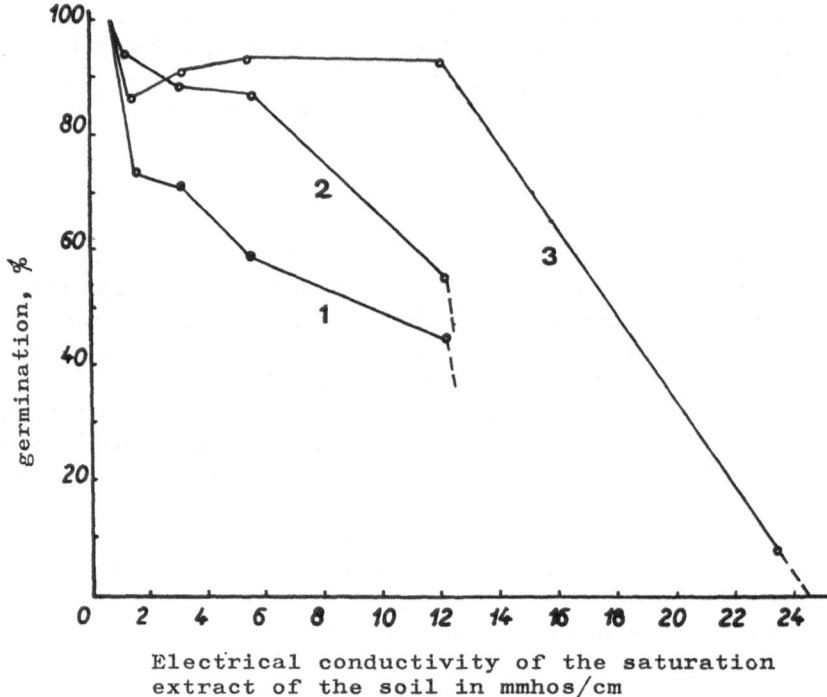

Fig. 1. Capacity of germination of the seeds of Graminaceae in relation to the salinity of the soil.

1 = durum wheat   2 = common wheat   3 = barley

mmhos/cm. On the other hand, barley appeared to be much more resistant. This species even showed a slight decrease in germination to lower values (1.81 mmhos/cm), improving thereafter and showing a constant behaviour up to the highest values of electrical conductivity.

The capacity of barley to germinate under conditions of excessive salinity was seen by the author already during several tests made in 1951–53 with salty soils of the Puglia Plateau (LOPEZ, 1954). In those tests, the object of which was to determine methods of cultivating that region, barley, oats and beans were sown and compared during those years when the research was executed. In the second year, even after additional, relief irrigation, the oats and the herb failed to give any yield while the barley had a production of 14 quintals per hectare.

Leguminous Plants

Leguminous plants proved less resistant. Table II and Figure 2 show the data obtained concerning the capacity of germination of the follow-

TABLE II

Germination of leguminous plants in saline soil

Solution added to the soil	Electrical conductivity of the saturation extract of the soil in mmhos/cm	Equivalent in soluble salts in g/kg of soil	Germination, %		
			Alfalfa	Alexandrine clover	Vetch
A	0.68	0.25	100	100	100
B	2.72	1.00	86.4	81.5	92.7
C	4.99	1.84	75.6	62.3	82.5
D	8.93	3.29	64.7	53.1	82.5
E	16.42	6.04	42.1	23.8	73.1
F	23.16	8.52	0.0	0.0	0.0

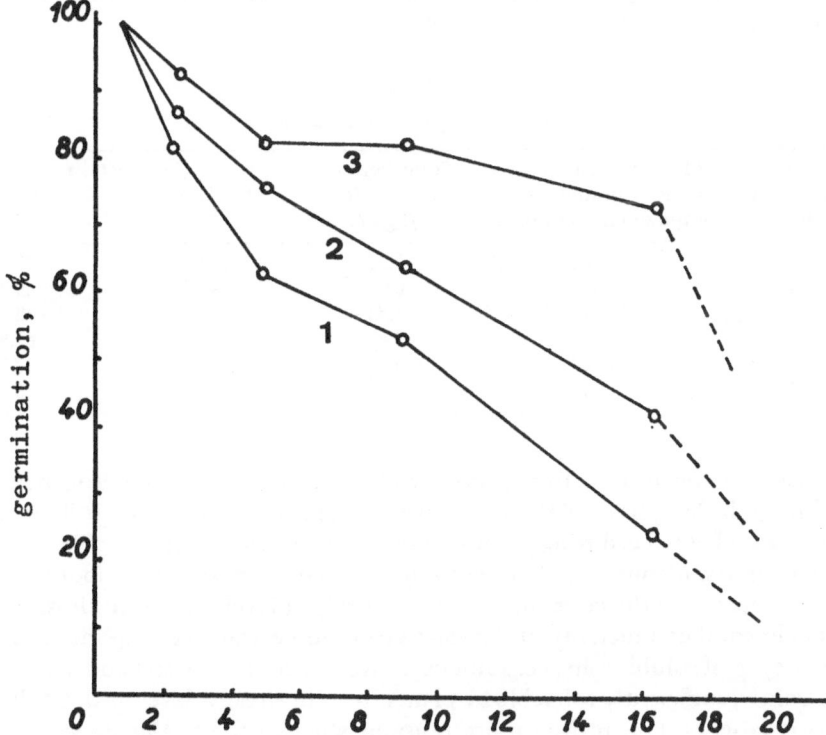

Electrical conductivity of the saturation extract of the soil in mmhos/cm

Fig. 2. Capacity of germination of the seeds of leguminous plants in relation to the salinity of the soil.

1 = alexandrine clover   2 = alfalfa   3 = vetch

ing three species considered: Alfalfa, Alexandrine clover and vetch.

They showed considerable sensitivity to salinity. As a matter of fact, their capacity of germination at the level of the electrical conductivity of the saturation extract amounting to 16.42 mmhos/cm, the maximum limit of toleration, declined as regards alfalfa to less than 50% as compared to the control and in respect of the Alexandrine clover to 23.8%. Vetch proved to be more resistant than the two aforementioned species and its capacity of germination at such salinity levels dropped to something less than ⅓.

Tomato

The two varieties of tomato proved extremely sensitive in saline environment.

Table III and Figure 3 show data concerning the capacity of germination of the seeds of this species in salty soil.

TABLE III

Germination of tomato in saline soil

Solution added to the soil	Electrical conductivity of the saturation extract of the soil in mmhos/cm	Equivalent in soluble salts in g/kg of soil	Germination, %	
			S. Marzano	P. 19
A	0.63	0.23	100	100
B	2.54	0.93	91.4	89.2
C	4.12	1.52	80.5	78.5
D	6.58	2.42	52.2	64.8
E	10.08	3.71	4.4	27.0
F	23.40	8.61	0.0	0.0

As a matter of fact, this species, while resistant in saline environment during the last phases of the vegetative cycle, proved extremely sensitive in the first phases, i.e. during germination and emergence. Considerable signs of weakness manifesting themselves in a decline of emergence to about 50% as compared to the control, are noted already at levels of 6.58 mmhos/cm of electrical conductivity of the soil's saturation extract, corresponding to 2.54 g/kg of soluble salts, i.e. a not excessive salt content in soil and, consequently, sufficiently tolerable by plants that are already developed. Of the two varieties, P. 19 proved more resistant whereas the S. Marzano variety, although showing a slight superiority against the P. 19 variety at the lower levels, is extremely sensitive when the electrical conductivity of the saturation extract exceeds 6 mmhos/cm.

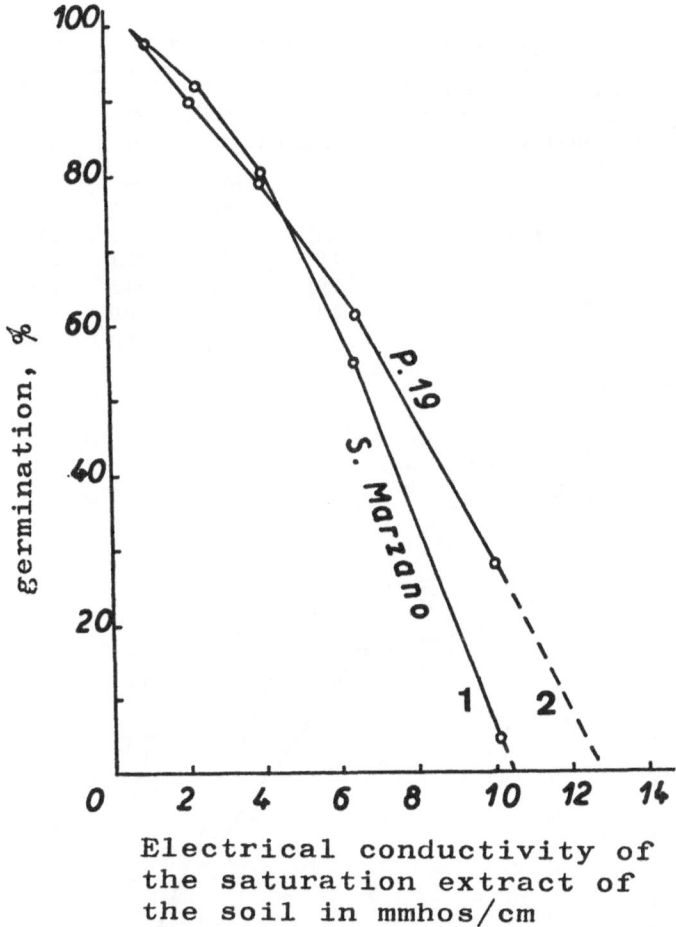

Fig. 3. Capacity of germination of the seeds of tomato in relation to the salinity of the soil.

1 = S. Marzano   2 = P. 19

Horticultural Plants

The seeds of the horticultural plants examined, viz. lettuce, broccoli, endive and beet, showed little sensitivity.

The results of the examination of the capacity of germination concerning these seeds are shown in Table IV and Figure 4.

Among these, the most resistant appeared broccoli which at a level of 13.60 mmhos/cm showed a capacity of germination of 64% as compared to the control, whereas at the same level, that of beet fell to about half and those of lettuce and endive to about ¾.

G. LOPEZ

TABLE IV

Germination of horticultural plants in saline soil

Solution added to the soil	Electrical conductivity of the saturation extract of the soil in mmhos/cm	Equivalent in soluble salts in g/kg of soil	Germination, %			
			Broc-coli	Let-tuce	En-dive	Beet
A	0.93	0.34	100	100	100	100
B	2.56	0.94	86.6	92.6	74.3	100
C	4.56	1.68	76.5	90.2	67.2	87.0
D	8.02	2.95	72.3	89.1	64.0	74.0
E	13.60	5.01	62.5	23.8	26.0	52.0
F	23.70	8.72	0.0	0.0	0.0	0.0

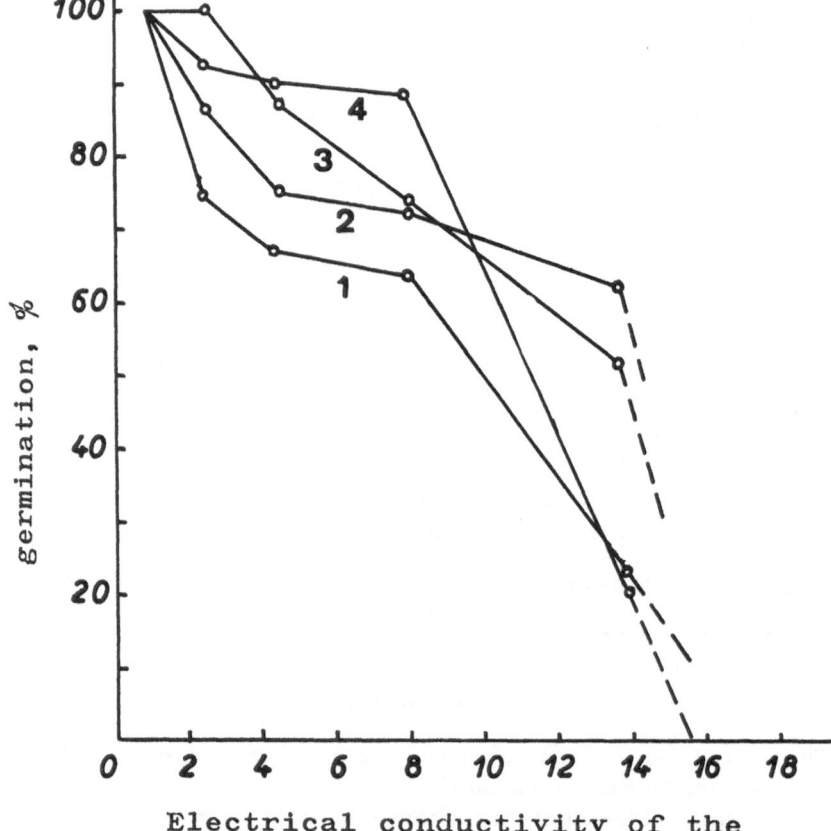

Electrical conductivity of the saturation extract of the soil in mmhos/cm

Fig. 4. Capacity of germination of the seeds of horticultural plants in relation to the salinity of the soil.

1 = endive   2 = broccoli   3 = beet   4 = lettuce

Effects on the Capacity of Germination

The influence of the salinity of the soil on the capacity of germination manifested itself not only in the reduction of the capacity of germination itself but also in a delay in the commencement of germination and emergence. In fact, whereas at the first levels of salinity the commencement of germination takes place simultaneously with the control, at levels exceeding 5 mmhos/cm electrical conductivity of the saturation extract, i.e. at about 2 g of soluble salts per kg of soil, the germinative phase begins as a rule 5 to 6 days later.

For example, in the case of barley emergence begins after 5 days at a level of 5.65 mmhos/cm while the few seeds emerging at the level of 23.72 mmhos /cm, do this 9 days after the control.

The same may be said of the other species considered: wheats begin emergence at the same salinity level 5 to 6 days after the control; alfalfa and Alexandrine clover after 6 days; vetch after 2 days; lettuce after 6 days; endive after 8 days and beet after 3 days.

CONCLUSIONS

Until now we considered the maximum limits of resistance of the plants examined. However, these limits indicate the degree of their adaptability and survival in an anomalous environment such as a soil imbued with salt. As we deal with seeds of cultivated plants such a criterion is of a very limited practical significance from the agronomic point of view.

A more cogent criterion from this point of view would therefore be to consider, for an appraisal, those limits of salinity within which the capacity of germination of the seeds approaches most that of a soil under normal conditions.

Therefore, if we consider the limits of salinity of the soil, still calculated according to the electrical conductivity of the saturation extract (CE), within which the capacity of germination of seeds is 85 to 90% as compared to that of the non-saline soil of the control, the studied plants may be divided into the following 3 groups:

CE (electric conductivity) = 0–4 mmhos/cm:
    little resistant plants: durum wheat, alfalfa, Alexandrine clover, *P. 19* tomato, broccoli and endive.

CE = 4–12 mmhos/cm: moderately resistant plants: common wheat, vetch, *S. Marzano* tomato, lettuce and beet.

CE = 12–18 mmhos/cm: resistant plants: barley.

Such subdivision should be considered limited to the variety of species examined and to the type of soil. As a matter of fact, it is not impossible that specific varieties within the same species may react differently just as in a soil whose structure and composition differs from that studied, the capacity of germination may have a different behaviour.

SUMMARY

The author, after having noticed that in a saline soil the growth of plants depends in many cases on how they react to salinity in the first phases of the vegetative cycle, has carried out tests of capacity of germination on soils imbued with salt at various degrees of salinity.

On the basis of the results obtained, he has subdivided the plants into the following three groups: (a) little resistant plants comprising: durum wheat, Alfalfa, Alexandrine clover, *P. 19* tomato, broccoli and endive; (b) moderately resistant plants: common wheat, vetch, *S. Marzano* tomato, lettuce and beet; (c) resistant plants: barley.

RÉSUMÉ

Puisque dans la Pouille, le long de la côte adriatique et ionienne, on pratique sur vaste échelle l'irrigation à l'eau plus ou moins saumâtre, le Centre Agricole Experimental de Bari a effectué des épreuves de la capacité de germination des graines des plantes d'emploi courant dans un terrain salé à différents degrés de salinité.

La limite maximum d'endurance dans la germination des graines, exception faite pour l'orge, se situe à une concentration saline de l'eau de 16 g/l; la conductante électrique de l'extrait de saturation du terrain est de 12,1–16,4 mmhos/cm, équivalent à un contenu salin de près de 4,5–6,3 g de sels par kg de terrain.

L'orge s'est revelé la plante la plus résistante parmi celles qu'on a examinées: sa capacité de germination, en effet, n'a baissé qu'à 93%. Les deux espèces de blé ont au contraire perdu près de la moitié de leur capacité de germination.

Parmi les legumineuses la vesce a fait preuve de grande résistance tandis que la luzerne et le trèfle se sont revelés peu résistants à ces limites.

Les tomates qui résistent fort bien, comme on le sait, à l'irrigation à l'eau salée, se sont au contraire revelées très sensibles pendant la période de germination et d'émergence. Leur capacité de germination a baissé jusqu'à 27% pour les tomates *P. 19* et jusqu'à 4,4% pour les tomates *S. Marzano*.

Très sensibles à ses limites se sont revelées aussi les plantes potagères parmi lesquelles le brocoli de rave a été le plus résistant.

RIASSUNTO

L'A. dopo aver rilevato come in un terreno salino la crescita delle piante dipende, in molti casi, da come queste reagiscono alla salinità nelle prime fasi del ciclo vegetativo, ha condotto delle prove di germinabilità su un terreno salinizzato a diversi gradi di salinità.

Dai risultati ottenuti egli ha suddiviso le piante nei seguenti tre gruppi: (a) piante poco resistenti e comprendenti: grano duro, erba medica, trifoglio alessandrino, pomodoro P. *19*, broccoletto di rapa sessantino e scarola verde; (b) piante mediamente resistenti: grano tenero, veccia, pomodoro *S. Marzano*, lattughino riccio e bietola da costa; (c) piante resistenti: orzo.

REFERENCES

ALLISON, L. E. 1964. Salinity in relation to irrigation. *Adv. in Agron.* 16: 129–180.

AYERS, A. D. 1952. Seed germination as affected by soil moisture and salinity. *Agron. J.* 44: 82–84.

AYERS, A. D. & H. E. HAYWARD. 1948. A method for measuring the effects of soil salinity on seed germination with observations of several crop plants. *Soil Sci. Soc. Amer. Proc.* 13: 224–226.

BOTTINI, O. & E. LISANTI. 1951. Ricerche sui terreni salsi del Tavoliere di Puglia. *Ann. Sper. Agr., Roma*, n.s. 5: 234–265.

DEWEY, D. R. 1960. Salt tolerance of twenty-five strains of Agropyron. *Agron. J.* 52: 631–635.

JACKSON, M. L. 1958. Soil chemical analysis, Prentice-Hall, Inc., Englewood Cliffs, N.J.

LOPEZ, G. 1954. Ricerche sul miglioramento dei terreni salsi del Tavoliere di Puglia. *Ann. Sper. Agr., Roma*, n.s. 8: 1561–1574.

REPP, G. I., McALLISTER, R. DeVERE & H. H. WIEBE. 1959. Salt resistence of protoplasm as a test for the salt tolerance of agricultural plants. *Agron. J.* 51: 311–314.

RICHARDS, L. A. et al. 1954. Diagnosis and improvement of saline and alkali soils. U.S. Dept. Agric. nr. 60.

Author's address: Professor Dr. Giacomo Lopez, Stazione Agraria Sperimentale, Via Ulpiani 1, Bari, Italy.

Utilization of Sea-water and Coastal Sandy Belts for Growing Crops in India

by

E. R. R. IYENGAR, T. KURIAN & A. TEWARI

India affords great potentialities for harnessing the enormous water reserve of the sea and at the same time the vast coastal sandy belts estimated to cover 20 million acres along the 3,500 miles long coastline, for growing crop plants. These sandy soils are mostly of marine origin. A few are wind deposits. These sandy areas are distributed along a climatic profile with very different climates comprising regions ranging from those with a well distributed rainfall to semi-arid tracts with long dry periods.

This Institute which is devoted to the studies on various projects for the recovery of chemicals from sea-water and seaweeds, has recently envisaged an investigation into the possibilities of using sea-water for plant growing. The research dealt chiefly with food crops, taking advantage of the changing salinity of sea-water during different seasons, sufficiently deep sandy soil with a reasonably low water table and of such plant species tolerant to sea-water salinity and the given environmental conditions.

The project chart (Chart A) indicates in brief the scope of research activities proposed to be undertaken in various disciplines of sea-water agriculture. The studies in this field are presently aimed at the evaluation of plant species which withstand fluctuating salinity of sea-water and at the selection of suitable coastal sandy areas for field experiments. This communication deals with the result of the preliminary experiments conducted so far.

GERMINATION OF CROP SEEDS

Many varieties of cereals, pulses, oil seeds, millets and leguminous and graminaceous fodder seeds were examined in order to determine the sea-water tolerance and their early seedling growth under laboratory conditions. The crop varieties were selected either for their high yielding or their salt resistance. The seeds were subject to germination experiments with different dilutions of sea-water ranging from 100–20,000 ppm in petri-dishes on filter papers. The emergence of radicles was taken as the criterion for the daily germination counts. Some of the cereals and millets and fodder crop seeds tolerated even a higher salinity than 20,000 ppm; the relative tolerance of the different varieties of these crops are given in Fig. 1–6. Bajra (*Pennisetum typhoides* S. & H.) varieties are more tolerant than groundnut, and paddy more than redgram. Among the leguminous fodder seeds *Medicago sativa* was better resistant than others and *Lolium perenne* amongst the grass seeds. The graminaceous species were higher tolerant than leguminous varieties tested. The rate of imbibition of solutions was higher in seeds of redgram and groundnut than in the seeds of paddy and bajra (Tables I, II & III). The uptake of solution was inversely proportional to the sea-water dilutions used, perhaps due to osmotic effect. But this phenomenon was not significantly different in various treatments presumably because the osmotic effect of the sea-water dilutions on seeds is superimposed by the specific ionic effect of Na^+ and Cl^- ions, thereby reducing or completely inhibiting germination at higher concentrations of sea-water dilutions. More information about antagonistic effects of the added salts to counteract the ill effects of Na^+ and Cl^- ions would be important in improving the germinable condition of seeds in sea-water.

POTCULTURE EXPERIMENTS WITH CROP PLANTS

Experiments on the establishment of certain plant species by fresh water irrigation or rainfall and subsequent irrigation with sea-water dilutions in order to study the effect on growth and quality of plants have been reported earlier (KURIAN et al., 1964). The adaptation of food crops, wheat and bajra to sea-water dilutions was attempted in pots (lined with asphalt) containing sandy soil. The plants were established with fresh water and occasional nutrient solution before sea-water treatments were administered. Application of 10,000 and 15,000 ppm of sea-water dilutions resulted in reduction of the growth of vegetative portions (Table IV). There was no adverse effect on floral initiation but the maturity of plants was hastened by a few days. It was felt necessary to amend the sea-water with deficient nutrient elements (N, P & K) by the addition of potassium and calcium salts for bajra. Such applications proved to improve the vegetative growth and yield by manyfolds.

Wheat – var. *Karchia*, a winter-season crop was experimented in pots by application of sea-water and its amendments. Low doses of potassium salts with constant nitrogen level improved the growth of plants in general and in particular the yield of grains over sea-water treatments (Table V). Similar observations on flowering and maturity of crops was noted as in bajra.

A balanced nutritive environment under saline conditions would go a

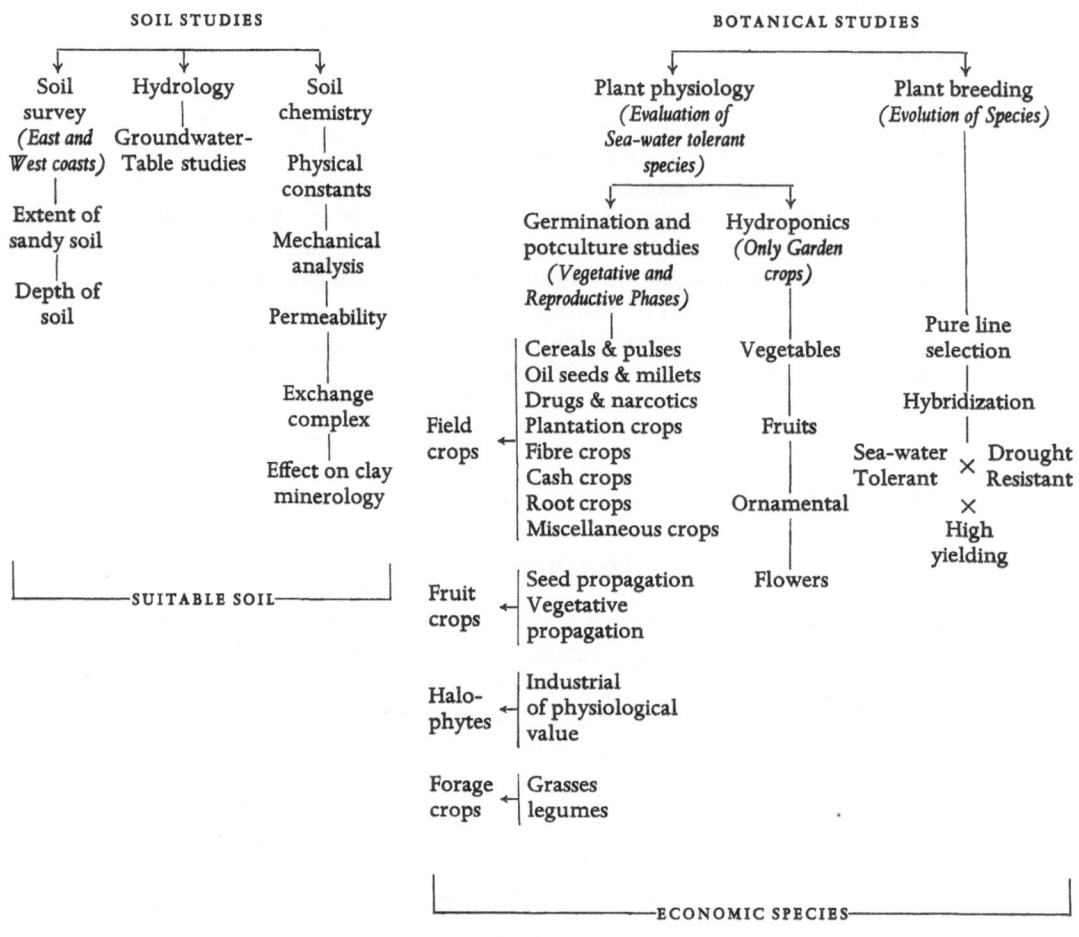

AGRICULTURE

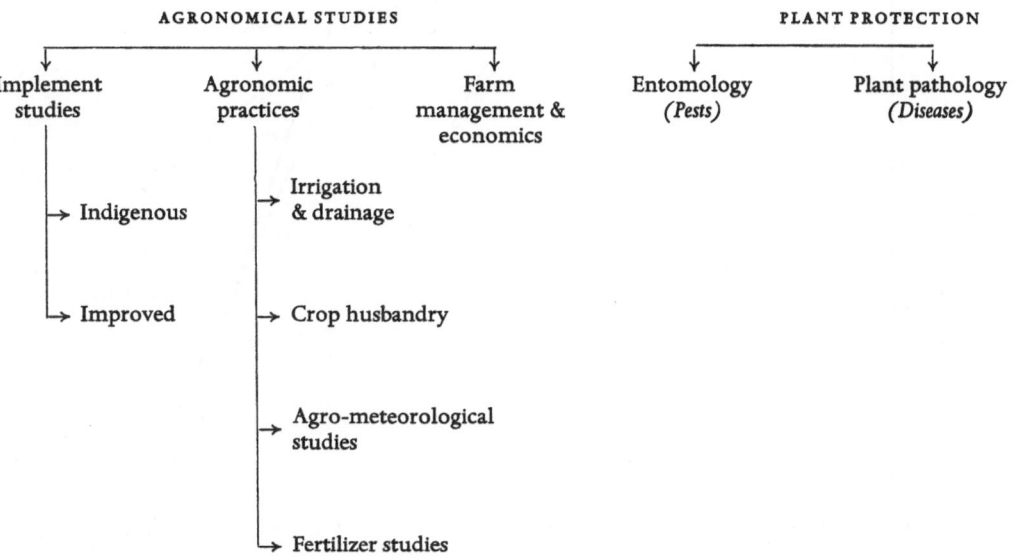

EXPERIMENTS ——
        ↓
FARMING

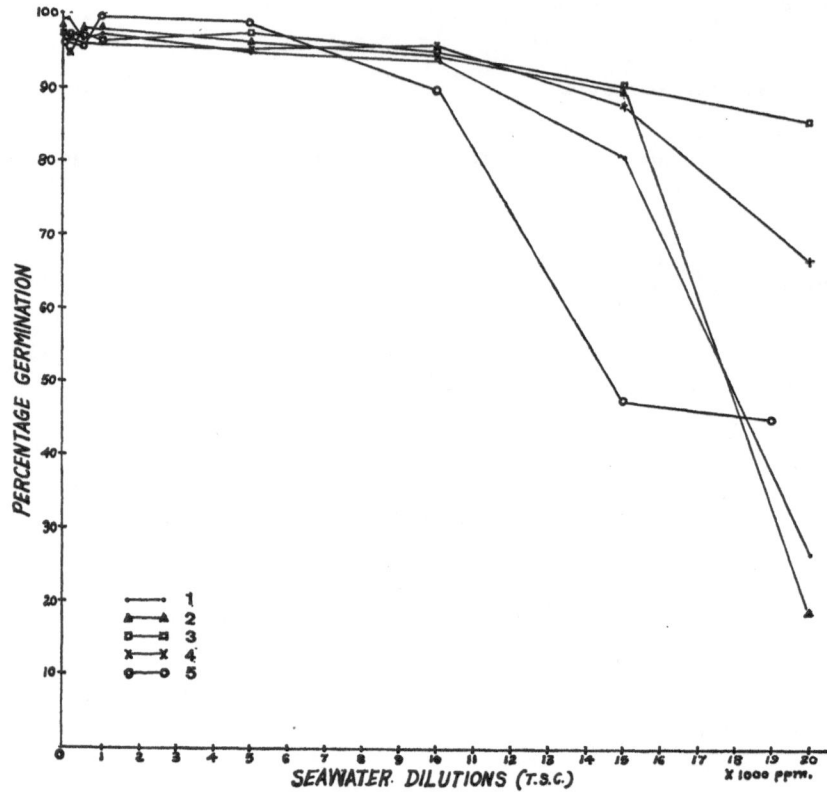

Fig. 1. Germination of Bajra (*Pennisetum typhoides* S. & H.).

1 = local collection 1   4 = local collection 2
2 = N-26-15-2   5 = Chadi
3 = Babapuri

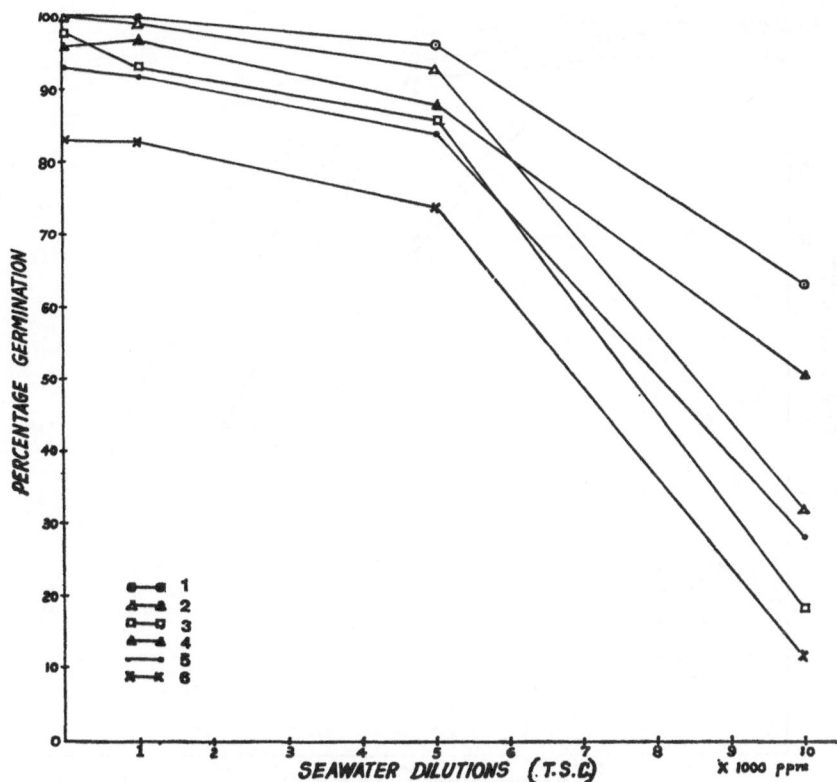

Fig. 2. Germination of ground-nut in sea-water.

1 = Punjab   4 = local collection 2
2 = AH-334   5 = local collection 1
3 = AK-12-24   6 = AH-32

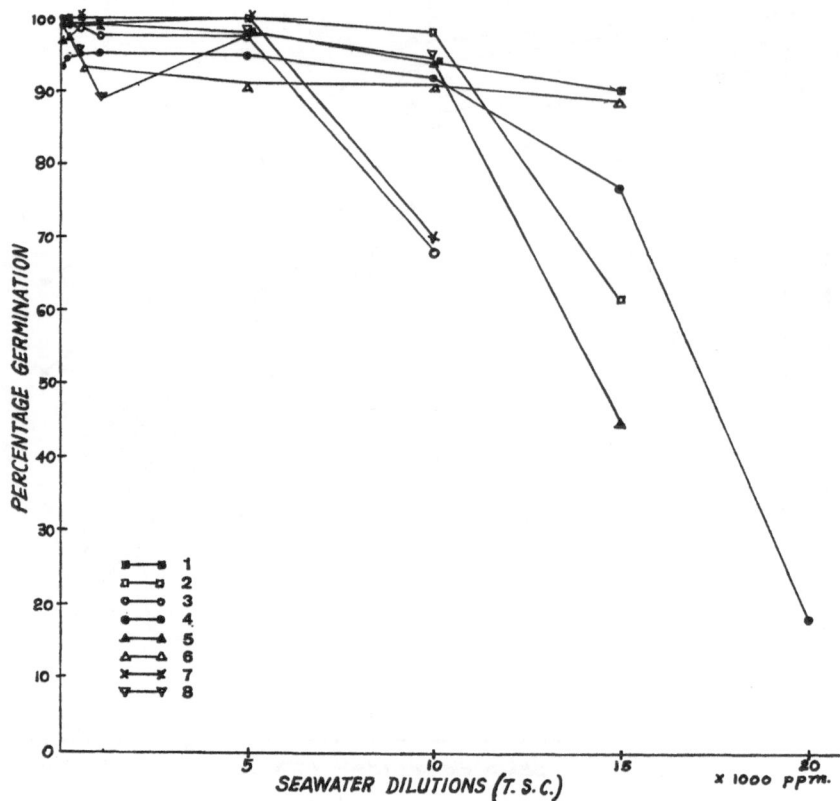

Fig. 3. Germination of Paddy in sea-water.

1 = C-87	5 = Kalarata
2 = C-10022	6 = C-1032
3 = NP-49	7 = C-75
4 = Sathi 34-36	8 = C-1327

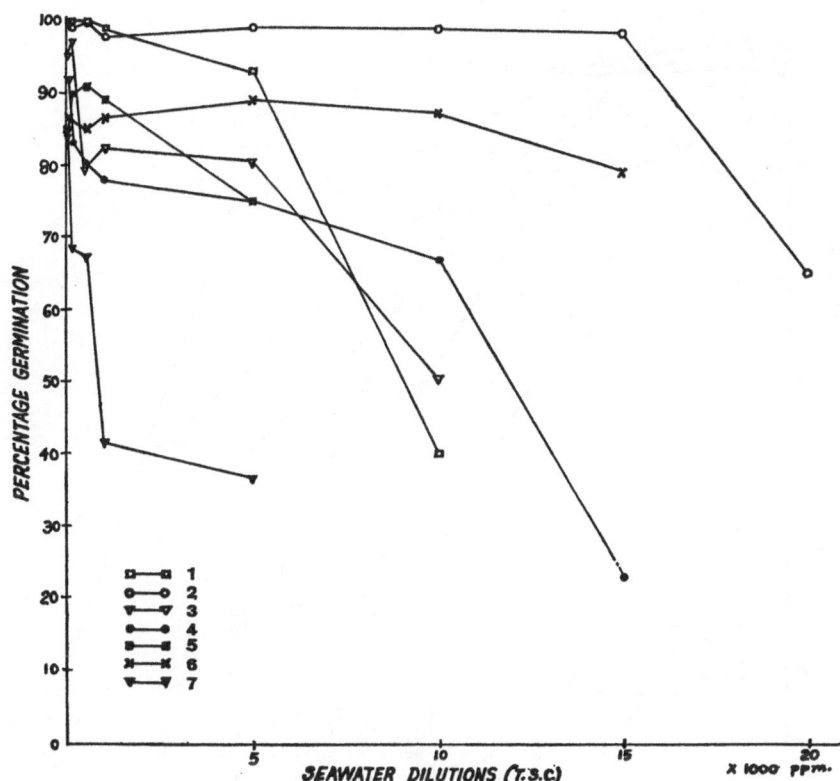

Fig. 4. Germination of pulses in sea-water.

1 = Bengal gram (Chitra)
2 = Green gram NP-23
3 = Red gram NP-24
4 = Black gram NP-6

5 = Lentil hybrid 1
6 = Red gram NP-80
7 = Red gram NP-16

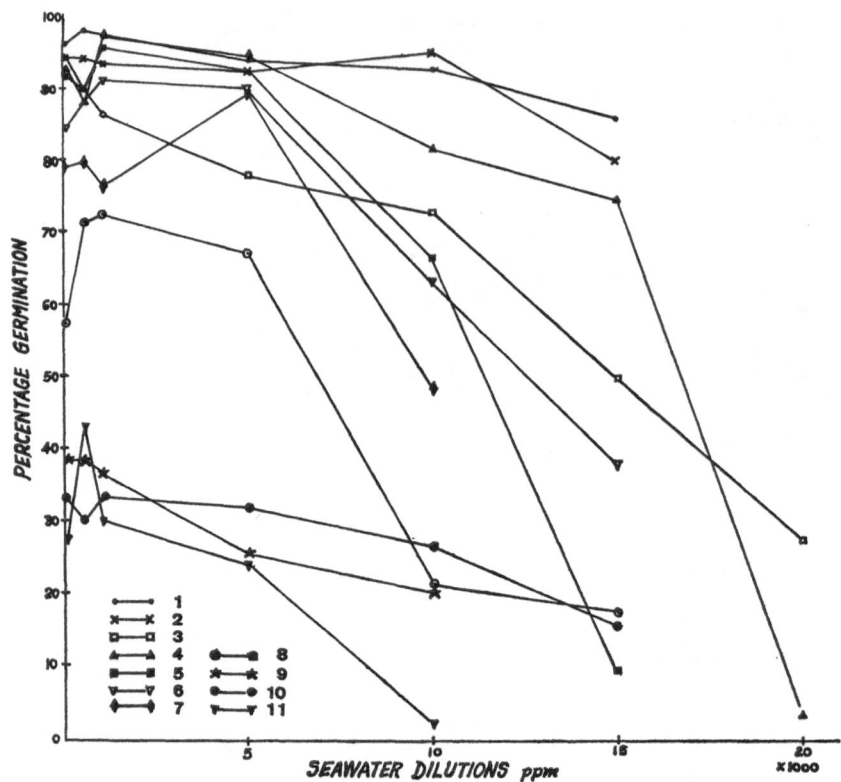

Fig. 5. Germination of leguminous fodder seeds in sea-water.

1 = *Medicago sativa* (perennial)
2 = *Medicago sativa* (local)
3 = *Mucuna cochinchinensis*
4 = *Medicago sativa* (Poona)
5 = *Phaseolus aconitifolius*
6 = *Mucuna* sp.

7 = *Lotus corniculatus*
8 = *Centrosema pubescens*
9 = Vernal alfalfa
10 = *Melilotus alba*
11 = *Trifolium alexandrinum*

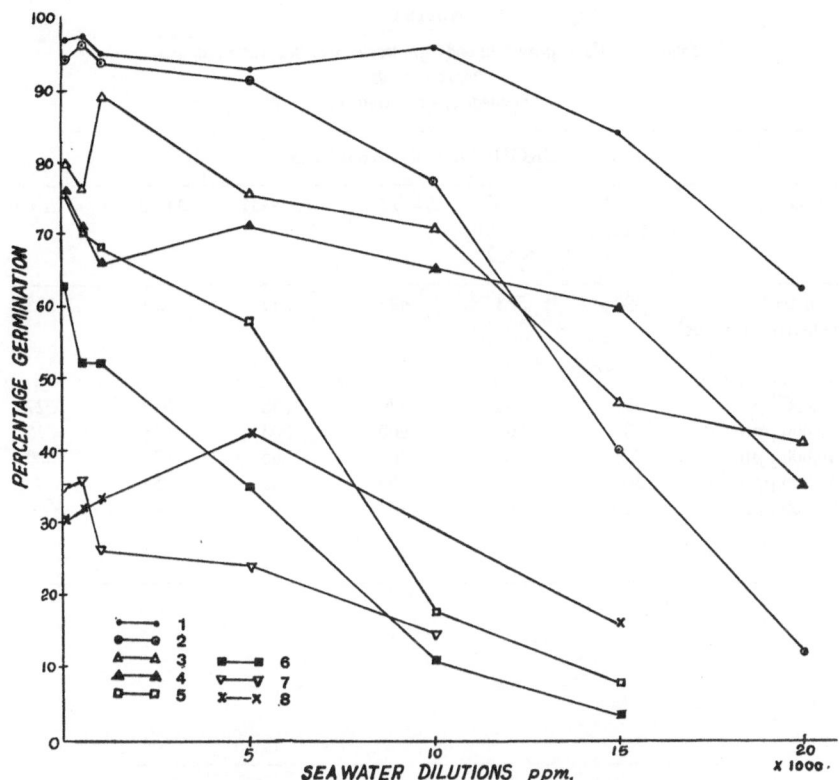

Fig. 6. Germination of graminaceous fodder seeds in sea-water.

1 = *Lolium perenne* P-312   5 = *Panicum antidotale*
2 = *Agropyron trachycaulum*   6 = *Sorghum sudanense*
3 = *Agropyron elongatum*   7 = *Sorghum halepense*
4 = *Echinochloa colonum*   8 = *Sorghum almum*

TABLE I

Solution uptake by groundnut and bajra treated with sea-water dilutions
mg/g of seeds
(Mean of 5 replications)

GROUNDNUT VARIETIES

Treatments	Local collection No. 1	Local collection No. 2	AK-12–24	AH-334	AH-32	Punjab
Control (Distrilled water)	390	523	496	340	520	535
Sea-water dilutions						
1,000 ppm	394	508	476	518	510	527
5,000 ppm	385	509	460	503	491	525
10,000 ppm	378	487	460	485	472	500
15,000 ppm	360	468	456	482	455	—
20,000 ppm	313	461	434	—	—	—

BAJRA VARIETIES

Treatments	Local collection No. 1	Local collection No. 2	N-26–15–2	Babapuri	Chadi
Control (Distilled water)	313	348	310	382	759
Sea-water dilutions					
100 ppm	313	356	324	395	396
500 ppm	314	349	307	402	380
1,000 ppm	313	341	301	389	379
5,000 ppm	303	336	290	388	365
10,000 ppm	303	312	301	365	355
15,000 ppm	284	308	289	367	356
20,000 ppm	283	305	250	364	352

TABLE II

Solution uptake of paddy varieties treated with sea-water dilutions; mg/g of seeds
(Mean of 5 replications)

Treatments	C-1032	C-1327	C-10022	C-87	C-75	Sathi 34–36	Kalarata	NP-49
Control Distilled water	263	237	210	251	201	218	207	257
Sea-water dilutions								
100 ppm	263	204	213	261	203	239	212	264
500 ppm	262	217	210	244	202	234	206	259
1,000 ppm	262	198	215	255	196	232	205	255
5,000 ppm	247	198	201	254	200	223	203	251
10,000 ppm	246	195	188	237	181	214	188	240
15,000 ppm	242	192	189	218	180	207	179	222

TABLE III

Solution uptake of pulses treated with sea-water dilutions; mg/g of seeds
(Mean of 5 replications)

Treatments	DIFFERENT PULSES EXPERIMENTED					
	Bengal gram chitra	Green gram NP-23	Black gram NP-6	Red gram NP-24	Red gram NP-16	Lentile Hybrid-1
Control Distilled water	633	466	524	1177	686	1011
Sea-water dilutions						
100 ppm	951	467	329	1170	1060	1078
500 ppm	978	494	296	1092	1078	1037
1,000 ppm	924	476	202	1102	1078	1014
5,000 ppm	916	405	202	1059	1046	960
10,000 ppm	896	312	198	1040	998	939
15,000 ppm	883	254	192	1014	943	897

TABLE IV

Growth behaviour of Bajra – var. Babapuri irrigated with
sea-water dilutions and its amendments
(Mean of 32 replications)

Treatments	Height of plants (cm)	Number of leaves	Dry matter production			Percentage Nitrogen in seeds (on oven dry basis)
			Shoot (g)	Root (g)	Grain yield (g)	
Control Tapwater + Hoagland solution	215.1	9.6	16.720	1.860	6.774	2.361
Sea-water dilutions						
10,000 ppm	137.0	6.5	8.840	1.345	1.821	1.140
10,000 ppm + Hoagland solution	191.9	9.3	14.020	1.874	7.626	2.282
10,000 ppm + K = 37.46 meq/l and H_3PO_4 = 19.27 meq/l	173.0	8.8	12.792	1.621	5.473	1.524
As above with Ca = 5.60 meq/l	172.0	9.1	15.223	2.012	5.174	2.527
15,000 ppm	139.5	6.6	9.022	1.297	1.422	1.449
L.S.D. at 5%	12.22	1.0	1.463	Not sig-	1.428	0.499
„ at 1%	16.39	1.41	1.964	nificant	1.709	0.671

long way in improving the crop production both in quantity and quality (HEIMANN, 1958). Needless to mention that extensive studies of this nature would be necessary to evaluate the type of fertilizer and the economic use of it in such irrigation under different climatic regions.

SOIL STUDIES

Low clay content, high permeability and less waterholding capacity of sandy soil explains the suitability for highly saline water or sea-water irrigation (NARAYANA et al., 1964). In such soil no salt accumulation can take place since any salt left over during one irrigation will be washed down to the deeper layers by the subsequent irrigations (KURIAN et al., 1964, BOYKO & BOYKO, 1959, BOYKO, 1966). Unlike heavy textured soils, the sandy soil does not come to dispersed state when interrupted by fresh water application or rain fall.

CONCLUSIONS

Sea-water tolerance of crop seeds at germination stage is specific to each of the species and some of the varieties. Inhibition of germination and retardation of plumule and radicle growth in higher concentrations of sea-

TABLE V

Yield characters of wheat – var. Karchia irrigated with sea-water dilutions and their amendments
(Mean of 12 replications)

Treatments	Length of earhead (cm)	Number of spikelet per earhead	Number of grains per earhead	Grain yield mg per plant	1,000 grain weight (g)	Weight of glumes per earhead (mg)
Control						
Tapwater	12.6	25.5	23.6	940	33.228	255
Control						
Tapwater + Hoagland solution	13.5	33.3	32.8	1839	38.213	282
Sea-water dilutions						
10,000 ppm	12.1	26.5	23.3	862	35.966	245
10,000 ppm + Hoagland solution	11.6	28.5	27.6	1668	34.719	210
10,000 ppm + K = 11.2, H_3PO_4 = 7.48 meq/l	12.3	26.8	25.1	1543	35.454	233
10,000 ppm + K = 5.61, H_3PO_4 = 3.74 meq/l	11.1	27.0	26.3	1625	33.719	185
15,000 ppm	12.7	25.8	24.1	749	32.306	266
15,000 ppm + Hoagland solution	11.9	26.0	23.8	1055	34.069	187
15,000 ppm + K = 13.87, H_3PO_4 = 11.22 meq/l	11.7	27.3	26.6	892	28.114	202
15,000 ppm + K = 6.94, H_3PO_4 = 5.60 meq/l	11.6	28.0	27.1	788	31.670	205
L.S.D. at 5%	0.86	4.23	2.74	28.39	1.104	10.22
,, at 1%	1.14	9.85	3.66	37.86	1.472	13.63

water dilutions can be attributed to the lesser imbibition rate of the solutions along with superimposed specific ionic effect.

Established plants in sandy soil, when irrigated with sea-water dilutions results in retardation of growth. Addition of salts of potassium and/or calcium in proper doses to create a nutritively balanced condition will prevent considerably the toxic effect of harmful ions of sea-water.

Nevertheless, the extent and the depth of sandy soil and the plant species adaptation to changing salinity and climate will govern the irrigation of sea-water on sandy soils.

ACKNOWLEDGEMENT

We are thankful to Dr. D. S. DATAR, Director of this Institute for his keen interest and encouragement during the course of investigations.

RÉSUMÉ

Grâce à son littoral de 3,500 milles et à ses vastes bandes sablonneuses cotières (environs 20 millions d'acres), l'Inde est un pays où il est possible d'utiliser l'eau de mer pour la cultivation des plantes.

Les conditions de base pour ce type d'irrigation sont les suivantes:

(1) La presence d'un sol sablonneux suffisamment épais, et d'une nappe phréatique raisonnablement profonde;

(2) L'existence de variétés économiques de plantes tolérant les variations de la teneur en sel de l'eau de mer.

Cet Institut a récemment effectué des recherches portant sur la possibilité d'utiliser l'eau de mer, notamment pour la production de denrées alimentaires. Les études préliminaires à ce sujet visent à établir les espèces pouvant tolérer des variations de salinité, et en même temps à sélectionner les étendues des côtes pour les épreuves sur le terrain.

On a examiné de nombreuses variétés de céréales, de légumineuses, de millet, de graines oléagineuses, ainsi que de graines fourragères de légumineuses et graminacées, afin d'étudier la germination et la croissance initiale de petites plantes. Les graines des différents variétés ont montré des tolérances diverses aux dilutions aqueuses à des époques données.

Il est étonnant de remarquer que nombre de céréales, graines de millet, et plantes fourragères peuvent prendre et pousser avec une réduction plus faible dans la croissance des petites plantes, à des concentrations allant de 10,000 à 20,000 mg/l (T.S.C.) (Teneur totale en sel).

L'emploi initial d'eau douce afin de permettre aux plantes de prendre, suivi par l'irrigation avec eau de mer diluée et des eaux auxquelles on a ajouté des sels de calcium et de potassium, ont montré les nombreux avantages qu'offrent ces méthodes par rapport au traitement avec eau de mer, notamment en vue du développement et du rendement des cultures soit au point de vue qualitatif qu'au point de vue quantitatif.

Sont actuellement en cours des études portant sur le traitement préliminaires des graines en vue d'en augmenter la tolérance à la salinité dans la phase de germination et de croissance, ainsi que d'autres études visant à sélectionner les variétés culturales qui conviennent aux régions caractérisées par des conditions climatiques différentes.

L'étude comparée de l'argile limoneuse du littoral et des terrains sablonneux, portant sur leurs propriétés physico-chimiques, par rapport à l'application de l'eau de mer et de ses dilutions a prouvé que les terrains sablonneux s'adaptent mieux des argiles limoneuses aux irrigations avec eau de mer.

RIASSUNTO

Grazie al suo litorale di 3,500 miglia e alle sue vaste fasce sabbiose costiere (circa 20 milioni di acri), l'India è un paese in cui si prospetta l'utilizzazione dell'acqua di mare per la coltivazione delle piante.

Le condizioni essenziali per tale irrigazione sono:

(1) la presenza di un terreno sabbioso di spessore sufficiente con una falda freatica ragionevolmente profonda;

(2) l'esistenza di specie redditizie di piante in grado di resistere alle oscillazioni del contenuto salino dell'acqua di mare.

Di recente questo istituto ha proceduto a ricerche concernenti la possibilità di utilizzare l'acqua di mare, particolarmente per la produzione di derrate alimentari. Gli studi preliminari a questo riguardo tendono a stabilire le specie che possono tollerare le variazioni di salinità, e a selezionare i tratti costieri adatti agli esperimenti di campo.

Sono state esaminate numerose varietà di cereali, leguminose, miglio, semi oleosi e sementi foraggere di leguminose e graminacee, per studiare la germinazione e la crescita iniziale delle pianticelle. I semi dei vari prodotti hanno rivelato tolleranze diverse alle diluizioni acquose in epoche determinate.

E' sorprendente notare come molti cereali, semi di miglio e di piante da foraggio abbiano mostrato la capacità di germinare ed attecchire con una minore riduzione nella crescita delle pianticelle, a concentrazioni comprese tra i 10,000 e i 20,000 mg/l (T.S.C.).

L'uso iniziale di acqua dolce per far attecchire le piante, e la successiva irrigazione con acqua di mare diluita e con acque corrette con sali di calcio e di potassio, hanno rivelato notevoli vantaggi di simili metodi rispetto a trattamenti con acqua di mare, per quanto riguarda lo sviluppo e la resa delle coltivazioni, sia dal punto di vista qualitativo che da quello quantitativo.

Sono ora in corso degli studi sul trattamento preventivo delle sementi per aumentare la tolleranza al grado di salinità nella fase di germinazione e di crescita iniziale delle pianticelle, e studi miranti a selezionare le specie di colture adatte per le regioni dei differenti ambienti climatici del paese.

Lo studio comparato dell'argilla limosa del litorale e dei terreni sabbiosi concernente le loro proprietà fisico-chimiche in relazione all'applicazione dell' acqua di mare e delle sue diluizioni, ha dimostrato che i terreni sabbiosi sono più adatti delle argille limose alle irrigazioni con acqua di mare.

REFERENCES

BOYKO, H. & E. BOYKO. 1959. Seawater irrigation: A new line of research on bioclimato-
logical plant soil complex. *Int. J. Bioclimatol. Biometeorol.*, 3 (II, B₁): 1–24.
BOYKO, H. 1966. Basic ecological principles of plant growing by irrigation with highly
saline or seawater, p. 131–200. *In* H. BOYKO [ed.] Salinity and Aridity – New Approaches
to Old Problems. Mon. Biol. 16. Dr. W. Junk, The Hague.
HEIMANN, H. 1958. Irrigation with saline water and the ionic environment. Potassium
Symposium Madrid, 173–220.
KURIAN, T., E. R. R. IYENGAR, M. R. NARAYANA & D. S. DATAR. 1964. Effect of seawater
dilutions and its amendments on tobacco. Paper presented at the symposium on
"The problems of Indian Arid Zone" held at Jodhpur, November 1964.
NARAYANA, M. R., V. C. MEHTA & D. S. DATAR. 1964. Effect of seawater and its dilutions on
some soil characteristics. Paper presented at the symposium on "The Problems of
Indian Arid Zone" held at Jodhpur, November 1964.

Authors' address: Central Salt and Marine Chemicals Research Institute, Bhavnagar,
India.

Nuclear Biology and Irrigation with Sea-water

by

JOSEPH STERNBERG

Perhaps the most distinctive characteristic of our modern scientific world is the replacement of the ivory tower of the solitary thinker by a team of investigators from an array of varied disciplines. The amount of information acquired during the past decades was so enormous that the dialogue between the scientists became very difficult, for lack of a common denominator. This is why we are witnessing the emergence of new liaison disciplines whose role is to find the unifying link between the too numerous branches of sciences, either through the agency of a common concept or of a common technique; it might be called a new brand of philosophy in our technocratic world. Nuclear science may enter this concept, for it touches almost every compartment of human activity, from archaeology to law and mental health, from gigantic engineering projects such as melting the polar icecaps, to explorations into the microcosmos, or search for microbial life in other planets. It was inevitable that the nuclear scientist should be attracted by this project, whose goal is the production of new food on land with the aid of sea-water.

This conference has already benefitted from the impact of nuclear energy, but only as a source of power in the application of desalination on a large scale. This is the realm of the engineers, and I am sure that significant advances will be obtained from this collaboration.

The aspect I would like to discuss here is different, namely, the biological and medical implications of large scale application of irrigation with sea-water.

For the biologist and especially for the physician, irrigation with sea-water is as revolutionary an advance as the discoveries of vitamins, hormones, or antibiotics: it will save millions of human beings from dying of malnutrition and the procession of miseries bound to hunger. However, this concept has a long way to go from its present state before practical results are achieved. It may have unexpected drawbacks that can be avoided if a careful study is made at every step from the engineering phase to the final determination of the nutritional value of edible plants grown in regions irrigated with salt water. This is comparable to another project that took two years of research and development before the first tangible results were seen, i.e., the preservation of food by irradiation. Born out of military need to preserve food in a warm climate, this project consisted in irradiating fresh or prepared food with large amounts of γ or β radiations in order to prevent microbial decay or premature sprouting. Started under auspicious conditions, the project succeeded rapidly in inhibiting sprouting in potatoes and maintenance for as long as two years at normal cellar temperatures. Many aspects had to be studied prior to its large-scale acceptance: package problems, i.e., replacement of the metallic can by a clear plastic package, cost of the procedure and availability at the large agricultural centers, and consumer's acceptance, i.e., attractive appearance and taste. The most important factor was, however, the wholesomeness of the irradiated food, an aspect that at one point endangered the entire project. Indeed, the animals fed with the irradiated food over a prolonged period developed severe nervous troubles, anemia and blindness; fortunately, it was learned later that these symptoms were due to a fault in the experimental design rather than to the irradiated food itself; nevertheless, the psychological impact of this drawback is still being felt. However, the excellent results obtained in the U.S. Army Laboratories in Natick, Mass., were crowned in 1963 by the F.D.A.'s approval for consumption of irradiated bacon, broilers and potatoes.

I have dwelt on this point long enough to stress the extreme caution needed before introducing a procedure affecting the food of a large segment of the population of the world. Some of the points presented here are potential lines of research that merit a high priority if human consumption is ultimately being contemplated. Even if this seems a remote goal at present, research should be carried out parallel with the engineering phases.

The radioactive atom can be used in two manners:

(1) as a carrier of radiation, whose effect on the living cell can be used for a definite purpose, and where the chemical nature of the atom is immaterial.

(2) as a means of identifying the atom through its radioactive tag; here, the chemical nature of the atom is important, as opposed to the former case where the physical nature of the radiation carried by the atom is essential.

The first aspect is being used in medicine for the treatment of various disorders, and on a large scale in agriculture to obtain mutants of usable plants by irradiation. This is another promising aspect for better food production, and has already yielded results in various countries of the world: I will only mention the significant increase in the grain yield of irradiated rice obtained by GOPAL-AYENGAR and his collaborators at the Atomic Energy

Research Establishment in Trombay, India. Production of mutants thriving at higher degrees of salination could constitute a problem to be studied with radiations, but this is a project for the future when the other phases of sea-water irrigation are well under way. For the moment, it seems that the tracer applications or radioactive atoms will be more useful in determining the parameters of research in this vast new field of activity.

The tracers can be introduced into a given system or they can be produced by temporary activation of the constituents of the compound examined. This latter procedure is called *Activation analysis* and will be discussed in greater detail.

ACTIVATION ANALYSIS IN THE CHEMICAL DETERMINATION OF TRACE ELEMENTS OF SEA-WATER

The procedure of activation analysis is a relatively simple and extremely sensitive method of identifying and determining very small elements in a sample; it consists in irradiation with neutrons or protons, which renders the components radioactive and permits their identification by measurement of the radiation. Its sensitivity is greater than that of any chemical or physical method, and it is already used in biology, medicine and even forensic medicine (determination of arsenic in hair, etc.).

Our knowledge of the oligo-elements of sea-water is either incomplete or in many cases subject to criticism, because of the lack of precision of the analytical procedure. Table I presents the list of trace elements detected in sea-water, with the year of their respective determination; obviously, some of these elements must be determined anew by more modern procedures.

An inventory of the chemical composition and especially of trace elements in sea-water becomes mandatory in the case of large scale applications of irrigation with sea-water. Indeed, it is well known that living organisms have the ability to concentrate some elements to a considerable extent: Table II gives the concentration factors for some elements present in sea-water, for the algae, the invertebrata and the vertebrata.

This phenomenon becomes very important in the case of irrigation with sea-water: indeed, bio-concentration is not limited to the living cells of sea-water, but is a universal characteristic of living matter. In some cases, this can be followed by concentrations of some trace elements to toxic levels, as is the case of selenium, which is present as a soluble salt in regions with alkaline soils in the USA (Kansas) as well as in other parts of the world. Many species of pasture plants concentrate this elements 3,000–10,000 times when growing on seleniferous soils (*Astragalus, Stanleya, Oonopsis, Xylorriza*): the cattle grazing on this pasture are stricken with a severe disorder called *alkali disease*, which is caused by the replacement of organic sulfur-containing vital compounds (proteins, enzymes) with inactive or toxic seleniated compounds. The selenium content of sea-water is far lower than that

TABLE I

Trace elements in sea-water

Element	ppb (μg/l)	Year of determination	Bio-concentration
Aluminium	5–10	1953	?
Antimony	0.2	1940	?
Arsenic	1.6–5.0	1952 activation analysis	++ (algae, medusae)
Barium	30–90	1936	?
Bismuth	0.2	1940	?
Cadmium	0.03–0.06	1952	− toxic (Zn antagonist)
Caesium	2	1938	see fallout paragraph
Cerium	0.4	1937	?
Chromium	1–2.5	1952	?
Cobalt	0.1	1940	+++ (vitamin B 12)
Copper	25–	1952	100–600
Boron		1938	++
Fluorine	1,4000	1933	+
Gallium	0.5	1940	?
Gold	0.05–0.5	1955	?
Iodine	50	1935	+++
Iron	1–50	1947	+++
Lanthanum	0.3	1937	?
Mercury	0.03	1937	? (toxic)
Manganese	1–10	1945	++
Molybdenum	12–16	1952	+
Nickel	1.4–2.6	1956	?
Radium	$0.07–0.58 \times 10^{-7}$	1952	?
Rubidium	120	1957 *	=
Scandium	0.04	1937	?
Selenium	4–6	1953	+++
Silver	0.1–0.3	1940	?
Thorium	0.01–0.001	1949	=
Tin	3–5	1952	?
Titanium	1–9	1952	?
Tungsten	0.09–0.10	1956	?
Vanadium	2.6–4.6	1956	+
Yttrium	0.3	1937	?
Zinc	9–21	1952	+++

The * signifies determination by activation analysis.

of the soil (4 μg/l), but all of it is in soluble compounds; a larger proportion of selenium salts coprecipitates with iron or manganese and falls to the bottom of the sea. It is conceivable that the selenium concentration in some of the plants grown in the regions irrigated with sea-water might reach dangerous levels if the amount of used water is enough to allow this bioconcentration.

Another example that will be studied in greater detail in the next pages is that of cesium and rubidium. Because of the considerable importance taken

TABLE II

Concentration factor of some elements by members of the marine biosphere (KRUMHOLZ et al., 1957)

Element	ppm (sea-water)	Algae	Invertebrata	Vertebrata
Na	10,000	1	0–0.5	0.07–1.0
S	900	10	5	2
K	380	25	0–10	5–20
Ca	400	10	10–1000	1–200
Sr	70	20	10–1,000	1–200
P	0.07	10,000	10,000	40,000–2,000,000
I	0.05	10,000	50–100	10–100
Fe	0.01	20,000	10,000–100,000	1,000–5,000
Zn	0.01	100	1,000–5,000	1,000–30,000
Cu	0.003	100	5,000	1,000
V	0.001	1,000	100	20

on by radioactive cesium as a component of fallout and radioactive wastes, it became mandatory to study the biological cycle of the correspondent stable elements in sea-water. SMALES & SALMON (1955) used activation analysis to determine cesium and rubidium in sea-water and in some algae: these authors found that algae contained relatively less rubidium (1/10–1/20 per equal volume) and cesium (1/5) than did equivalent volumes of sea-water.

These few examples show the importance of accurate determination of the composition of sea-water, in the case of irrigation with sea-water and the ensuing use of the cultivated plants for animal feed.

SOIL STUDIES USING RADIOACTIVE TRACERS

Tracers were used long before the discovery of radioactive elements in studies on the microcirculation of water. SCHOELLER (1959) made an excellent review of the problem of radioactive tracers, chiefly in the study of subterranean waters. The cardinal aspect in irrigation with sea-water is the verification of H. BOYKO's hypothesis of "subterranean dew," upon which is based cultivation on sand and sandy loam. This new concept assumes that salt covers each grain of sand with a thin film, thus leaving a vapor chamber in the microscopic cavities between the grains; the vapor is then condensed into water, supplying the feeder roots of the plant with fresh water. BOYKO even calculated the film of salt on the surface of sand particles, about 1 mμ thickness, and concluded that such a continuous film is practically impossible (1964, 1966).

These findings agree with our knowledge of the ion exchanger properties of sandy loams and sand, as well as clay; working with Cesium 137, WIJK & BRAAMS (1960) found that a slightly alkaline clay absorbed about 50%

of the element on particles less than 2 μ diameter, i.e., only 2% of the total soil mass. SPITSYN et al. (1958) showed that sandy loams and clays absorbed 90% of a solution of Cs 137, while sand absorbed 50% of the same solution. Since cesium is a relative of the main salty components of sea-water, Na and K, similar experiments with Na 22 and K 42 are essential to furnish evidence of the BOYKO-theory. Comparable investigations are being carried on in our laboratory using artificial sea-water to which tracer amounts of Na 22 have been added; the solution is passed through a column of sand and sandy loam, and the rate of absorption is measured. In a further experiment, plants will be grown and the amount of Na 22 taken up by the plant will be determined. If no significant amount of sodium is detected in the plant, then the hypothesis of subterranean dew receives significant experimental evidence. The experiment can be completed by simultaneous addition of tritiated water to the labeled sea-water and studies in function of time of the water uptake of the plant.

METABOLIC STUDIES ON PLANTS GROWN ON SANDS IRRIGATED WITH SEA-WATER

This area of investigation is essential if the species grown with saline water or sea-water are to be consumed by animals or humans. Thus far, of the about 200 species grown by BOYKO & BOYKO (1959) in the desert garden of Eilath, some are used for fodder (e.g. Acacia) and others for human consumption (e.g. mulberry and pomegranate); in their experiment with sea-water the BOYKOS (1964, 1966) used even sugarbeet and barley, where the ratio edible/inedible part is far larger than in the former species. It is too early to project lines of interest in this vast field of investigation; these will be decided when the species are chosen for large scale culture and eventual animal or human consumption. However, as a general indication, it seems that studies on the concentration of some undesirable components in the edible part should be given priority; the importance of selenium was shown before, and it should be remembered that some elements could exert some antagonistic effects on the functions requiring a metallic activator as is the case of molybdenum, a copper antagonist and a prosthetic part of the enzyme xanthine-oxidase, or cadmium, the antagonist of zinc, which exerts a toxic effect on the renal glomeruli.

PUBLIC HEALTH PROBLEMS CONNECTED WITH SEA-WATER IRRIGATION

Only the aspects incidental to the use of radiation or radioactive isotopes will be considered here. The first aspect to be considered is the possibility of transmission of bacterial or viral diseases to plants irrigated with sea-water. This is unlikely, since sea-water contains few microbial or viral species that could in turn spread disease to the land environment.

The second aspect is perhaps the most important point to be considered when large scale irrigation becomes feasible and when plants grown with sea-water are destined for animal or human consumption. In the previous paragraphs, the possible bio-concentration of some stable oligo elements was discussed in detail, emphasis being placed on a toxic level for some of them. The example of Cesium with regard to the activation analysis procedure was cited with the purpose of stressing the importance of bio-concentration of radioactive elements present in sea-water. At present, the concentration of fission products in the ocean is very small and does not present any cause for concern, because of the enormous dilution factor.

The radioactive elements can be released into the sea and ocean from the following sources:

(1) *fallout* from atmospheric nuclear weapon tests, either regional or global.

(2) *radioactive wastes* from industrial uses of nuclear energy: thermo-nuclear power stations, research reactors, etc.

(3) *accidental release* of radioactive material from nuclear-powered engines used for propulsion of ships.

The recent estimates of the USA Atomic Energy Commission give a total of 17.9 Mc (Megacuries) (equivalent to tons of radium per Mc) Strontium 90 injected into the stratosphere up to January 1963, date of the last moratorium on nuclear-weapon testing. Of this amount, 6.2 Mc remained in the stratosphere and continue to fall regularly on the global surface, and approximately the same amount has already been deposited; taking this last figure as the total deposition, 3.6 Mc of Sr 90 were introduced into the ocean, or 17.5 mc/km^2 for the Northern Hemisphere and 5 mc/km^2 for the Southern Hemisphere. Assuming a thickness for the mixed water of 100 meters, it follows that the concentration of Sr 90 averages 0.17 $\mu\mu$c/l for the Northern Hemisphere and 0.05 $\mu\mu$c/l for the Southern Hemisphere. The Japanese authors MIYAKE & SARUHASHI (1960) estimate that about 20% of the total sea surface of the Northern Hemisphere (3 × 10^7 km^2) is contaminated with fallout and contains an average of 1.5–2 $\mu\mu$c/l. MIYAKE & SIGIURA (1955) analyzed sea-water collected 300 miles Northeast of Bikini 7 months after the nuclear explosions were carried out on that island. The results give an over-all radioactivity of 715 $\mu\mu$c/l, at least 350 times higher than the average level; the main radioactive constituents were the following.

It is interesting to note that Cerium 144, a rare earth produced by fission of Uranium 235, is found in a higher proportion than Strontium 90; this was interpreted by the Japanese authors as being an incorporation of the fallout component by plankton present in surface waters in the Pacific. This fact throws a different light on the problem of radiocontamination with sea-water, since it follows that irrigation water taken from the surface is bound to have a different pattern of concentration of radioactive elements than that taken from the depths of the ocean. The fission of Uranium 235 yields a

TABLE III

Nuclide	Activity ($\mu\mu$c/l)	% of total
Y 91	260	37%
Ce 144, Ce 141	400	5%
Ru 106, Ru 103, Rh 106	26	3.8%
Zr 95	15	2.2%
Sr 90, Sr 89	14	2.0%

TABLE IV

Fission products of Uranium 235 (slow neutrons) from 1 kg

At 10 seconds after fission		At 20 days		At one year	
Total activity		9,800,000 curies		31,000 curies	
Nuclide	% of total	Nuclide	% of total	Nuclide	% of total
Rubidium		Lanthanum 140	13.9%	Cerium 144–	
Cesium	20.0%	Barium 140	12.0%	Praseodym 144	52.8%
		Praseodym 143	12.0%	Niobium 95	14.7%
Bromine		Cerium 141	9.7%	Zirconium 95	7.2%
Iodine	27.7%	Xenon 133	6.3%	Promethium 147	5.7%
		Zirconium 95	5.9%	Yttrium 91	3.8%
Xenon		Yttrium 91	5.6%	Strontium 89	2.7%
Krypton	32.4%	Iodine 131	5.6%	Ruthenium 106–	
		Strontium 89	5.0%	Rhodium 106	4.9%
Molybdenum		Neodym 147	5.0%	Strontium 90–	
Tellurium	5.2%	Ruthenium 103	4.4%	Yttrium 90	3.7%
		Niobium 95	4.2%	Cesium 137–	
Antimony	4.8%	Cerium 144	2.3%	Barium 137	2.9%
		Praseodym 144	2.6%	Ruthenium 103	0.8%
Yttrium		Molybdenum 90	1.3%	Rhodium 103	0.8%
Lanthanum					
Cerium	4.5%				
Strontium					
Barium	2.6%				

wide range of elements of different chemical nature, and of various levels of radioactivity, as is shown in table IV.

The proportion of rare earths and non-metabolites is considerable, but the concentration of these elements by plankton changes the perspective entirely. To these elements, one must add the radioactive atoms produced by the slow neutrons released during fission, which could be practically any element colliding with the neutrons. In fishes caught in the North Pacific in 1954, Japanese workers detected 3.8–7.7 mμc/g tissue of Zn 65, Fe 55, Co 60 and Mn 54.

Thus, the diversity of fission elements present in ocean water from the fallout of atmospheric weapon-testing is increased by the preferential uptake of some elements by plankton. Since this concerns microscopic organisms, one must take into account a possible radiocontamination of the irrigated plants with water drawn from the surface of the ocean. Also, the possibility of vertical contamination of the deeper regions of the sea must be considered, since the dead organisms will fall to the bottom and increase the radioactivity level of the water.

With the last moratorium of nuclear weapons testing and with the hope of a long-lasting settlement of this problem, the possibility of radiocontamination of the ocean water by fallout decreases steadily and will be reduced to insignificant values in a decade, when the projects of irrigation might become a reality. But in the meantime, the industrial application of nuclear energy makes giant strides, especially in the field of power production and propulsion; a conservative estimate is that towards the end of the century about 6% of the total power production of the world will be generated by nuclear energy; by 1970, approximately 263×10^9 kwh/year will be installed in the world, or about 1.3% of the total world consumption. This is being produced by about 80 power reactors, one quarter of them in the U.S.A. and almost the same proportion in Great Britain.

Propulsion by nuclear energy is based on the same power reactor as that used for generating electricity, with the difference that the former is mobile, and therefore, subjected to more hazards than the latter. For the time being, nuclear propulsion is mainly utilized for military purposes: about 33 submarines of the U.S.A. fleet are powered by nuclear energy, and more than double this figure is foreseen by 1970. Surface ships range from huge aircraft carriers to icebreakers, for military and civilian use. The advantage of nuclear propulsion is obvious for long range carriers; for instance, the uranium load of 312 kg U 235 contained in about 8,000 kg natural uranium in the core of the reactor of the commercial ship "Savannah" would enable this vessel to circle the world 13 times without refueling,

The future outlook is very optimistic, but for the danger of radio-contamination of the environment; indeed, the reactors are liable to have the same accidents as any other human enterprises, the mobile ones far more so than the stable ones. The Windscale accident in 1957 was followed by large scale contamination of the environment, with a diffusion of radioactive iodine over an area of 200 square miles in the midwest coast of England; an equivalent of 20 kg radium was generated within 24 hours of the accident; 2/3 of this amount was deposited on the ground within an area of 100 miles to the south, the remaining third was carried across the Channel by the wind and traveled as far as Poland and Norway. The radioactivity of the cloud diminished considerably after crossing of the Channel, from 1,030 $\mu\mu$c/day over the Mersey to 1.6 $\mu\mu$c in Norway; probably, a large proportion fell into the sea-water.

The tragic sinking of the nuclear powered submarine Thresher is still in everybody's memory; what must be stressed is the impending possibility of nuclear accidents in the oceans, with all the implications of large scale radiocontamination of the environment. Statistics have shown that in spite of all the precautions taken to avoid marine accidents, collisions involving nuclear-powered ships will happen at the rate of about 1–2 per year. A nuclear-powered ship the size of the Thresher could have, after a long period of function, an amount of radioactivity of about 360 Mc (equivalent to 360 tons of radium), which might be released into a region and thereafter dilute uniformly with the water; the level of radioactivity is then a function of the volume in which the wastes are diluted, barring any bio-concentration; taking 100 $\mu\mu$c/l as the maximum permissible concentration of radioactive material, it appears that the amount of radioactivity diluted in the entire Mediterranean and Black Sea will result in 55 $\mu\mu$c/l, but 10,200 $\mu\mu$c/l in the Baltic Sea and 2,240,000 $\mu\mu$c/l in San Francisco Bay. In essence, the problem raised by sea-water irrigation is in a way the opposite of that posed by treatment of radioactive wastes; while the latter aims to dilute the radioactive element until a permissible concentration is obtained, the former reverts the process by favoring bio-concentration through irrigation. It remains to see what amounts of water are employed and what is the radioecological importance of the diluted fission products. The best known in this field are Strontium 90 and Cesium 137; the latter has been frequently mentioned in previous paragraphs, and should be enlarged upon here.

Cesium 137 is an element generated by the fission of Uranium 235, with a yield of 6.3%; this means that 1 kg Uranium 235 will produce 63 g Cesium 137, which will predominate in the fission products because of its long half-life (t/2 = 30 years); its strong γ emission (0.662 Mew) together with its metabolic relationship with potassium make this radiocontaminant one of the most dangerous for humans.

As seen in the former tables, the corresponding stable cesium is a trace constituent of every living matter: plants contain cesium in concentrations between 0.1 ppm (algae) and 89 ppm (Verbena); invertebrata contain more than vertebrata (138 ppm. vs. 32 ppm). The concentration factor for the radioactive Cs 137 ranges from 1,000–25,000 for plankton to 400–1,000 for submerged vascular plants and far less for floating or emergent plants. In fishes the isotope reaches a concentration factor of 3,000–10,000; transfer to the food web seems to be very rapid: it takes two hours for the isotope to be fixed by the algae, a fact that might be of importance in the decontamination of sea-water by a layer of algae prior to irrigation. Of considerable importance are the findings of the Russian workers GULIAKIN and YUDINT-SEVA (1957, 1958) that cereals grown in a water-medium containing the same amount of isotope as in clayey sand or loam accumulate 1,500 times more Cs 137 than do cereals grown on clay; also, PENDLETON & UHLER (1960) noted that plants grown in flooded conditions concentrated 450 times more Cs 137 than the same plants grown in normal soil.

A very recent paper published by the members of the Biology Service of

Euratom illustrates this point; the Ispra reactor is located on the eastern shore of Lake Maggiore, whose only outlet is a tributary of the Po river. The waters serve to cool the reactor as well as to irrigate the lowland rice cultures on a region larger than 1,000 squares miles. During the years 1963 and 1964, the level of Cs 137, Sr 90, Ru 106 and Ce 144 increased in rice until harvest; in grains, the hull contained most of the Ce 144 and Sr 90, while Cs 137 was chiefly in the endosperm. This latter radiocontaminant appeared to be readily available to rice from flooded soils. All radiocontaminants had higher levels in the perennial meadows bordering the flooded region. The concentration of the radiocontaminants in the animals of the same region reflected their habitat and dietary habits. Thus, amongst the insects, the adult dragonfly had a significantly high level of Cerium 144, while another aquatic insect (*Ranatra*) exhibited high levels of Manganese 54, reflecting the corresponding high level of stable manganese. Tadpoles had 2–5 times more radiocontaminants than adult frogs, and molluscs higher levels than fishes; again, high levels of Manganese 54 were detected in molluscs, for the same reasons as listed before.

This investigation revealed that the present levels found in the eco-system are well below the dangerous limit, but also that an accidental release of high level wastes from the neighbouring reactor will be followed by a large scale contamination of the entire food web depending on the water system in a fertile agricultural area.

At present, the amount of Cs 137 in vegetable foods is well below the permissible level: a kilogram of food contains between 0.050 and 0.110 mμc Cs 137, the highest level being found in the cereals. However, there arises the problem of irradiation of the total body of man if larger quantities of Cs 137 are ingested. This is already a problem in the Northern regions, where the peculiar ecological situation brought the following complication: the large nuclear explosions set off by the Soviet Union in 1961 produced a fallout pattern with a maximum intensity in the Arctic regions. The fallout elements were washed out by the rain and snow and absorbed rapidly by the lichens, plants that obtain almost all their water supply from the atmosphere. The caribous and reindeers, in turn grazed on the contaminated plants and their body burden of Cs 137 increased considerably; finally, the Eskimos, Laps and Indians, whose staple diet was reindeer and caribou meat, have at present a body burden 30–50 times higher than those living under the same conditions but whose food habits are different. This will be dealt with in detail elsewhere, but serves to show that in the complex situation of modern technological advances, every technical or biological development is followed by a chain reaction of effects on the environment, sometimes unpredictable, as was the case in the epidemics of congenital malformations produced by the apparently harmless sedative thalidomide; it is the role of the scientist to encourage the application of global scale projects that will be of help in feeding the population of the future, but it also is his duty to foresee every possible danger and to obviate it.

J. STERNBERG

CONCLUSION

The radioactive atom can be utilized in many ways in studies on the
problems arising from the project of irrigation with sea-water; it can trace
the fate of the macro- or microconstituents of sea-water into the soil, the
plants and the animals that will eventually feed on the plants grown with
salt water. It will show whether some constituents present in minute
quantities in salt water reach toxic levels in the plants or whether the
nutritional value of the plants grown in a salt medium remains similar to
that of the plants grown in soil and irrigated with fresh water. Finally, the
presence of radioactive elements released by the industrial use of nuclear
energy is a factor that must be taken into account, and the possible bio-
concentration of these elements must be analyzed and prevented.

REFERENCES

The references mentioned in the text, as well as the pertinent material discussed in this
paper, are found in the following volumes:

Arid Zone Programme, UNESCO. 1954. *Reviews of Research on Problems of Utilization of Saline
    Water* Vol. 4 (UNESCO, Paris).
Arid Zone Program, UNESCO. 1959. *Recent Progress in Hydrology of Arid Zone* Vol. 12 (UNES-
    CO, Paris).
BOYKO, HUGO. 1966. Basic ecological principles of plantgrowing by irrigation with highly
    saline or sea water. p. 131–200. *In* H. BOYKO [Ed.] Salinity and Aridity – New Approaches
    to Old Problems. Dr. W. Junk, Publishers, The Hague.
BOYKO, HUGO & ELISABETH BOYKO. 1964. Principles and Experiments regarding direct
    irrigation with highly saline and seawater without desalination. *Trans. N.Y. Acad. Sci.*
    Ser. II, 26, suppl. to No. 8, p. 1067–1102.
BOYKO, HUGO & ELISABETH BOYKO. 1959. Seawater irrigation – a new line of research on a
    bioclimatic Plant-Soil-Complex. *Int. J. Biochim. Biometeor.* 3, (II), Sec. B₁: 1–24.
BOYKO, HUGO & ELISABETH BOYKO. 1966. Experiments of plantgrowing under irrigation
    with saline waters from 2000 mg/l T.D.S. up to seawater of oceanic concentration,
    without desalination. p. 214–284. *In* H. BOYKO [Ed.] Salinity and Aridity – New Ap-
    proaches to Old Problems. Dr. W. Junk, Publishers, The Hague.
BROECKER, W. 1963. Radioisotopes and Large-Scale Oceanic Mixing, in M. N. HILL (Ed.),
    The Sea. Interscience Publishers, New York, pp. 88–108.
HARVEY, H. W. 1957. The Chemistry and Fertility of Sea-Waters. University Press, Cam-
    bridge.
MIYAKE, Y. 1963. Artificial Radioactivity in the Sea, in M. N. HILL (Ed.), The Sea. Inter-
    science Publishers, New York, pp. 78–87.
SCHULTZ, V. & A. W. KLEMENT JR. 1963. Radioecology; *Proceedings of the First National Sym-
    posium on Radioecology Colorado, 1961.* Reinhold Company, New York.
United Nations Report of the Scientific Committee on the Effects of Ionizing Radiations
    1962; General Assembly, 17th Session, Document 16-A/5216.

Author's address: Joseph Sternberg, M.D., Professor of Physiology and Nuclear Medicine, Faculty of Medicine, University of Montreal, Montreal, Canada.

Problems of Irrigation with Brackish Water in Italy

by

Luigi Cavazza

INTRODUCTION

In regions suffering, if only seasonally, from an insufficient hydric regime, agriculturists have always endeavoured to find additional water for irrigation. In Italy, as in all other countries with similar conditions, this interest in irrigation has continuously increased after the second world war for the coincidence of various causes. The main causes were the generally higher standard of living, the need for more efficient human work, the availability of larger capital, the much more perfected techniques in agriculture and irrigation, the huge market expansion and organization of agricultural production, etc.

Under the drier climate of Southern Italy irrigation is more needed than in the North, but it is also more difficult and expensive there, for the rivers are usually shorter, poorer and more typically torrential. In some regions there is even a complete lack of true water streams, particularly where the geological structure of the soil is of carstic type, as for instance in Puglia. Under these conditions, it was tried already in old times to use water from brackish layers and recently, even to utilize it with a higher salinity than accepted in the past. One tried also to draw it from deeper layers, and to use it also for plants once considered traditionally as crops for dry farming only.

In Italy, brackish water for irrigation is mostly drawn from wells. In the last 15 years the expansion of this kind of irrigation in certain Italian coastal zones was greatly assisted by Government grants to land transformation works; but in this field, namely the use of saline irrigation, the initiative was always left to private promotors. For this reason, because brackish underground waters are usually not utilized by communities for crop production but autonomously by small [1] or medium sized farms with very intensive

1. The average size of 21 farms studied by the "Ente Irrigazione di Puglia" along the Bari coastline was 2.04 ha.

cultivation methods and an above average good soil, there prevails in these zones a perfected and often very old empirical irrigation tradition. We are, however, confronted there with a general lack of research, of exact technical data (especially from the agronomical viewpoint), and of legal norms.

Irrigation with brackish water is practiced in Italy mostly in the coastal zones, on a discontinuous area and with a frequency decreasing as we progress to the South and less freshwater surface streams are available. The most interested regions are Puglia and Sicily. The width of the coastal zone is very variable according to the orographic and geologic nature of the places: generally only a couple of kilometers from the sea-shore, more seldom up to 10 or even 20 km (for instance in the "Salento"). Well frequency is usually in an inverse ratio to the distance from the sea-shore or, more exact, to the depth of the water layer, with the exception of a narrow littoral band often of less than 100 m width, sometimes, however, up to 2–3 km which is not intensively exploited. The frequency of the wells along the coast is also highly variable; on the Bari coastline, for instance, there are often up to 25 wells per km length.

The surface irrigated in Italy with brackish water is not known, nor the number of wells and the extension of the brackish layers. This surface may be roughly estimated to a few tens of thousand hectares, i.e. less than 1% of the total irrigated Italian land, and thus less than 5% of that irrigated from normal wells of fountain-heads.

More exact data, however, are available for certain zones, first of all for the Puglia, where irrigation with brackish water always was and still is most essential and where its problems have been most earnestly studied. According to the latest results of research conducted during the last twenty years by the Institute for Irrigation in Puglia and Lucania, most of Puglia lands hold, at most varying and often remarkable depths, brackish layers sometimes even one on top of others (Tavoliere di Foggia). They are nearly always, and especially the deepest layer, exposed to the danger of salinity. It is estimated that this concerns 63 of the approximately 68 m³/s constituting the total outflow of Puglia water layers. Of these waters, wholly fed by the regional meteoric precipitations, some 26% emerge, mostly near the sea coast, in surface springs, and of these 17.5 m³/s one quarter is already exploited for irrigation, while another quarter would still be available. The most remarkable of these springs are those of the Idume, of the Chidro, of the Tara (giving together 4 m³/s approximately) already totally exploited; further the many small springs along the coastline (the so-called "Christ waters"), and the sundry undersea springs (e.g. the conspicuous "Ring of S. Cataldo" in the Taranto Gulf). Approximately 16% of the total outflow of the Puglia layers is used for irrigation by means of wells (the water of some 2,500 of these is more or less brackish), while the remaining 58% flows at present into the sea. It is calculated that of this only additional 14 m³/s can be used, owing to the necessary precautions against excessive further salination. Thus in Puglia up to 50% of the total under-ground outflow is considered to be usable.

In Sicily, hydrogeological research is carried out by the Institute for Agricultural Reform, and water quality is investigated by the Catania Institute of Agricultural Chemistry. 4,098 springs have been counted by the Government Hydrographic Service, many of them of carstic type, with a total estimated yield of 33 m³/s. More than half of the irrigated Sicilian lands are using brackish wells, and the waters of some surface streams (the Caltagirone, the Dittaino, the Stretto torrent, the Salso, etc.) are brackish also.

The practical sources of brackish water for irrigation in Italy may be classified in four different groups, namely:

(a) carstic type layers,
(b) phreatic or artesian layers somehow connected with the sea,
(c) waters beneath the dunes,
(d) surface water streams variously salinated.

The use of layers whose salinity is of pure terrestrial origin is negligible in Italy.

We shall deal briefly with the typical problems of each of the above groups.

THE EXPLOITATION OF CARSTIC LAYERS AND THE PROBLEM OF THEIR SALINATION

The mechanism of the exploitation of carstic layers has been the best studied in Italy. The hydrogeological aspects and the connected practical problems, already outlined by DABELL (1955), have recently been thoroughly investi-gated in an ample set of studies by ZORZI & REINA (1955–1964) at the Institute for Irrigation in Puglia and Lucania; the many and detailed results of these studies remarkably contribute to the quantitative knowledge of the Puglia underground hydrology.

The carstic layers consist of waters contained in the net of cracks (average porosity approximately 20%) characteristic of the compact stratified calca-reous formations of the Mesozoic (their thickness ascertained in Puglia is up to 4,500 m), or of waters imbueing the calcareous tuff of a later period that often accompanies them along the coast. This crack system normally communicates with the sea; where, as it sometimes happens, especially in the Salento, the mesozoic limestone is covered near the coast by less per-meable or impermeable formations (like clays), the linkage of the carstic layer with the sea takes place below these formations. The structural singularities of this hydrological system explain its characteristic function-ing, the many problems connected with its exploitation, and the difficul-ties in studying it.

The statics and the dynamics of these layers, which have already been studied by many authors, may be briefly schematized as follows, on the basis of the GHYBEN-HERTZBERG theory, by now widely accepted.

A certain part of the meteoric waters (e.g. approx. 2/3 on most of the Puglia basins, equivalent to approx 4l/s × km² of collecting basin) percolates through the frequently very shallow soil that covers the mesozoic limestone. After that it drops more or less vertically in the crack system, at an average speed measured in Puglia of approx 3 m/d (ZORZI & REINA), until it reaches the underlying salt waters of marine origin and is communicating with the sea. The percolated fresh water, being lighter (specific weight measured in Puglia 1.0028) floats on the salt one (specific weight of Adriatic water 1.0280). Under conditions of static balance and absolutely no mixing, it stays higher than the average sea level in a relative quantity theoretically given by h = d/(D–d), where d is the specific weight of fresh water and D the specific weight of sea-water. Therefore, under conditions in Puglia, the freshwater layer keeps above sea level for approx 1/40 of its own thickness.

In fact the boundary between salt water and fresh water is not well defined. ZORZI's research has shown that for approx 4/5 of the theoretical thickness of the freshwater layer, starting from the top, the salinity decreases very little, while deeper down it suddenly begins to increase fast until it reaches sea-water concentration. The presence of this gradual transition zone alters in the freshwater layer somehow the theoretical relationship between the portion above and the one below sea level, which is no longer 1/40 but drops to approx 1/30. This agrees fairly well with former observations in sundry mediterranean countries (DABELL, 1955).

The higher hydrostatic level of the carstic layer in respect to the sea causes a lateral water flux towards the coast, valuated in the Bari littoral at approx. 300 l/s per km of coast. This flux, as against the vertical one, occurs wholly within a water saturated medium and at extremely variable velocities according to the geometric characteristics of the crack system and to the water volume to be drained; in the Salento, for instance, ZORZI & REINA have measured a velocity of approximately 20 m/d. In fact, the declivity of the carstic layer is in Puglia from 0.3 to 0.5‰ (in the Salento) and from 1 to 2‰ (in the Bari province).

The regime of this layer is influenced by that of the rains. These, concentrated in Puglia for about 70% in the autumn-winter season, bring an enrichment of the layer, while from march-april until to september the layer flows undisturbed toward the sea.

If we draw up fresh water from this layer by a well, thus lowering its level, the floating balance of this water on the salty one is broken, and the salty water will tend to re-ascend until the relationship between the portions of fresh water over and under the sea level is brought again to approximately 1/30; in other words the salty water tends to re-ascend approximately 30 times the lowering of the level in the well. It is clear that, the deeper the well reaches under sea level, the more easily will it be invaded from below by the cone of salt water and the more salty its water will become. The maximum level depression, beyond which there will be salt water invasion, is given, according to ZORZI, by S–P × 0.03125, where S is the distance of the undisturbed layer above the sea level, and P is the distance of the bottom of the well from the same sea level in the cases when the well descends below that level. ZORZI, however, advises to limit such depression to a half, i.e. not to exceed the safety depression of the layer. If we consider that in cylindric wells the yield increases with the depth but in the same ratio increases the danger of salination, it is clear that there exists an optimum depth permitting a maximum yield of water not exceedingly salty.

This movement of liquid masses, however, takes the more time the higher the resistance is met by the liquid flux and, furthermore, since it does not take place in a homogeneous medium, as is assumed in first approximation by the theoretical schemes. As a first result, this explains what was remarked by ZORZI & REINA, namely that the daily level fluctuations in the wells drawn by pumps do not noticeably affect the average salinity of the water drawn; only the average variations during the whole irrigation season become relevant, also more or less in different cases; often the most dangerous increase in salinity happens about september, when the first autumn rains have started enriching the water layer. In a similar way the influence of the tides and sea-winds is noticed only in the wells very near the shore, for which the empirical tradition already spoke of layer oscillations of the order of 1/2 meter.

As the crack system in the limestone is not homogeneous, the level depression in the well is not followed by a symmetric water influx in respect to the well axis. It is well known that water flows into each well through cracks often easy to identify, and its salinity may widely differ according to their geometrical characteristics, their orientation and their connections. A prevalence of wide and pervious cracks with an horizontal course, for instance, facilitates freshwater influx, whereas vertical and deep cracks tend to bring salty water. For this reason DABELL (1955) advises to seal such dangerous cracks, and also to make a larger use of wells with filtering tunnels. These, in most cases in other mediterranean countries, permitted a remarkable increase in water yield without requiring a dangerous lowering of the liquid weight; only exceptionally it may happen that a filtering tunnel proceeds in the same direction of a main crack, and therefore does not increase the water yield.

As the carstic layer descends toward the sea and thus becomes thinner as it nears the coast, the danger of its salination increases, with the additional possibility of a lateral intrusion of sea-water. It is for this reason, besides the troubles in cultivation caused by the cold and salty winter winds, that there exists already in fact a coastal band free of wells. Depending on the proclivity of the water layer and on the drawing system employed, this band has a width from 100 m up to 1 km; according to DABELL it should rather be extended to some 1–2 km in the Bari region, and according to ZORZI to 3 km in the Salento; it could be done without this where the mesozoic limestone is covered along the coast by impermeable formations several tens of meters thick (ZORZI).

The irregular cracking of the Puglia's limestone also explains the difficulty found by BOTTINI & LISANTI (1955) in establishing a clear and simple relationship between the salinity of the water drawn and the distance of the well from the sea. On the one hand, PANTANELLI maintains that its salinity is in an inverse proportion to the distance from the sea, and actually does not reach farther than 1–2 km in the Bari region, or 4–5 km in the Lecce region. On the other hand, BOTTINI & LISANTI (1955), in a relevant study of the coast up to 3 km from the sea near Bari, found that wells functioning within the first 600 m from the sea had a more similar salinity, which is as a rule rather high ($4.6‰ \pm 1.5$), whereas the more distant wells, though showing an average salinity not much different from the near ones, had significantly larger irregularities ($5‰ \pm 2.8$). This leads to think that wells drawing from a rather thin fresh layer are affected by the underlaying salt water strongly, quickly and about in equal measure, whereas for those farer away, even if they draw from a thicker layer, the speed of salination would be essentially affected, besides the pumping regime, by the crack system that the salt water must go through, by a longer way.

One of the most striking proofs of the irregularity of the carstic underground system is given by the distribution of the carstic springs and of their yields. It is calculated that the Tara springs group by itself almost totally drains the Tara basin; likewise the Chidro spring, east of Taranto, drains most of the Manduria basin. As against these underground systems converging into great springs, the flow of the layer along the Bari littoral is very much dispersed.

Another debated problem is how salination of wells is affected by the use of pumps as against the traditional "norie" (paternoster-pumps). These, activated by an equine animal, were the only system used in the past and permitted to lift water from no deeper than about 15 m. It is generally agreed since the least 10 years and, according to DABELL (1955), also in other mediterranean countries, that the pump causes much stronger mixing of fresh water with the underlying salt water. This is the case because of the different kind of action and of the larger yield. Unpublished data of the Ente Irrigazione give for 12 wells operated by "norie" an average yield of 89 ± 16 l/min, and for 7 wells operated by pumps of 186 ± 62 l/min. This was one reason for the diminution of pump diffusion in the last ten years.

According to DABELL (1955), however, the difference between the two systems consists in the fact that with the "noria" the water table is seldom lowered below half its height on the bottom of the well, i.e. its depression

as a rule does not exceed 50–70 cm. The pump instead often almost completely empties the well, thus stimulating a much stronger ascent of the
underlying cone of salt water and consequent salination of the freshwater
layer.

But recently REINA compared the regime of two wells, one operated by
motor pump and the other by "noria", and found that, at least in normal
cases, "noria"'s superiority is practically negligible, being normally limited
to the first hours of operation (the salinity of the water drawn increases by
an average of 0.50–0.25‰ from the beginning to the end of a working day).
Anyhow this effect would disappear after 30–40 days of irrigation; and this
would make the pump definitely preferable.

According to DABELL (1955) the carstic layer could be still better utilized, by
studying the possibility of re-charging it during the rainy season, by water
collecting at the foot of hilly formations, or by conveying uplifted water in
distribution channels more or less parallel to the coastline.

In order to preserve the lateral coastal barrier, DABELL (1955) advises,
besides the method used today of leaving a coastal band of sufficient width
undisturbed, to enrich the fresh layer in special cases with water brought
from elsewhere, or to make an artificial barrier with impermeable materials as concrete, bituminous emulsions, bentonite, silica gels, calcium
acrilates, etc.; or, at last, to try the sealing effect of explosives on tufaceous
rocks.

Altogether, from the studies related above, we receive a rather satisfactory schematic picture of the dynamics of the carstic layers and in particular, quantitatively, of those of Puglia. But there are still many aspects
requiring deeper studies and verification. Many estimates available today
depend on parameters and coefficients assumed as valid but taken from
various authors who obtained them under different conditions; many
valuations (of surfaces, yields, etc.) are unfortunately subject to heavy
determination errors which, when brought into the calculation, often
unvoluntarily spoil the final results.

Certain inferences are drawn from too few cases and require confirmation and consideration on the variability of the results. For instance, it
seems that a deeper, also theoretical, study of the processes of salt diffusion
in water might lead to a better understanding of the salination of layers at
rest or at least of the transition zone from salty to fresh waters. More data
on hydrostatic loads may give better details on the lines of flow and on the
water amounts in the layers. Deeper comparison between the effects of
"norie" and pumps, and further study of the relationship between meteoric precipitation and the layers will surely clarify the occurrence of salination caused by pumping.

Undoubtedly the use of carstic layers requires much stronger precautions
than that of normal layers, in order to prevent intolerable salination.
Already the Ente for the Irrigation of Puglia and Lucania has succeeded in

having the underground waters of the whole Puglia put under administrative protection; thus research and use of these waters must be authorized by the "Genio Civile." According to REINA, the alteration of the sea bottom near the coast should also be regulated, to prevent excessive dispersion into the sea of the carstic waters.

The problem of coordinating the control of all those interested in one and the same underground water complex in more contiguous provinces is more complicated. In each case a suitable general exploitation plan must be drawn if public control has to make sense. ZORZI, finally, remarks that too detailed norms cannot be prescribed because the irregularity of the crack system mentioned above does not allow to foresee, beforehand and on a large scale, the interference of nearby wells drawing from the same layer. Therefore the exploitation program must be a gradual one and must be followed closely by continuous controls in order to detect any new manifestation and to make it possible to intervene with timely regulations as soon as necessary.

OTHER LAYERS WITH SALINITY AFFECTED BY THE SEA

A position sufficiently typical of the zones near river mouths, where the permeability of the alluvial cover decreases as it approaches the coast, has been studied first by DE MARCHI, IPPOLITO & COTECCHIA (1958), and recently in more detail by ZORZI & REINA in the "Tavoliere di Foggia," regarding the salinity of the water layers. In these conditions there may exist layers retained under pressure in permeable strata (sands and conglomerates) relatively deep (10–100 m, 30–40 in the lower Tavoliere di Puglia), covered for instance by silt and yellowish sandy clays and lying on the impermeable foundation of bluish pliocenic clays. If the material containing the water layer becomes ever less permeable as it nears the sea, it may happen that the water succeeds more easily in reaching the surface by filtering through the covering silt, than in discharging into the sea. In fact it was found that in a wide zone, near the surface, often at less than 5 m depth, there exists a phreatic layer fed by the water rising from the artesian layer; from this phreatic layer, then, the study of the piezometrics lets us infer that there must be, through the overlying soil, a continuation of the ascending flow and its dispersion into the atmosphere by evapotranspiration. This process, of much variable intensity according to the depth of the phreatic layer and also to other factors, has been valued, on the basis of the balance of the layers, at an average of 124,000 m^3/year $\times$ km^2, that is 124 mm/year, corresponding to a continuous flow of approx 4 l/s $\times$ km^2 and for the whole interested extension equal to 2 m^3/s.

The residual horizontal flow is very slight, and practically absent in the upper and less permeable strata of the alluvial cover. This makes the diffusion of sea-salt into the water layer very easy, which therefore, near the sea and up to 2–3 km from the coast, is brackish, with a salinity decreasing as the depth and the distance from the sea increase.

This scheme elaborated by ZORZI & REINA explains rather well the distribution of waters of different salinity in the upper artesian layer of the Tavoliere di Foggia, the pressure distribution in the layer at various depths, the existence of salty soils on the surface especially where the phreatic layer is more brackish, and also – as we shall soon see – their qualitative composition. The order of magnitude of the evapotranspiration volume required by this scheme is quite plausible, but it would be very useful to verify it, especially taking into account the laws regulating the flow of the water in the soil covering the water layer.

An important practical consequence of this scheme is that, in order to avoid the conspicuous water losses into the atmosphere, it would be enough to lower the hydrostatic pressure in the main artesian layer by drilling wells. This, according to the authors, would permit to use for irrigation the water which is now lost, and at the same time it would bring down the salination of the upper phreatic layer and of the soils where most of the salts coming from the deeper strata are now accumulating.

THE WATERS UNDER THE DUNES

The existence of freshwater layers under the coastal dunes has been known for a very long time and also the mechanism of their formation has been explained; they have not been studied in Italy of late. These layers are very frequent along the Italian coasts, but they are seldom used for agriculture. This can be done only when it is possible to draw small water volumes by means of "norie" or, more often, by hand or beam buckets, for very intensive vegetable cultivation, mostly only for nurseries, transplantations and critical culture periods; or when mostly vegetables can be grown directly on dune sands properly leveled and settled, so that the cultivated soil profile (about 70–100 cm) should allow the roots to feed directly from the surface of the freshwater layer. Fine examples of this technique exist in various Italian regions, e.g. in Puglia (coast between Manfredonia and Barletta), Abruzzo, Veneto, Toscana, etc.

The existence of these layers beneath the dunes has sometimes brought confusion in the study of the underground hydrology of the coastal zones. These layers are in fact very limited, as they consist of coastal bands of a few hundred meters width and of little depth. In section they float on the seawater, they have lenticular shape and their statics follow the GHYBEN-HERZBERG scheme explained already when speaking of the carstic layers; from these the layers beneath the dunes differ because the permeable material is more uniform, more porous (often above 40%), and by far more permeable. For these reasons these layers have usually a limited size, and are more easily salinated by the underlying water; they are also more affected by the tides and by the stirring caused by mechanical drawing means; in these cases the use of pumps is therefore much more dangerous.

It is not known if anybody has studied how relevant (at least in the spring) may be here the condensation of atmospheric water vapour (the air being usually very damp near the sea) in the lower sand layers kept cooler than the air by the underlayer. Such a condensation is also being made much easier by the high porosity of the sand (BOYKO & BOYKO, 1964).

In the dune sands, owing to their physical characteristics, the absorption of rain-water is more complete, and the losses by evaporation are much smaller.

BRACKISH SURFACE WATER STREAMS

These phenomena have not been much studied in Italy, but they are also comparatively less frequent. Sometimes they originate from carstic springs and they are then usually very short (a few kilometers) near the coast as e.g. the Tara in Taranto province; their problems are included in the more general one of the carstic layers.

Some other times, especially in Sicily, surface streams become brackish by crossing salty soils or even formations linked to rocksalt layers (e.g. the Salso River).

Important, but little studied so far, is the problem of salinity caused by sundry pollutions. In some branches of the Po Delta and in some channels linked thereto the salinity has reached up to $3\%_0$, as a result of the immission not only of waters drained from salty soils and perhaps also of sea infiltration, but also of salty waters coming from metaniferous deposits. We have no exact data nor published studies on this subject to date, but it is certainly very important, if we consider which agricultural zones are affected by it.

THE COMPOSITION OF THE FIXED RESIDUE IN THE BRACKISH WATERS

The composition of the salts dissolved in all the brackish waters is, in a first approximation, similar to that of sea-water. But, as O. BOTTINI has remarked, a finer analysis shows that these waters cannot be considered simply diluted sea-water. We shall thus describe the more typical cases from the data of BOTTINI & LISANTI (1955), further from those of the Ente per l'Irrigazione for Puglia and Lucania, and from those of GIOVANNINI, FICHERA & GIORGI (1958) for Sicily.

In the carstic phenomenon, the water during its descent toward the aquifer absorbs a small quantity (an average of $0.5\%_0$ in the Salento) of salts from the soil, mostly $Ca^{\cdot\cdot}$, SO_4^{-} and CO_3^{-} ions. But further on it absorbs salts from the sea and as the salinity increases (we find often $4-7\%_0$ along the Bari coast, somewhat less in the Lecce region farther away from the sea), the ionic composition becomes always more similar to that of sea-water, with Cl' anions prevailing and cations in the following decreasing order: Na, Ca, Mg, K; in most brackish waters NaCl constitutes 60–80% of the total fixed residue; no problems ever arose concerning boron content or sodium carbonate residue.

In the case of groundwater beneath the dunes, rain has a short descent through the sand and in this medium the surface phenomena and therefore the ionic exchange are negligible. In consequence there is only salination from the sea by mixing (convection) or by simple diffusion.

More complicated is the salination of water layers near the river mouths, as already explained for the Tavoliere di Foggia and for other similar cases (Piana di Sibari, etc.). The waters of the artesian layer, as long as they are more than 10 km from the sea and above the sea level, have a very low salinity decidedly of terrestrial origin (more SO_4/Cl, Ca/Cl, Mg/Cl, CO_3/Cl, K/Na, Mg/Na, and less Na/Ca, Mg/Ca than in sea-water); but in advancing toward the coast the salinity of the deep layer increases and its composition approaches that of sea-water. Near the coast these changes gradually involve the upper layer as well. However the waters nearer the surface and farther from the sea, as already remarked by Bottini & Lisanti (1955), are brackish (1.99–3.28‰ of salty residue) but much richer in sulphate and in calcium than sea-water. This agrees with the scheme advanced by Zorzi & Reina, foreseeing a strong evapotranspiration and therefore salt accumulation in the upper phreatic water layer.

As to the composition of the surface brackish streams, it usually mirrors the origin of their salinity. Neglecting the obvious characteristics of the carstic springs, in Sicily the brackish surface streams contain often more calcium than the brackish waters of marine origin S.P.[1] often about 50%, and S.A.R. mostly less or not far from 10%); this is perhaps also due to the fact that often these streams pass through chalky formations. Altogether the composition of the fixed residue of these waters is much variable, but usually satisfactory; dangerous boron excess was never found, and sodium carbonate extremely seldom.

For the waters of the Po di Volano and of the linked channels we do not yet have any published data, but their composition probably mirrors that of the water from the sea and from metaniferous deposits.

SALINITY CONTROL OF SOILS IRRIGATED WITH BRACKISH WATER

One of the foremost problems is that of preventing the progressive accumulation of salt in the soil. The classical rules are: abundant irrigation (the so-called filtration co-efficients express the fraction of excess water required in function of the salinity of the irrigation water and of the drainage water) and shortened turns, so as to take into account not only the minera-

1. S.P. = sodium percentage; S.A.R. = sodium absorption ratio.

logical component of the telluric water, but also the osmotic one; indirect measures are all those providing a good soil structure and therefore permeability, and a good drainage.

Unfortunately no careful research on this subject has been carried out in Italy. At the Palermo Institute of Agronomy it has been tried to irrigate with the waters of the Salso River for five years, studying various irrigation volumes and turns, soil washing and amending; but with decidedly negative results on clay soils.

Quite different is the position in the Puglia littoral, which has been repeatedly studied by PANTANELLI, BOTTINI, LISANTI and FICCO. There the availability of brackish waters coming from the carstic layers is very often associated with the presence of soils of the red earth type, usually laying on the same compact and cracked mesozoic limestone in whose depth the water layer circulates. These soils have usually excellent porosity and good permeability (infiltration velocity of the order of 10^{-1}–10^{-3} cm/s) and good natural drainage. This drainage is often made still easier by moderate thickness of the agricultural soil lying directly on the cracked rock or, in the case of allochthonic red earths, on porous and permeable calcareous tuff; all these conditions make any drainage net or work unnecessary. The rains, concentrated for about 70% in the autumn-winter season, very efficiently wash out the soil, besides refilling the water layer.

Under these circumstances irrigated cultivation is usually practiced every second year, with various rotations of which the most frequent is: tomato (for paste) – cabbage (transplanted among the tomatos in July) – wheat (or oats or barley). Thus the salt accumulated in the soil during an irrigation season may be washed out during the following two winters. It has been tried to irrigate every year, leaving the washing of the soil to one rainy season only, but often with unsatisfactory results; though FICCO succeeded in getting good yields by irrigating four years in continuation.

The specific irrigation volumes practiced under these conditions are very limited (the average for 18 farms studied by the Ente for Irrigation in Puglia and Lucania is 244 ± 56 m³/ha); these volumes, used in farms whose average soil thickness was 38 cm, were not much different from those used in the deeper alluvial soils of the "lame" (alluvial valleys) or in those lying on porous tuff. The average irrigation turn was 6.7 ± 1.1 days with a slight and not general shortening in the hottest period of the season. With a number of 13–17 turns we reach a seasonal total of approx 3200–4100 m³/ha. The corresponding amount of salt brought into the soil, with an average salinity of 5‰, would thus be 16–20.7 t/ha. With an hydric field capacity, e.g., of 32% of the dry weight of the soil and a wilting coefficient of approx 16% (values frequent in the Bari red earths), we can calculate that if this salt would all accumulate in the soil, say in the first 40 cm, it would cause an increase in the osmotic pressure respectively of 7 and 15 atm, or, as total water potential, respectively of little more than 7 and 30 atm in respect to the hydric field capacity and the wilting coefficient. These values do not allow an intensive vegetable cultivation, even of salt resistant varieties. But in fact, already BOTTINI & LISANTI (1955) did not succeed in evidencing relevantly different salinity in the Bari soils irrigated with brackish waters as against those not irrigated; after them FICCO found that most of the salt brought into the soil with the irrigation water is brought away by the same water (proportions of from 26 to 83% of the total brought salt were found to have been washed out by the end of the irrigation season).

This clearly proves that the irrigation volumes used, even if small, are so applied as to ensure a good drainage and thus to prevent excessive salt accumulation. The washing effectiveness of these irrigation volumes depends probably on the fact that, due to the short turns, the water is given when the soil is still well far from the wilting coefficient (an essential condition for the tomato cultivation in Puglia), more exactly, conceding the existence of a slight percolation and with the assumed data, when less than 5% of the water (referred to the dry soil has been dispersed by evapotranspiration. Under these humidity conditions the hydric conductivity of the soil would be rather high and would allow a faster and more uniform water distribution in the soil mass and toward the deep strata. Thus even the salt displacement by diffusion would be strongly accelerated and dangerous concentrations near the roots and even near the evaporating soil surface would be avoided. In fact, slat surface efflorescences in the earth ridges more exposed to evaporation are rare and, if they occur, very slight.

The usual irrigation system is that of infiltration from furrows. This is particularly suitable for tomato culture and, especially in this type of soil which hinders lateral water diffusion, compels to little irrigation volumes not easily controllable. With this system the water tends to imbue the soil somewhat irregularly, notwithstanding the uniformity in the direction of the length of the furrows, which are kept no longer than 5 m. The water thus tends to descend in depth under the furrow, effecting a measure of percolation.

The same technique as described above is applied independently from the thickness of the soil, because the existence of underlying cracked or porous rock does not necessarily increase the percolation; in fact the descent of the water from the soil into a system of pores or of larger cracks requires that at the separation surface the water tension be near zero, i.e. that the soil in touch with the rock be saturated, thus holding more water. On the other hand the limestone rock is not finely and regularly cracked and allows local stagnations. On the contrary, the case of soil lying on porous tuff, though favouring the same condition at the separation surface, is more similar to that of the deep soils, permitting capillary water ascent toward the soil.

That portion of the soil on which the plants are growing and which is always in connection with the irrigation system, (strips between furrows with double plant rows) is less exposed to leaching; on the contrary it is subject to salt concentration due to absorption by the roots and, at the sides of the strips, to evaporation into the atmosphere. Notwithstanding the easy salt diffusion mentioned above, these causes sufficiently explain the salt accumulation that FICCO found in the soil at the end of the season, and which requires the more complete and effective washing away during two consecutive irrigation seasons. That this cannot be eliminated is proved by the fact that in three years tests by FICCO the washing away obtained in consecutive irrigation seasons gradually decreased from 83 to 39 and then to 26%. DABELL (1955) on this point suggested to try at winter's end an abundant washing away with the same water of the carstic layer which has at that time the lowest salinity; but he himself warned against the danger of depleting the layer with consequent faster salination, which could annul any gain.

For the perennial (polyannial) cultures (citrus, grapes, artichokes) the washing of the soil is made possible either by its structure (sandy in the Taranto and Sicily citrus groves) or by the little watering required (Puglia curtain vineyards and artichokes). In conclusion, while the general outline of the irrigation technique empirically elaborated in Puglia is clear enough, it must be admitted that the hydric and saline balance in the soil, its dynamics and its relationships with its characteristics are still very uncertain for lack of exact experimental data permitting to quantitatively verify the various aspects of the outlined scheme.

THE CHEMICAL AND PHYSICO-CHEMICAL CHANGES
EFFECTED IN THE SOIL BY BRACKISH WATER

As already pointed out when dealing with the Italian brackish waters, those studied in Sicily by GIOVANNINI, DI GIORGI & FICHERA (1958) have low percentages of sodium compared to the other cations (S.P. values mostly around 50%), and S.A.R. seldom above 10. In Puglia instead, during the irrigation season, with salinities around 4–5‰ corresponding to an electrical conductivity of 6000–7000 micromhos/cm, S.P. is mostly 10–20% and S.A.R. between 60 and 80. It is clear that the Puglia brackish waters easily reach a dangerous composition for certain soils, essentially because of their low content in calcium. In fact, as shown by PANTANELLI and BOTTINI, the usefulness of these waters depends essentially on the lime content of the soils. In 38 samples analyzed by BOTTINI & LISANTI (1955) this lime content averaged 13.8% with often very high peaks (maximum found 46.6%) which constitutes a precious store of calcium ions able to counterbalance the sodium excess in the waters, so that in practice the exchange capacity of the Bari red earths (of the order of 30–32 me/100 g of earth) is mostly saturated by the calcium ion (sometimes, according to LOPEZ, up to 80–90%). It must further be added that the same composition of the colloidal fraction of the red earth makes it less liable to structure alterations by excess sodium, due to its high content of aluminium and iron sesquioxides and to the prevalently illitic and in lesser measure caolinitic nature of its siallitic minerals.

In practice, some sampling done by FICCO along the adriatic coast between Bari and Brindisi near Egnazia after four years of consecutive irrigation did not show any rise in salinity but a slight pH rise from an initial 7.8 to 8.2 while, as already mentioned, there was also a decrease in the fraction of salt washed away by irrigation.

All this hints to a certain degradation of the initial soil structure, against which little could be helped by the manure spread in the modest annual amount of 15 t/ha. This amount is unfortunately normal in the Bari littoral where organic fertilizer is scarce and precious.

It is worth noting that in the Puglia red earths, normally rich in total potash and well provided with exchangeable potash, the sodium brought in by the brackish waters presumably helps the displacement of potassium from the absorbing complex, thus assisting the nutrition of the plants. But we do not yet have experimental data permitting to evaluate the practical relevance of this process, nor its longterm effects on the chemical fertility of the soil. This soil in fact, in Puglia (and against the old saying that he who fertilizes with salt makes rich the father and poor the son), still and always remains rich in exchangeable potash. The frequent lack of calcium in the Taranto pleistocenic red sands is compensated by the lower salinity of the local waters (1.90‰ for the Tara river) and by their better quality (S.P. 46% and S.A.R. 4.8), and in addition also by the favourable texture of the soils.

BRACKISH WATERS AND CULTIVATED CROPS

Of the crops irrigated with brackish waters, the most traditional and diffused crop along the Bari littoral is the tomato for paste; in the last 20 years in Puglia there has been a remarkable extension of the artichoke; where salinity is lower and soil conditions more favourable, citrus is irrigated (in Taranto province and, in Sicily, e.g., at Bagheria and Casteldaccia), pepper, eggplant, etc., or, with a few waterings per year, grapes for table consumption and other cultures once considered dry, like the olive tree. Researchers in many countries have already amply investigated how different crops are affected in quantity as well as in quality by brackish water irrigation. In Italy nothing much was done after the worthy work by PANTANELLI which also summed up the former research by PASSERINI, GALLI and BIGNANI.

How different crops are affected by salinity may depend on their different resistance against the osmotic potential of the telluric water, or on effects of ionic balance of the solution forming in the soil. It is thus clear that the resistance of different species to salinity cannot depend only on the salinity of the water used, but on all affecting conditions, like the chemical and physical structure of the soil, the climatic environmental factors and the cultural technique. For citrus, for instance, the use of slightly brackish water (1.5–3‰) gives excellent yield in the Taranto region and in parts of Sicily (Bagheria and Casteldaccia) because those soils are decidedly sandy; the use of more salty water along the Bari littoral is possible only for the tomatoes (for paste) and often older seedlings must be transplanted, because they are more salt-resistant than the young ones. For some cultures, like artichokes and table grapes, brackish water can be used because only 2–3 summer waterings are required, bringing later and smaller quantities of salt. For all cultures irrigation by sprinkling has to be much more limited, because of the dangerous concentration of salt on the leaves.

This variability of agronomic conditions explains the existing uncertainty, and not only in Italy, in setting the limit between fresh and brackish waters and above all the maximum limit of the practical usability of a certain water, or the one of salt resistance of single species. In fact, if we indicate the salinity by the fixed residue (g/l or ‰) as it is still the use in Italy, waters are generally considered brackish that have a residue above 2‰ (approx. concerning to an electrical conductivity of 3000 micromhos/cm). But a study commission named by the Ente for Irrigation in Puglia and Lucania set the limit to 3‰ (4500 micromhos/cm), and according to PANTANELLI no water should be called brackish that does not contain at least 5 g of salt per litre (above 7000 micromhos/cm). We have already mentioned that the values mostly found during the irrigation season in Puglia are 4–5‰ (6000–7000 micromhos/cm) reaching often up to 8‰, which BOTTINI considers being the maximum tolerated by tomatoes. Seldom and temporarily only even water with a salinity up to 10 and 12‰ could be used, i.e. one third of sea-water concentration. In other countries, as the United States of America, waters of such high salinity would not even be considered.

In fact, it is often difficult to judge about the fitness of brackish water for irrigation because the problem is put in uncertain terms. First of all, when speaking of water from carstic layers the salinity of the well at the beginning of the season should not be confused with that prevailing during the pumping, and the variations during the year should also be taken into account. But even when considering waters of well defined salinity, it is of little practical use to ascertain the maximum salinity beyond which the plant cannot survive, supposing that it can be ascertained with enough precision; what is important is to establish the possibility of an economically profitable cultivation. For each crop and in well defined agro-ecological conditions, the production curve should be established in function of the water salinity, so as to allow later and from time to time, on the basis of economic parameters (sundry prices, etc.) to find out which maximum salinity, with the optimal irrigation volume, still ensures a profitable production.

If the problem is put on these terms, it seems possible to establish scales of resistance of the various crops to brackish water, different with the changing of any of the mentioned parameters, as the soil type, the water quality, the prices level, etc. PANTANELLI rightly remarked that plants tolerate greatly different degrees of salinity when the roots draw directly from a saline solution or when they only find themselves in a soil moistened by the same solution; besides, in the soil the limits rise little if it is sandy, much if it is clayish, little if it is poor in iron carbonate and in free aluminium, much if it rich in these components; According to PANTANELLI, it is always the vegetables that profit most from brackish waters irrigation.

He has even found that, provided the salinity does not exceed some 3.3‰, it is actually essential for getting the largest and best yields of tomato, eggplant, pepper, cabbage, fennel, asparagus and artichoke. FICCO, using much more saline water (up to 7–8‰), confirmed the possibility of using brackish water on many vegetables and field crops, but got decidedly negative results on eggplants, pepper and celery, in agreement with the consensus among practitioners.

The production results, in fact, must not be only expressed in terms of simple yield. Unpublished results of a trial carried out at the Bari Institute of Agronomy, for instance, show that in the average on 4 pepper varieties irrigation with brackish water (the same used by PANTANELLI), though not causing statistically relevant differences as against the use of fresh water (potable water of the river Sele), induced a remarkable drop in the average weight of the fruits (-19.6%) and a rise in their number ($+22.8\%$), so that brackish water gave an equal and even apparently larger yield; also the shortening of the turn from 7 to 3.5 days in the average raised the weight of the fruits by 8.3%, while the number of fruits per plant was not affected. These results, which deserve further confirmation and would allow to interpret PANTANELLI's remarks, suggest the hypothesis that brackish waters stimulate a higher differentiation of flower buds or the development speed of the flowers themselves, while salinities even not too high would decidedly damage the weight of the fruits, besides the well known inverse relationship between this weight and the number of fruits per plant. In conclusion, for all species the salinity must not exceed a certain not very high value. Proof thereof are the average poor yield of tomato in the Bari province, the low size of the plants which are therefore raised without supports, and the difficulty in growing tomatoes for table consumption or to be peeled, like the St. Marzano variety, requiring larger irrigation.

At this point it is worth noting that various authors, including PANTANELLI, have noticed a reduction of the evapotranspiration coefficient as a result of brackish water irrigation. But beside the little economic relevance

of these considerations and notwithstanding the great variability of these
coefficients, it must be stated that, given the average irrigation volumes
used in the Bari region and with an average tomato yield of approx. 25 t/ha
the actual evapotranspiration coefficients are of the same order of magnitude (approx. 120 l/kg of fresh fruit) as those found for fresh water by other
authors (MARIMPIETRI and BALDONI). Thus the usefulness of brackish water
irrigation under the practical conditions in which it can be economically
applied, is very doubtful.

The low specific water consumptions prevailing in the Bari orchards
(some 109 mm/month and thus decidedly lower than the potential evapo-
transpiration in the same zone) result in proportionally low yields. This is
not inconsistent with the higher drought resistance found by PANTANELLI
as a result of brackish water irrigation, but it limits its practical relevance.

In the litterature on brackish water irrigation it is often neglected to
consider how the following crops may be affected by the soil salination at
the end of the season, even if it be temporary. This is usually important
during the autumn, when the rains have not yet completed their washing
action. For this reason, in the Bari region, cabbage is conveniently raised
after the tomatoes; when once, instead of cabbage, peas and spinach were
grown in alternate rows, there appeared a most evident difference between
the two. The peas, by stunted growth and diffuse chlorosis, show exactly
that they had been sown where the watering furrows had salinated the
soil.

As already mentioned, plants are affected not only by the brackish water
concentration, but also by its composition, or better by the soil composition after interaction with the water. Otherwise it could not be explained
that in hydroponic cultures, even when particularly halophytic plants are
grown, the salt concentration cannot be pushed somewhat nearer than 2/3
of that of sea-water, corresponding to some 15 atmospheres of osmotic
pressure.

The importance of calcium in the water and much more in the soil in
balancing sodium and even magnesium in the Puglia conditions has been
well shown first by PANTANELLI and later by BOTTINI. An interesting problem was raised by HEIMANN in connection with the relationship Na/K,
whose importance would be basic for brackish water irrigation. But recently FICHERA & DI GIORGI (1964), working in the laboratory on a very salt-
sensitive plant as the bean, could not discover that this theory may be
applied to the current practical conditions. The fact that on the irrigated
Bari red earths potash fertilizers did not usually get remarkable results lets
us suspect that the rich exchangeable potassium content of these soils fulfils
by itself the task pointed out by HEIMANN.

The complex action of brackish irrigation does not show in the crop
quantity only. PANTANELLI has remarked that plants thus treated give fruits
more tasty, of better colour and consistence, richer in mineral salts and

sugar, with heavier dry residue and more resistant peel; the plants themselves hold better against drought and also against some parasites (*Fusarium*). Most technicians and farmers are of the same opinion, though experimental data are lacking on the matter.

In fact, just the agronomic aspects of brackish irrigation have been most neglected and should be more deeply studied. For instance, the following widely held beliefs should be properly tested: that brackish irrigation makes eggplant more bitter and fibrous, pepper sharper, cucumber more bitter, that the "cruciferae" tend to blossom earlier and that this tendency would even be transmitted to the following seed, that broad bean raised on soil previously irrigated with brackish water becomes better resistant to the parasite *Orobanche*, etc.

REMARKS ON SOME ECONOMIC ASPECTS OF THE PROBLEM

The economic side of brackish water irrigation must be briefly considered in order to see the problem in the frame of practical reality. The provision of water should be treated separately from its use in the field.

For brackish surface streams no special problems occur. There against, with brackish underground layers, especially with those of the carstic type, the limitations and the precautions required by the danger of salination of the layer may deeply affect the profitability of their use.

Up to 30–40 years ago, in Puglia the carstic layer could be exploited only where it was no deeper than 10–15 m below the field level, that being the maximum lifting power of the "norie" (paternoster-pumps); the wells were usually dug only 1–1.5 m below the water surface. With the introduction of pumps, water could be economically lifted up to some 50 m; in the last ten years also this limit was more and more frequently passed and now, for some uses, 100 m and sometimes even 150 m and more are reached.

At the same time technical progress made drilling cheaper, and better knowledge of the underground hydrology in Puglia lessened the boring risks, so that at present it is often worth-while to try it down to several hundred meters in the search of artesian layers for irrigation and, of course, also for drinking water.

Besides, the expansion of the irrigated surface was strongly stimulated by the increasing use of brackish waters for formerly non-irrigated crops (especially curtained vineyards), if only with low seasonal volumes and eventually only with emergency interventions. Sometimes, whenever possible, also very salty water is used, diluting it with fresh water coming from elsewhere, which is more expensive or available in lesser quantities (e.g. water from the drinking water net-work, or rain-water from cisterns, or sewage water); this is often done in the littoral vegetable gardens in Bari and in some Taranto citrus groves.

As the use of deeper waters became more profitable, technicians of the

Ente for Irrigation in Puglia and Lucania were induced to plan a certain exploitation of the Gargano carstic layer, limited to the lower fields or for drinking purpose; but, in the Tavoliere di Puglia, also of the deep carstic layer which is found, being under pressure at the depth of several hundred meters under the powerful cover of the blue pliocenic clays. Little more than ten years ago, this deep layer was not even thought of.

It also cannot be ruled out that the use of the brackish waters of the carstic layers, now typical of individual autonomous farms, may in time and under certain conditions be entrusted to collective net-works, in order to achieve a better co-ordination. This would help also to regulate the use of the layer and to better the quality of the water.

In the case of brackish waters, any setting of an optimum seasonal irrigation volume is complicated, simply for the different resistance to salinity of the various crops. Since the problem has not yet been sufficiently studied, we can only sketch the general outlines. In all events it can easily be seen that the most convenient irrigation value for brackish water, if agronomically less efficient, cannot be, as a rule, but inferior to that of fresh water or of only slightly brackish water. It is therefore doubtful whether under these conditions such crops as fodder could be expanded, the more as in Southern Italy these crops are already economically problematic even where irrigation is easier. In the past, the tomato was dominant where brackish irrigation was practiced; now its cultivation has become much less profitable and is being replaced by other vegetables like artichokes or, where salinity is low or can be lowered by dilution, by pepper, pumpkin, etc. In the Bari littoral tomato cultures are also often replaced by curtained vineyards for table-grapes, the latter requiring only 2–4 summer waterings and enjoying a much better market.

In general, one is under the impression that the traditional and technically so refined cultivation of tomatoes for paste, which for centuries had permitted the use of the most brackish waters, struggles today against the changed economic conditions and risks, and may follow the same destiny as the dry cultivation of tomatoes.

In many zones, for instance in the Taranto region, this dry cultivation has within some ten years already practically disappeared, essentially because of the revolutionary rise in the wages of agricultural manpower.

SUMMARY

In Italy saline water used for irrigation comes from the following sources: (a) carstic aquifers, (b) common phreatic and artesian aquifers having some relationship with sea-water, (c) underdunal aquifers, (d) variously salinized stream water. The use of water from aquifers having a salinity essentially originated from soil is negligible.

The main problems which arise in practice from the use of saline water are discussed.

It is necessary to reduce the salinization of the water to a minimum. This involved geo-hydrological problems concerning aquifer utilisation or surface water pollution. Good progresses were made during the last 15 years in certain Italian regions, particularly Apulia, on underground hydrology, water resource evaluation, and aquifer utilisation referring to brackish water. The influence of the following factors on the salinization of carstic aquifers has been examined: distance from the sea, tides and winds, depth of well or gallery, lifting device (pump or "noria" (paternoster-pump)), water table depression in the well, time during the irrigating season, general features of the fissure system in the carstic rocks, etc.

Research on phreatic and artesian aquifers near river mouths in the Tavoliere of Foggia provided interesting viewpoints on the hydrologic balance and water salinization problems. A progressive reduction of permeability of the rocks in artesian aquifers, while approaching the coast, results in an upwards water flow which is dispersed in the atmosphere by evapotranspiration. In proximity of the coast the water flow seaward becomes negligible to a point that spreading of salt is permitted from the sea to the aquifers. In order to eliminate this loss of water and to prevent salinization of upper aquifers and soils, it has been suggested to reduce the artesian pressure by water pumping.

Utilisation of fresh water under dunes has not been sufficiently investigated; the same holds for surface flowing saline waters, such as those found in Sicilia streams or the ones of the Po delta, which are polluted by methane deposit waters.

Deeper knowledge of the problems hitherto mentioned are wanted. An other important group of problems concerns the danger of damages from water salinity to agricultural soils. Saline water analysis has been broadly developed in Apulia and Sicily. An overimposition of salinity of marine origin to one of soil origin is evident; results fit satisfactorily to the scheme derived from geo-hydrological survey. Generally speaking, in carstic aquifers, sodium percentage increases (S.P. more often 60–80%) as well as sodium absorption ratio (S.A.R. more often 10–20), when salinity increases. Analysis of surface waters yields generally much better results in Sicily.

An irrigation trial with saline water seems to give discouraging results on clay soils in Sicily. In Apulia, on the contrary, irrigation with brackish water has very ancient tradition which has become possible because of very favorable soil properties (high calcium content, good permeability, suitable colloid composition, etc.) and because of the concentration of rainfall in autumn and winter, and the alternation of a year with an irrigated summer crop, with one without irrigation. Small watering amounts are applied (ca 250 m³/ha), however, with a high frequency (6–7 days), so that leaching of a large proportion of salts brought by brackish waters is permitted during the irrigation period; the remaining salt amount being normally leached by two consecutive rain seasons.

A third group of problems is concerned with saline water-plant relationship. Behaviour and salt resistance of different crops have been observed, in relation to soil type and cropping technique. When moderately salinized water (3‰) was used, an increase in fruit number per plot, but a decrease in fruit size was observed on pepper; with the same water, increase in yield and quality improvement have been indicated, while with increasing water salinity yield decreases and quality becomes lower. There is a strong need of broader research work on the agronomic aspects of saline water use.

Economic problems of the brackish water use are mentioned.

RÉSUMÉ

Les eaux salées utilisées en Italie proviennent essentiellement de (a) nappes karstiques (b) nappes phréatiques ou artésiennes communes ayant quelque rapport avec l'eau de mer (c) nappe sexistantes sous des dunes (d) cours d'eau de surface plus ou moins salés. L'emploi d'eaux de nappes dont le sel provient principalement du sol, est négligeable.

Les principaux problèmes créés dans la pratique par l'emploi d'eaux salées, peuvent être résumés de la façon suivante:

Il est nécessaire de réduire au minimum le degré de sel contenu dans les eaux utilisées. Il s'agit de problèmes concernant les aspects hydro-géologiques de l'utilisation des nappes d'eau ou inhérents à la pollution des cours d'eau superficiels. De nombreux progrès ont été effectués aux cours des quinze dernières années dans certaines régions de l'Italie et en particulier dans les Pouilles, quant à la connaissance de l'hydrologie souterraine, à l'évaluation des ressources hydriques et aux problèmes de l'utilisation des nappes.

L'on a en outre pris en considération les influences que la distance de la mer, l'action des marées et des vents, la profondeur des puits ou des galléries filtrantes, le système de soulèvement (norias ou pompes), le niveau de la surface de l'eau dans les puits, la saison et la subdivision de l'irrigation dans la dite saison, les caractéristiques générales du système de fissures du terrain karstique, etc. ont sur le degré de sel contenu dans les nappes karstiques.

L'étude des nappes artésiennes ou phréatiques dont l'emplacement est voisin de bouches fluviales dans le Tavoliere de Foggia a permis de faire d'intéressantes découvertes quant au bilan hydrologique et au degré de sel de ces eaux. La manifestation d'un flux vers le haut qui se disperserait par la suite dans l'atmosphère par évapo-transpiration, a été attribuée à la réduction progressive de la perméabilité du matériel qui contient la nappe artésienne au fur et à mesure qu'elle s'approche de la mer; près de la côte l'on a pratiquement un arrêt du flux et cela permettrait l'invasion de sel de la mer dans la nappe. Pour éviter d'importantes pertes d'eau et afin de réduire le degré de sel contenu dans les eaux superficielles et dans les sols, l'on a suggéré de réduire la pression de la nappe artésienne en pompant l'eau dans des puits.

L'utilisation des nappes sous les dunes et des eaux salées à la surface du sol a été peu étudiée. (Par exemple: les cours d'eau siciliens ou ceux du delta du Po, polluées entre autre à cause de l'entrée des eaux salées des gisements de métane).

L'on souligne qu'il serait opportun de procéder à une étude et à un contrôle plus approfondis des problèmes mentionnés ci-dessus.

Un deuxième groupe de problèmes importants est celui qui concerne les dommages que l'eau salée utilisée pour l'irrigation porte au sol cultivé. L'étude la plus approfondie de la composition du résidu constant des eaux salées italiennes a été menée dans les Pouilles et en Sicile. Cette étude nous a permis de voir que très souvent l'eau est salée en raison d'une superposition, variable en mesure, d'eau de mer et de sels provenant du sol; cela représente souvent une base pour l'interprétation des résultats obtenus au cours des études géo-hydrologiques. En général, dans les nappes karstiques, au fur et à mesure que le degré de sel augmente, l'on voit s'accroître le pourcentage de sodium par rapport aux autres cations (très souvent 60–80%) ainsi que le S.A.R. (très souvent de 10 à 20); quant à la composition des eaux salées de surface en Sicile, elle est bien meilleure. Les essais d'irrigation à l'aide d'eaux salées sur des terrains argileux en Sicile semblent donner des résultats négatifs. Dans les Pouilles, l'ancienne irrigation traditionnelle et perfectionnée, au moyen d'eaux salées est possible grâce aux propriétés très favorables du terrain (contenu de calcium élevé, perméabilité, nature des colloïdes, etc.), à la concentration des pluies en automne et en hiver et à l'alternation d'une année d'irrigation à une année de culture sèche. Tous les 6 ou 7 jours l'on irrigue avec des volumes réduits d'eau (ca 250 m³/ha), de sorte qu'une bonne partie du sel apportée par les eaux salées est éliminée au cours de la saison même.

Le troisième groupe de problèmes concerne les rapports de l'eau salée avec les plantes cultivées. L'on a observé leur résistance au sel ainsi que le comportement des différentes espèces, en prenant en considération aussi le type du sol et la technique de culture. Avec des eaux modérément salées (3 ⁰/₀₀ environ) l'on a observé que le nombre des fruits augmentait mais que leur poids diminuait. Quand la quantité de sel est réduite, la production augmente et la qualité s'améliore, mais si le degré de sel s'accroissait l'on aurait des résultats négatifs même sur le plan qualitatif. L'on signale qu'il serait bon de procéder à un approfondissement des recherches surtout en ce qui concerne les aspects agronomiques.

L'on discute certains aspects économiques de l'irrigation à l'aide de l'eau salée.

RIASSUNTO

In Italia si utilizzano acque salmastre provenienti essenzialmente da (a) falde carsiche, (b) comuni falde freatiche od artesiane aventi qualche relazione con le masse marine, (c) falde sottodunali, (d) corsi d'acqua superficiali variamente salinizzati. Trascurabile è l'impiego di acque di falda con salinità di origine prevalentemente terrestre.

I principali problemi che sorgono in pratica con l'impiego di acque salmastre possono venire così riassunti.

Occorre ridurre al minimo la salinizzazione delle acque utilizzate. Si tratta di problemi riguardanti gli aspetti idrogeologici dell'utilizzazione delle falde o riguardanti l'inquinamento dei corsi d'acqua superficiali. Notevoli progressi sono stati compiuti negli ultimi 15 anni in certe zone italiane e precisamente in Puglia, nella conoscenza dell'idrologia sotterranea, nella stima delle rispettive risorse idriche e nei problemi dell'utilizzazione delle falde.

Si sono inoltre prese in considerazione le influenze sulla salinizzazione delle falde carsiche della distanza dal mare, dell' azione delle maree e dei venti, della profondità dei pozzi o delle gallerie filtranti, dell'impianto di sollevamento (norie e pompe), della depressione del pelo liquido nei pozzi, del procedere della stagione irrigua, delle caratteristiche generali del sistema di fessure della roccia carsica ecc.

Lo studio di falde artesiane e freatiche in vicinanza di foci fluviali nel Tavoliere di Foggia ha fornito interessanti vedute sul bilancio idrologico e sulla salinizzazione di queste acque. Alla progressiva riduzione di permeabilità del materiale in cui è contenuta la falda artesiana, col procedere verso il mare, si è attribuita la manifestazione di un flusso verso l'alto che si disperderebbe poi nell'atmosfera per evapotraspirazione; nei pressi della costa, la pratica cessazione del flusso verso il mare permetterebbe l'invasione di sale dal mare nella falda. Per evitare cospicue perdite di acqua e ridurre la salinizzazione delle acque più superficiali e dei terreni, si é suggerito di abbassare la pressione della falda artesiana mediante emungimento da pozzi.

Poco studiata risulta l'utilizzazione delle falde sottodunali e quella delle acque salmastre fluenti in superficie (per es. quelle dei corsi d'acqua siciliani o quelle del delta padano, inquinate anche per l'immissione delle acque salse dei giacimenti metaniferi).

Si mette in evidenza l'opportunità di ulteriori approfondimenti delle conoscenze e di verifiche per tutti i problemi menzionati.

Secondo gruppo importante di problemi è quello riguardante i danni della salinità dell'acqua irrigua al terreno agrario. Lo studio della composizione del residuo fisso nelle acque salmastre italiane è stato più ampiamente condotto in Puglia ed in Sicilia. Esso dimostra molto spesso la sovrapposizione in varia misura di una salinità di origine marina ad una di origine terrestre e permette spesso di interpretare i risultati degli studi geo-idrologici. In generale nelle acque delle falde carsiche, col crescere della salinità cresce pure molto la percentuale di sodio rispetto agli altri cationi (più spesso del 60–80%) e l'S.A.R. (molto spesso tra 10 e 20); molto migliore risulta in Sicilia la composizione delle acque salmastre fluenti in superficie. Prove di irrigazione con acque salmastre su terreni argillosi sembrano dare risultati negativi in Sicilia. In Puglia l'antica e perfezionata tradizione irrigua con acque salmastre è resa possibile per le proprietà molto favorevoli del terreno (elevato contenuto in calcio, permeabilità, natura dei colloidi ecc.), per la concentrazione delle pioggie in autunno-inverno e per l'alternanza di un'annata irrigua ed una in coltura asciutta. I piccoli volumi di adacquamento sono applicati con turno frequente che permette il dilavamento di una buona parte del sale nel corso della stessa stagione irrigua.

Un terzo gruppo di problemi riguarda i rapporti dell'acqua salmastra con le piante in coltura. Sono stati osservati la resistenza alla salinità ed il comportamento delle diverse specie, anche in relazione al tipo di terreno ed alla tecnica colturale. Con acque non molto salate (3 $^0/_{00}$ circa) si è rilevato un aumento del numero dei frutti, ma una loro riduzione in peso; con debole salinità sono stati segnalati aumenti di produzione e miglioramenti qualitativi, ma si avrebbero peggioramenti anche qualitativi su certe specie col crescere della salinità. Si segnala la necessità di un approfondimento delle ricerche soprattutto sotto gli aspetti agronomici.

Si discutono alcuni aspetti economici dell'impiego irriguo delle acque salmastre.

REFERENCES

BIGNAMI, P. 1933. L'irrigazione con acque salate. *Atti Cong. Naz. Acque, Bari.*; *Agric. Colon.* 29: 411.

BOTTINI, O. 1959. Tradizione e ricerca in Italia nell'uso delle acque salmastre per irrigazione. *Atti Acc. Econ. Agr. Georgof.* s. 7, 6: 15 p.

BOTTINI, O. 1961. Tradition et recherche en Italie dans l'utilisation des eaux saumâtres pour l'irrigation. p. 251. *In* Salinity problems in the arid zone. *Proc. Teheran Symposium, 1958 UNESCO.*

BOTTINI, O. 1963. L'irrigazione con acque salmastre, Volumi e turni. *L'Italia Agric.* 100: 555–561.

BOTTINI, O. & E. LISANTI. 1955. Ricerche e considerazioni sull'irrigazione con acque salmastre praticata lungo il litorale barese. *Ann. Sper. Agr.*, n.s. 9: 401–436; *Ann. Fac. Agr. Bari.* 9: 19–60.

BOYKO, H. & E. BOYKO. 1964. Principles and Experiments regarding direct irrigation with highly saline and sea water without desalination. *New York Acad. Sci., Trans.* Ser. II, 26. *Suppl.*, p. 1087–1102.

CERRUTI, A. 1948. Sulle sorgenti sottomarine dei mari tarantini e sulla loro eventuale utilizzazione per l'irrigazione nel Salento. *Risv. Agric.* 7: 118.

COTECCHIA, V. 1955. Influenza dell'acqua marina sulle falde acquifere in zone costiere, con particolare riferimento alle ricerche d'acqua sotterranea in Puglia. *Geotecnica.* 2: 105.

DABELL, J. P. 1955. Some observations on the use of brackish water for irrigation along the Bari-Brindisi coastal plain. F.A.O. 55-9-5602, (stencilled).

DE GENNARO, E. 1954. Attività dell'E.R.A.S. *In* Le ricerche e le utilizzazioni delle acque sotterranee nell'Italia meridionale e nelle Isole. *Atti II Cong. int. Irrig. Bonif., Algeri.*

DE MARCHI, G., F. IPPOLITO & V. COTECCHIA. 1958. Indagine sulle acque sotterranee del Tavoliere di Puglia. *Cassa p. Mezzog. Docum.* n. 2.

DI PRIMA, S. 1962. Problemi, ricerche e rimedi per le acque salse irrigue e terreni salmastri. *Terra Pugl.* II (2): 21–29.

FICCO, N. 1960, 1961. Un triennio di sperimentazione irrigua con acque salmastre. Ente Svil. Irrig. Trasf. Fond. Puglia Luc. Bari, 80 p.; *Genio rur.* 24: 753–766.

FICCO, N. 1961. Aspetti della utilizzazione irrigua delle acque salmastre nel Mezzogiorno. *Atti Conv. int. Prob. Irrig. Bac. Mediterr.* 259–262, Cons. Gen. Bon. Foggia.

FICCO, N. 1963. Irrigazione con acque salmastre, p. 297–317. *In* Tecnica dell'Irrigazione. *Cassa p. Mezzog.*

FICHERA, P. & M. C. DI GIORGI. 1964. Nuovi orientamenti sull'utilizzazione delle acque saline. I. L'influenza del rapporto potassio-sodio su colture salino-sensibili (*Phaseolus vulgaris* L.). *Agrochim.* 8: 179–188.

GIOVANNINI, E. 1957. Contributo alla conoscenza delle acque di irrigazione. Nota I. Le acque correnti di superficie e le acque di pozzo nella piana di Catania e zone limitrofe. *Ann. Sper. Agr.* II: 1319–1354.

GIOVANNINI, E., M. DI GIORGI & P. FICHERA. 1958. Contributo alla conoscenza delle acque di irrigazione. Nota III. Su alcuni aspetti applicativi delle acque di uso irriguo della Sicilia Orientale. *Tecnica agric.* 10: 559–568.

LOPEZ, G. 1968. L'Irrigazione con acque salmastre in Puglia (in press).

MILELLA, G. 1948. La sorgente subacquea Anello di S. Cataldo in Taranto. *Risv. agric.* 7: III.

PANTANELLI, E. 1928. Composizione delle acque freatiche del Tavoliere. *Rel. Uff. Irrig. E.A.A.P., Bari.* 2: 101–118.

PANTANELLI, E. 1937, 1929, 1941. Irrigazione con acque salmastre. *Mem. Staz. Agr. Sper. Bari,* 26, 72 p. 1937; *Risv. Agric.,* 37–41 & 83–90 1929; *Agric. Colon.* 35: 340–342. 1941.

PANTANELLI, E. 1943. Risorse idriche del Tavoliere. *Staz. Agr. Sper. Bari, Mem. 35.*

PANTANELLI, E. 1947. Le risorse idriche della provincia di Bari. *Risv. Agric.* 1, 6 p.

PANTANELLI, E. 1947. Le risorse idriche della provincia di Matera. *Risv. Agric.* 6: 451–460.

PANTANELLI, E. & A. BIANCHEDI. 1929. Irrigazione con acque salmastre. *Econ. della Capitan.* p. 67.

PASSERINI, N. & P. GALLI. 1927. Sperimenti intorno l'azione del cloruro sodico contenuto nell'acqua di irrigazione su di alcune piante coltivate. *Atti Acc. Georg.,* s.5, 24: 191; *Boll. R. Ist. Sup. Agr. Pisa,* 3.

PASSERINI, N. 1932. Innaffiatura con acque di mare. *Italia agr.* 69: 10.

REINA, C. 1957. Acque dolci e salate del sottosuolo di Puglia. *Atti XVII Cong. Geogr. Ital. Bari.*

REINA, C. 1961. Le sorgenti carsiche salmastre del Chidro. *Ass. int. Hydrogeol. Mem. 4.*

REINA, C. 1964. La noria o la pompa per lo sfruttamento delle acque sotterranee salmastre del litorale pugliese? *Puglia Agric.* 4: 347–359.

ULPIANI, C. & O. BORDIGA. 1913, 1927. Esperienze di irrigazione con acque salmastre; *Comm. Reale Irr.,* 2 Relaz. p. 193 Roma 1913; and *In* Opera omnia of C. ULPIANI 2: 689, F.lli Marescalchi, Casalmonferrato, 1927.

ZORZI, L. 1957. Possibilità di emungimento da falde dolci sostenute da acque salmastre. *Atti IX Cong. Unione Geod. Geof. Int. Taranto,* 2: 318–327.

ZORZI, L. 1963. Considerations on Italian legislation about ground water research and utilisation; *In* Science et Tech. p. les régions peu développées. C.N.R. Roma p. 65–70.

ZORZI, L. 1964. L'emungimento quale fattore di conservazione del potenziale idrico sotterraneo. *Atti VI Cong. Int. Genio rur. Losanna.*

ZORZI, L. 1968. Le risorse idriche sotterranee della regione pugliese. Indirizzi per un coordinato piano di utilizzazione. (In press).

ZORZI, L. & C. REINA. 1955. Sulla necessità di controllare e disciplinare le utilizzazioni di acque sotterranee nella penisola Salentina. *Atti VIII Cong. Naz. Irrig. Ital. Milano.*

ZORZI, L. & C. REINA. 1956. Sulla presunta idrografia sotterranea profonda della Capitanata. *Geotecnica*, 1: 11 p.

ZORZI, L. & C. REINA. 1957. Valutazione e sfruttamento delle risorse idriche sotterranee della Conca di Brindisi. *Giorn. Genio Civ.* 10.

ZORZI, L. & C. REINA. 1960. Nuove vedute sul movimento delle acque sotterranee nel bacino artesiano del Tavoliere di Puglia. *L'Acqua*, 1: 12 p.

ZORZI, L. & C. REINA. 1960. Bonifica idraulica superficiale con emungimenti da falde acquifere profonde. *Atti IV Cong. Int. Bonif. Irrig. Madrid.*

ZORZI, L. & C. REINA. 1962. Idrogeologia della provincia di Taranto. *Gior. Genio Civ.*, 2 : 19 p.

ZORZI, L. & C. REINA. 1963. Le acque sotterranee nel Salento. *Civ. d. Scambi* 81–84.

ZORZI, L. & C. REINA. 1964. Interpretazione idrogeologica della salinità delle acque sotterranee in alcuni bacini esoerici ed endoerici del Mediterraneo. *Sci. Tecn. Agr.* 4: 15–28.

Author's address: Professor Dr. Luigi Cavazza, Istituto di Agronomia, Università di Bari, Via Amendola 165, Bari, Italy.

On Saline Irrigation Problems in Sicily
Remarks to Professor Cavazza's Report

by

GIAN PIETRO BALLATORE

I should like to clarify that part of Professor CAVAZZA's report which deals with the activities of the Agronomy Institute in the University of Palermo. This Institute, under my direction, has interested itself – among other subjects – in the utilization, for irrigation purposes, of the waters of the River Salso which crosses Sicily from North to South. The water from this river becomes saline because of the hydric contribution of affluents originating from zones which belong, from the geological point of view, to gypsum-sulphuric formations also rich in rock-salt deposits.

The morphology of South Sicily would allow the building of a large reservoir, of 40 million m³ capacity, in order to convey to it the winter flows of the river which are then less rich in salt. Five years of daily recording of the river's flow and of its chemical composition has led to the conclusion that the waters of the projected reservoir would have a salinity ranging from 4,000 micromhos/cm at 25°C at the beginning of the irrigation season to 5,330 at the end of the season. As the irrigable soils downstream of the said reservoir are of a more or less clayey nature, some doubts arose as to the utilization of such waters. It was considered opportune, therefore, to carry out a series of studies and, to that end, a 6 ha experimental field was established, situated in an elevated zone in the proximity of the river. At one end of this field a crownshaped reservoir of suitable capacity was built and this is filled annually with the waters of the river. By means of appropriate devices, the salinity of the water in the small reservoir is brought up to about the same amount that would prevail in the large one to be built on the axis of the River Salso.

The experimental field has already been in operation for the last five years and we are of the opinion that the conclusions which are being drawn therefrom may also be valid for other zones of Sicily, where clayey soils are being irrigated with more or less saline waters collected into small artificial lakes in the hills.

Studies on the reaction to saline water on the part of several vegetable species have generally confirmed various facts of which we were already aware from the literature; however, for some of the species it has been possible to single out varieties having a different degree of tolerance towards salinity. Some of these species have been inserted into typical rotations of horticultural and zootechnical farms and have been treated with various irrigational techniques: gravity irrigation, shower irrigation, normal watering volume, washing watering volume, etc. Unfortunately all the species which underwent examination showed intolerance and yield limitations which increased with the years, although in different degrees according to the irrigation methods experimented with.

Analytical research done on drainage waters, soil soluble salts and on the evolution of the absorption system of the soil justifies the negative effects which were observed in regard to vegetation. Indeed, in spite of the efficiency of the drainage network, the field soil shows a tendency to transform itself into a magnesian Solontchak. On the other hand, even the interruption of irrigation for two years out of five (leading to the washing action resulting from a yearly rainfall of 400 mm, such as is recorded in that zone), does not bring soil salinity back to its initial degree.

Other research carried out on the effects of mineral and organic-mineral fertilizers at different dosage levels and ion ratios among nutritive elements did not lead to any appreciable results; some fertilization formulas, at least during the first years, do contribute towards decreasing the effect of saline accumulation but do not succeed in averting the grave danger of involution of the soil's fertility.

However, in the past two years, research has been carried out on lysimeters and experimental plots, with the object of improving the soil's structure by means of correctives, or of modifying the ionic imbalance induced by saline water by means of chemicals whose dosages are calculated on the basis of interchangeable sodium values and on the soil's capacity for cationic interchange. The great amount of data which have been collected and are now being processed give reason to believe that some interesting results will emerge.

As a matter of fact, in regard to the zones under our examination, the present knowledge of salinity balance enables us to affirm that in order to avert any danger of involution of the soil it is not possible to irrigate with a total of more than 1,200 m³ saline water per hectare; and by applying the rules of Mediterranean arid-culture, it is possible to anticipate by the end of summer the sowing of valuable crops (table-tomatoes, cauliflower, artichokes, peas, fennel, etc.), without having to await the autumn rainfall, thereby gaining time.

Because of these limitations, the Land Improvement Trust, operating in the same zone, is trying to give a new direction to the study of this problem involving the utilization of the River Salso's waters for irrigation. It is proposed to single out the sources of saline water which feed some of the

Salso's affluent streams in the downstream stretch of its course. The geological and hydraulic findings available thus far, give grounds for hope as to the feasibility of deviating or, at any rate, neutralizing the saline flows of the said sources, the sodium chloride content of which extend to $200-250\%_0$ values. According to preliminary calculations, by such a procedure, the salinity of the Salso's waters would be reduced to NaCl values of $0.6-0.7\%_0$, thereby rendering their use for irrigation purposes a simpler and less troublesome proposition.

Author's address: Professor Dr. Gian Pietro Ballatore, Istituto di Agronomia, Università di Palermo, Palermo, Italy.

PART II: FIELD TRIALS WITH SALINE IRRIGATION

Plantgrowing with Sea-water and Other Saline Waters in Israel and Other Countries

by

HUGO BOYKO & ELISABETH BOYKO

The main purpose of these experiments was to obtain an experimental answer to our ecological observations that highly permeable soils, particularly sand and gravel, greatly enhance the salt tolerance of plants in general. After the successful growth of about 200 species, mostly non-halophytes, in the Desert Garden of Eilath planted in 1949 (E. BOYKO, 1952, BOYKO & BOYKO, 1968) our experiments tried to answer the following questions:

1. It is possible to grow plants on sand or gravel by irrigation with sea-water, and which are the concentration limits for different species?
2. Is there any salt accumulation in the root layer, and is there any danger of such a salt accumulation in deep layers of highly permeable soils, like sand?
3. What is the growing behaviour of those plant species with which the experiments were made?

The difference between sea-water and saline underground water in their impact on living beings is mainly a *qualitative* one, and not, or to a lesser degree only, a quantitative one with regard to the Total Salt Concentration (T.S.C.). The explanation is that the various types of sea-water have very different concentrations, but they are all sea-water proper because their relative chemical composition, i.e. the percentage of their components, is alike in all of them. This specific composition presents such a well balanced ionic environment for the plants that their capacity for salt tolerance against irrigation with sea-water or diluted sea-water respectively seems to be multiplied in almost all species compared to their tolerance against irrigation with saline water of the same concentration but of another composition.

If we have at our disposal a particularly highly concentrated sea-water, e.g. that of the East Mediterranean Sea or of the Red Sea we can then experiment with most of the other sea-water types of the globe including

those of the ocean simply by means of dilution. The most striking features
of all sea-water types are:

a. The very *high chloride content*. All sea-waters are composed of about 75%
sodium chloride and about 10% other chlorides, mainly magnesium chloride and a relatively high percentage of calcium- and potash compounds as
antagonists.

b. Sea-water contains *all nutritive elements* the plants need, including also –
and this seems to be of particular importance – the necessary trace elements.

c. *Microorganisms*, living or dead, are to be found in considerable amounts
but no adequate study of their possibly decisive importance was made so
far in connection with this research.

These three specific features of sea-water are the reason why the biological effects achieved by experimenting with sea-water *cannot* be compared
with experiments made with NaCl solutions or even with synthetic sea-
water.

We used four types of sea-water in our experiments. They are:

1. *East Mediterranean Sea-water* taken from the shore south of Tel Aviv. It
contained according to the analyses carried out in the chemical laboratory
of the Ministry of Agriculture, Israel:

$$
\begin{aligned}
CaSO_4 &- 2{,}965 \text{ p.p.m.} \\
MgSO_4 &- 1{,}713 \quad ,, \\
MgCl_2 &- 3{,}530 \quad ,, \\
KCl &- 870 \quad ,, \\
NaCl &- 31{,}700 \quad ,, \\
\text{Various} &- \underline{2{,}822 \quad ,,} \\
& 43{,}600 \text{ p.p.m.}
\end{aligned}
$$

The actual concentration fluctuated between this and a higher one due
to evaporation from the partly open tank at our disposal wherein we had
to keep the sea-water until the next filling. The highest concentration on
record was 53,000 mg/l T.S.C.

The other sea-water types were simply made by appropriate dilutions
with tap water, namely 75, 50 and 25%. Control experiments were, of course,
carried out with fresh-water. Thus we used:

2. sea-water of oceanic concentration;
3. sea-water of the North Sea type
4. sea-water of the type of the Caspian Sea.

The types of sea-water used by us and their amplitude of concentration
are to be seen from the following Table:

Sea-water types used	Total Salt Content in part per million (ppm)	Sodium chloride (NaCl) in part per million (ppm)
1. East Mediterranean or Red Sea	40,000–53,000	30,000–39,750
2. Oceanic Concentration	30,000–40,000	22,500–30,000
3. North Sea (near Holland)	20,000–27,000	15,000–20,250
4. Caspian Sea or Aral Sea	10,000–13,250	7,500– 9,940
Fresh water for the control experiments with each species	± 400	± 200 Cl

Lacking any budget for these experiments we constructed cheap but exact lysimeters from empty asphalt barrels in order to have a continuous control, quantitatively and qualitatively, on water and salt percolation.

The ten species experimented with were:

1. *Agropyrum junceum* P.B.   (Mon. Biol. XVI, p. 238–248)
2. *Calotropis procera* R.Br.   (,,   ,,   ,,   ,, 248–254)
3. *Juncus maritimus* Lam.   (,,   ,,   ,,   ,, 254–259)
4. *Juncus punctorius* L.   (,,   ,,   ,,   ,, 259–261)
5. *Rotboellia fasciculata* Desf. (,,   ,,   ,,   ,, 261–266)
6. *Ammophila arenaria* L.K.   (,,   ,,   ,,   ,, 276–277)
7. *Agave fourcroydes* Lem.   (,,   ,,   ,,   ,, 271–275)
8. *Agave sisalana* Perr.   (,,   ,,   ,,   ,, 271–275)
9. Barley (Beduin strain)   (,,   ,,   ,,   ,, 267–271)
10. Sugar beet   (,,   ,,   ,,   ,, 277–279)

Of these the species Nr. 1–5 have been subjected to continuous measurements and the development of the stems, leaves, and fruitstands of each individual has been recorded separately. The species Nr. 6–10 were regarded as orientation experiments.

The pages in brackets mentioned after each single species refer to the detailed description of these experiments in the volume complementing this book, i.e. "Salinity and Aridity – New Approaches to Old Problems" (H. Boyko & E. Boyko, 1966) which appeared as vol. XVI of the Monographiae Biologicae. There detailed report is given on the development of the plants during the years of treatment with sea-water as well as a description of the new methods in order to enable verification or execution of similar experiments elsewhere.

All plants showed a many times multiplied salt tolerance compared with plants grown on normal agricultural soil.

It seems that in all species heliotropical growth was retarded due to irrigation with sea-water of concentrations higher than 10,000–13,000 ppm T.S.C., while horizontal and geotropical growth do not seem to be significantly influenced. Under irrigation with 10,000–13,000 ppm T.S.C., however, neither of the species *Calotropis procera, Juncus maritimus, Juncus punctorius* and *Agropyrum junceum* showed any significant retardation of heliotropic growth, and the (horizontally) creeping stems of *Rotboellia fasciculata* had even greatly elongated internodes under irrigation with 20,000 ppm T.S.C. sea-water compared with the individuals irrigated with fresh water or with the 10,000 ppm type.

Vitality or vigour as expressed by drought resistance seems to be strongly influenced insofar as it grows with higher concentrations up to 20,000–30,000 mg/l T.S.C. according to the species. The lowest drought resistance recorded during our "Dying Experiments" (carried out simply by stopping any irrigation) was determined in those plant individuals, which were used in the parallel control experiments and, for this purpose, formerly irrigated with fresh water.

Analyses made in the laboratory as well as on soil profiles in field investigations proved that no salt accumulation is to be feared in sand cultures, if the sand is deep enough to allow an adequate continuous drainage. On the other hand, plants which are able to accumulate salt in their vegetative organs, and which are regularly harvested are assisting in the desalination of the soil, a process which we called "biological soil desalination."

As Chairman of the International Commission of Applied Ecology of IUBS the senior author tried, since 1950, successfully to induce other countries to experiment with saline water and, since 1957, with sea-water, in similar lines. Sponsored partly by UNESCO, a number of such successful experiments with highly saline water and with sea-water could be performed since then along a climatic profile from the Tropics to Northern Europe.

Thus, for instance, the sea-water experiments, carried out in India (at the Central Salt & Marine Chemicals Research Institute, Bhavnagar) led, among other results, to the growing of a wheat variety under irrigation with sea-water from the Indian Ocean diluted to a concentration of 20,000 ppm. The yield obtained was about the same as that obtained from the same variety raised in the control experiments.

Another plant species, *Carthamus tinctorius* L., an edible oilplant, was successfully grown even with undiluted ocean water, of a total salt concentration of 36,000 ppm.

In Orinon (Spain) dozens of different species of vegetables, cereals and flowers were grown with the waters from the Atlantic Ocean; in these experiments the high tide was used to fill the reservoirs and direct irrigation was extremely successful. From this place stems also the very interesting historical information that we were not the first to grow plants with sea-water. Old documents prove it for a sand island near Bilboa more than 250 years ago and good harvests were recorded during a period of famine (E. ESTEBAN-GOMEZ, 1965, 1968).

In Sweden, near Stockholm, a 60% increase in the productivity of natural coastal pastures was achieved by using for irrigation the water of the Baltic Sea (HELGREN, 1959), which has "only" 6000–8000 ppm T.S.C., but is, qualitatively pure sea-water, its components being the same and in the same proportion as those in waters from the ocean.

M. P. D. MEYERING's (1966) experiments with the excellent fodderplant *Agropyrum junceum – boreo-atlanticum* Simon & Guin. belong, too, into this series. His irrigation water was taken from the North Sea at Spiekeroog and had a T.S.C. of 32,000 ppm. The best results he achieved were with 15,000–20,000 ppm T.S.C.

Apart from the Desert Garden at Eilat, quite a number of successful experiments have been carried out during the last two decades with highly saline underground waters. The most important ones – from the scientific point of view – are probably those executed in Tunis under the auspices of UNESCO (VAN HOORN et al., 1968) and in Southern Italy with the centre in Bari. There, the scientists at the Agricultural Research Station of the Government as well as of the University of Bari are since long successfully engaged in such experiments (FICCO, 1968; LOPEZ, 1966).

At some places these experiments have led already to economic applications as for instance in the Settlements of Wadi Araba and in the Juncus-esparto plantations on the sand dunes and sand flats between Yotvata and Timna about 35 kilometers north of Eilat; here this valuable plant-raw-material for the production of cellulose for first class printing paper is grown commercially under irrigation with waters containing 2000 ppm sodium chloride.

The success of all these experiments and their potential economic and social significance for the worldwide fight against hunger induced the 104 participants from 24 countries in the International UNESCO-WAAS-Italy Symposium (Rome, September 1965) on this subject to accept unanimously a resolution recommending all international and national Agencies to give strong support to further research in this field.
This Resolution reads:

The participants in the International Symposium on irrigation with saline water and sea-water with and without desalination, organized by the World Academy of Art and Science, the Accademia Nazionale di Agricoltura and the Consiglio Nazionale delle Ricerche, and sponsored by UNESCO, firmly believe that
results already achieved by irrigation with highly saline water indicate clearly that those arid areas at least where such a water supply is available, can be rendered capable of crop production, and they therefore strongly recommend to international and national organizations concerned with human welfare that financial provision continue to be made for the necessary expansion of research and field trials.
The following fields are seen as of particular importance:
1. Ecological, physiological, genetic, biochemical and agrotechnical stn-

dies on promising economic plants under direct irrigation with highly
saline and sea-water under various climatic conditions;

2. Research in the laboratories and in the field on *biological desalination* of
soil and water;

. 3. Microbiological studies connected with these lines of research.

Following this resolution, the Council of the disseminated World University of WAAS (World Academy of Art and Science) acknowledged this
field of research (i.e. Productivizing the Deserts) as of particular importance
for human welfare and it is to be hoped that soon a transnational teamwork of Institutions of Higher Learning from various countries will be
organized in order to accelerate the application of the feasible results in
this new research field.

SUMMARY

Successful experiments were carried out with ten ecologically greatly
differing species under direct sea-water irrigation on dune sand as medium
from July 1957–October 1963 in two series, including a "dying experiment"
without irrigation during the last two years. Four types of natural sea-
water were used: (1) East Mediterranean Sea-water; (2) Ocean concentration; (3) North Sea type and (4) Caspian Sea type. Fresh water (tap water)
was taken for the control experiments.

The ten species experimented with are: 1) *Agropyrum junceum* P.B.; 2) *Calotropis procera* R.Br.; 3) *Juncus maritimus* Lam.; 4) *Juncus punctorius* L.; 5) *Rotboellia
fasciculata* Desf.; 6) *Ammophila arenaria* Link; 7) *Agave sisalana* Perr.; 8) *Agave fourcroydes* Lem.; 9) Barley (Beduin strain); and 10) Sugar beet.

Of these the species Nos. 1–5 have been subjected to continuous measurements and the development of each individual has been recorded separately. The species Nos. 6–10 were regarded as orientation experiments.

As Chairman of the International Commission for Applied Ecology of
IUBS the senior author tried, since 1950, successfully to induce other
countries to work with saline water and, since 1957, with sea-water on similar
lines. Sponsored partly by UNESCO a number of successful experiments
with highly saline and sea-water were the consequence along a climatic
profile from the Tropics to the Baltic Sea.

On the basis of a resolution of the International UNESCO-WAAS-Italy
Symposium on saline irrigation (Rome, 1965) the World Academy of Art
and Science is now at work to organize a transnational teamwork in this
new field of research within the framework of its disseminated World
University.

RÉSUMÉ

Des épreuves ont été faites avec dix espèces très différentes du point de vue écologique, espèces soumises à irrigation directe d'eau de mer, avec du sable de dunes comme moyen à partir de Juillet 1957–Octobre 1963, en deux séries, y compris un "dying experiment" sans aucune irrigation pendant la dernière année. Voici les types d'eau naturelle de mer qui ont été employés: (1) Eau de la Mer Méditerranée de l'Est; (2) Concentration Océanique; (3) Type de la Mer du Nord; (4) Type de la Mer Caspienne. De l'eau fraiche (eau du robinet) fut employée pour des épreuves de contrôle.

Voici les dix espèces que l'on a essayées: 1) *Agropyrum junceum* P.B.; 2) *Calotropis procera* R.Br.; 3) *Juncus maritimus* Lam.; 4) *Juncus punctorius* L.; 5) *Rotboellia fasciculata* Desf.; 6) *Ammophila arenaria* Link; 7) *Agave fourcroydes* Lem.; 8) *Agave sisalana* Perr.; 9) Orge (type Bedouin); 10) Betterave à sucre.

Les numéros 1–5 des espèces sus-mentionnées ont été soumises à des contrôles continus et le développement de chacune d'entre elles enrégistré séparément; les espèces n. 6–10 furent considérées comme des épreuves d'orientation.

Comme Président de la Commission Internationale de l'Ecologie Appliquée, de l'IUBS, l'Auteur a réussi avec succès à convaincre dès 1950 d'autres pays à travailler sur des voies parallèles. Grâce à l'aide de l'Unesco un nombre d'épreuves ont été faites avec de l'eau de mer le long de la ligne climatique à partir des Tropiques jusqu'à la Mer Baltique.

Sur base d'une résolution adaptée par le Symposium International de l'UNESCO-WAAS-Italie sur irrigation avec de l'eau salée (Rome 1965), la World Academy of Art and Science est maintenant en train d'organiser une coopération transnationale qui s'occupera de cette nouvelle ligne de recherche dans le cadre de la "World University."

RIASSUNTO

Dal luglio 1957 all'ottobre 1963 sono state fatte due serie di esperimenti, incluso uno "fino alla morte" senza irrigazione durante l'ultimo anno, con dieci specie ecologicamente molto diverse, sottoposte ad irrigazione diretta con acqua di mare usando la sabbia delle dune come substrato.

Sono stati usati quattro tipi di acqua naturale di mare: (1) Acqua del Mediterraneo Orientale; (2) Acqua a concentrazione oceanica; (3) tipo del Mare del Nord; (4) tipo del Mar Caspio.

Per gli esperimenti di controllo è stata usata acqua fresca (acqua di rubinetto).

Le dieci specie esperimentate sono: 1) *Agropyrum junceum* P.M.; 2) *Calotropis procera* R.Br.; 3) *Juncus maritimus* Lam.; 4) *Juncus punctorius* L.; 5) *Rotboellia fasciculata* Desf.; 6) *Ammophila arenaria* Link.; 7) *Agave fourcroydes* Lem.; 8) *Agave sisalana* Perr.; Orzo (razza beduina); 10) Barbabietola da zucchero.

Fra queste, le specie 1-5 sono state sottoposte a continue misurazioni e lo sviluppo di ognuna di esse è stato registrato separatamente. Le specie 6-10 sono state considerate quali esperimenti di orientamento.

In qualità di presidente della Commissione Internazionale di Ecologia Applicata della I.U.B.S., l'autore principale tentò, fin dal 1950, di indurre altri paesi a lavorare su tali indirizzi. Con l'aiuto dell'UNESCO è stata fatta una serie di esperimenti con acqua marina in una gamma di climi che va dai Tropici al Mar Baltico.

REFERENCES

BOYKO, ELISABETH. 1952. The Building of a Desert Garden. *J. Roy. hortic. Soc.* London 76: 1-8.

BOYKO, ELISABETH & HUGO BOYKO. 1968. The Desert Garden of Eilat. *In* H. BOYKO [Ed.] Saline Irrigation for Agriculture and Forestry, Dr. W. Junk, N.V. The Hague.

BOYKO, HUGO (ed.]. 1966. Salinity and Aridity – New Approaches to Old Problems. Mon. Biol. XVI. pp. VIII + 408; Dr. W. Junk, N.V. The Hague.

(*in it*): 1) BOYKO, HUGO: Basic ecological *principles* of plant growing by irrigation with highly saline or sea water p. 131-200.

2) BOYKO, HUGO & BOYKO, ELISABETH: *Experiment* of Plant growing under irrigation with saline waters from 2000 mg/liter T.D.S. up to sea water of oceanic concentration, without desalination, p. 214-282.

An extensive bibliography is to be found in both papers.

FICCO, NICOLA. 1968. Irrigation with saline water in Puglia, p. 161-167. *In* H. BOYKO [Ed.] Saline Irrigation for Agriculture and Forestry, Dr. W. Junk N.V. The Hague.

ESTEBAN-GOMEZ, D. ISIDRO. 1965. Sestao, origen en el mundo del riego directo de cultivos agricolas con agua de mar, p. 1-32. Centro Coordinador de Bibliotecas de Vizcaya. Sestao, 1965.

ESTEBAN-GOMEZ, D. ISIDRO. 1968. Agricultural cultivation by means of direct irrigation with undesalinated sea-water p. 93-106. *In* H. BOYKO. Saline Irrigation for Agriculture and Forestry, Dr. W. Junk, N.V. The Hague.

HEIMANN, HUGO. 1966. Plant growth under saline conditions and the Balance of the Ionic Environment p. 201-213. Mon. Biol. XVI, Dr. W. Junk, N.V. The Hague.

HELGREN, GUNNAR. 1959. Irrigation with Seawater in Sweden. Report Conf. on supplementary irrigation, Commission VI. Int. Soc. of Soil Science.

LOPEZ, G. 1966. Irrigation with saline water in Puglia. p. 294-313. Mon. Biol. XVI. Dr. W. Junk, N.V. The Hague.

MEIYERING, M. P. D. 1968. Diluted seawater as a suitable medium for animals and plants, with special reference to *Cladocera* and *Agropyrum junceum boreo-atlanticum* SIMON. & GUIN. p. 331-356. Mon. Biol. XVI. Dr. W. Junk, N.V. The Hague.

VAN HOORN, J. W., CH. OLLAT, R. COMBREMONT & G. NOVIKOFF. 1968. p. 168-186. *In* H. BOYKO. Saline Irrigation for Agriculture and Forestry. Dr. W. Junk, N.V. The Hague.

Authors' address: World Academy of Art and Science, 1 Ruppin Street, Rehovoth, Israel.

Agricultural Cultivation by means of Direct Irrigation with Undesalinated Sea-water

by

Isidro Esteban-Gomez

Since the earliest times, the problem of utilizing sea-water had occupied the human mind. We may trace it back to the days of the Greek philosopher Aristotle, whose experiments and data on this subject are recorded. In a less remote past, we find the question dealt with in Francis Bacon Lord Verulam's book "Silva Sylvarum," published in London in 1623, and in a work by Louis Ferdinand, Count of Marsilli, a member of the Royal Academy of Sciences of Paris, the latest edition of which appeared in Amsterdam in 1725. To this short list we may add the curious sandstone filter used on the Canaries for purifying salt water – a device which goes back to the first inhabitants of those islands, the "Guanches."

While these attempts to desalinate sea-water, undertaken in the far past, show how ancient man's interest in this problem is, the results obtained and the experiments undertaken were based on the simple filtration of sea-water through different kinds of soil or sand. It took several centuries until these empirical experiments produced concrete results, and only in our times, since the last world war, the first practical and more or less profitable results on the difficult road to the desalination of sea-water became evident.

The extraordinary progress of science and technology in the last few years finally gives ground for hoping that we may place the inexhaustible water reserves of the ocean at the disposal of man and of the many drought-stricken parts of the earth, and that we shall develop efficient and cheap formulas for their exploitation.

In addition to the possibilities of sea-water desalination, there is also need for a new approach, simpler, less complicated and easier accessible to the poorer nations: the direct use of sea-water, at any rate for agricultural purposes, in order to approach the problem of the desalination of water and soil from a new viewpoint – the one which has given its name to this Conference: "New Approaches to Salinity Problems," and to which we participants have dedicated our efforts now for a number of years. I have the privilege to describe here one of these new approaches.

THE MIXED HYDROPONICS SYSTEM AND ITS IMPLEMENTATION ON THE ORINON BEACH (SANTANDER)

In 1958 we received an experimental plot on the Orinon beach in the province of Santander, and started there with our experiments in April 1959.

The beach of Orinon lies on the North coast of Spain, halfway between Bilbao and Santander, on the shore of the Cantabrian Sea, at a longitude of 43°24′ N. and a latitude of 3°19′30″ W. It is a sandy stretch with a length of 1 km and a similar width at low water, closed between two cup-shaped masses of rock, Mount Cerredo (645 m high) and Mount Candina (487 m high), which form its eastern and western limits; the coast is precipitous.

The flora and fauna of the beach is that which one would expect on an open sea-shore far from industrial centres and large rivers, and in that respect is, with the exception of certain climatological characteristics, similar to most other beaches in the world. The East-West boundary of the area farthest away from the sea consists of a dune formation two to three meters high, which has been formed by abundant and highly variegated spontaneous vegetation, including such species as *Agropyrum*, *Eryngium maritimum*, *Euphorbia paralias*, *Juncus*, *Genista horrida*, *Polypodium vulgare*, etc. Beyond these dunes lies the beach proper, which is of marine origin and has been formed by erosion caused by the sea and the predominant North-East wind. Its sand is reddish and has a very fine and homogeneous grain, composed as follows:

TABLE I

Mechanical Analysis of the Sand of the Orinon Beach

Size of grains:	in %
Fine sand (less than 0.02 mm)	82
Medium sand (0.02–0.2 mm)	17.5
Coarse sand (0.2–2 mm)	0.3
Coarse elements larger than 2 mm	0.2
	100.0
Calcium Carbonate	39.31
Highly soluble salts	0.50
Chlorine (as chlorides)	0.052
Organic matter	0.40
Reaction (pH)	6–6.5
Silica	59.24

GEOLOGICAL DATA FOR ORINON BEACH

Orinon, like the whole of the province of Santander to which it belongs, is of Secondary or Mesozoic formation, with a predominance of Cretacean layers.

Geologically speaking, the beach itself, which is enclosed between Mounts Cerredo and Candina, is a recent marine alluvial formation. The Urgoniensian subdivision of the mountains themselves, from the sea inwards, belongs to the Lower Cretacean. Farther inland, they belong to the Wealdensic substratum which forms a transition between the Jurassic and the Cretacean.

Above the plain formed by the village of Orinon, which is delimited by the base of Mt. Candina and the estuary of the river of Orinin (Rio Aguera) along a West-East line, with the sides of the angle meeting in the South at Pontarron, a continental alluvial plain has been formed along the bed of the Aguera. This plain consists of a clayey soil and is used as farmland by the inhabitants of the village. In the direction of the sea, this clay plain extends up to the dunes and beach of Orinon, where it is replaced by an intermediate zone of mixed clay and sand about 150 m wide. This area too is cultivated and bears much earlier crops than the clay land. It also has attractive stands of sea pine and eucalyptus.

AVERAGE SALINITY OF SEA-WATER AT ORINON BEACH

As already stated, the beach lies on the open sea and barely receives any fresh water, particularly in the summer months, the more so as the nearby Aguera river carries little water and does not affect the salt content of the sea-water, which dominates at high tide and has oceanic characteristics, as may be seen from the following tables:

TABLE II

Tides and Sea-water Conditions

Date	Tide	Place	pH	Temp. of sea-water	Density	Salts ‰
17.9.58	High	Point F.	5.5	21°C	1.029	36.86
26.8.60	Low	Centre of beach		21.5°C		
31.8.61	High	Point H.		22°C	1.025	31.12
10.9.64	Spring	Point F.		19.25°C	1.025	33.23

Depth: 0–1 m
Summer Months: June–September incl.

Average temperature of sea-water during summer	20.93°C
Average salinity of sea-water during summer	34.73‰

TABLE III

Composition of average salt contents as sampled during the summer months at Orinon

Components	% of sample	g/l
sodium chloride	77.8	27.019
magnesium chloride	10.9	3.785
magnesium sulfate	4.7	1.632
calcium sulfate	3.6	1.250
potassium sulfate	2.5	0.868
calcium carbonate	0.3	0.104
other components	0.2	0.069
	Total salts ‰	34.727

This sample also represents the water used for irrigation at our experimental station, where the main cultivation activities were carried out only in summer.

OTHER OCEANOGRAPHIC DATA ON ORINON BEACH

TABLE IV

Orinon Bay: Depth at Lowest Ebb

Point of reference	Depth in m	Bottom
Cape Cabollero, West end of Bay	8.7	Rock
„ , Centre of Cape	7.6	Sand
Village of Sonabia	8.–	„
„ opp. centre of Bay	8.4	„
Centre of Orinon Beach	1.7	„
Cercanda Island, East of Bay	10.–	„
Sand Beach at East of Bay	0.7	„

TABLE V

Tide Level at Orinon [1]

1. Data taken for the port of Patron Santander, nearest port to Orinon, taking into account the local tide difference (Flood: —0.1 m; Ebb: 0.0 m).

Year	Highest flood m	Lowest ebb m	Level at Orinon m
1959	4.20	0.10	4.10
1960	4.59	0.54	4.05
1961	4.30	0.00	4.30
1962	4.40	0.00	4.40
1963	4.78	0.33	4.45
1964	4.30	0.10	4.20
1965	4.30	0.10	4.20
	Average height of tide: 4.24 m		

This average height is the one which has been taken into account at Orinon in pumping the sea-water up for irrigating the plantation, for at many beaches, high tide allows several meters in raising the water for irrigation, which is a considerable advantage of the use of sea-water. It should, moreover, be noted that here at Orinon, when there were high tides and winds from the sea, particularly in spring and winter, the sea has at times flooded and destroyed the fields of the agricultural station, but, for reasons to be explained later, without causing the soil to become saline.

LOCAL CLIMATE AT ORINON BEACH DURING SUMMER

The climatological classification of Orinon Beach is "moderate maritime." The data provided below refer exclusively to four months of the summer season (June to September, which are the warmest months here). During these months the station is in full production, and they are therefore the only ones which directly affect the production of crops by our method of Mixed Hydroponics. The fact that the climatological data for other seasons are different does not affect the results of other, complementary investigations – oceanographical, mineralogical, etc. – which were conducted during the remainder of the year, at a time when it was also necessary to protect the installations and the land against the rigours of the weather caused by the nearness of the sea.

Temperature data

Average temperature during the cultivation season 23.3°C
with maximum temperatures up to 37–38°C

Humidity data

Average relative humidity in cultivation season 80.7%
Tending to drop to 40% during heat waves.

Data on a heat wave at Orinon Beach, for purposes of comparison with conditions during hamsin weather in Israeli Negev or a sirocco or irifi in the Spanish Sahara: "Suffocating heat, with persistent gales reaching velocity peaks of 66–77 km/h which uprooted some trees, flattened many of the cultivated crops on the ordinary fields of Orinon, and caused some damage to buildings in the village."
Maximum temperature: 40°C,[1] rising from 22°C.
Median temperature for the day: 30°C, from a median temperature for the previous day of 20.5°C.

[1]. These data were recorded during the heat wave at our Experimental Station at Orinon on September 16, 1961.

Relative humidity of air during heat wave: 40%, dropping from 80%.

Such heat waves are normally accompanied by a haze and considerable reduction of visibility, with the air above the beach containing great quantities of fine sand and dust up to a height of 3–4 m. This obviously endangers the crops, unless preparations for counter-measures are made.

Damage to crops grown at our Experimental Station due to sand storms may be estimated at no more than 2% if occurring during the normal vegetative period, and with the application of our emergency measures for such cases, which have been thoroughly studied in view of their extraordinary interest for desert zones, plantations will survive the adversity with no greater damage than has been indicated.

Wind

On the part of the Orinon Beach where our plantations are located, there always are fairly strong winds or breezes (No. 4 on the Beaufort scale, or 19–26 km/h). Mostly they blow from the North-East, with occasional exceptions in summer, chiefly during heat waves of the type described above, when the wind blows from the South or South-West.

TABLE VI

Monthly Rainfall at Orinon during Summer June–September (4 months) (in mm)

Year	June	July	August	Sept.	Total
1959	72.9	64.7	96.1	171.–	404.7
1960	111.9	74.–	57.5	182.3	425.7
1961	119.–	45.8	38.8	80.3	283.9
1962	35.6	36.8	12.5	49.6	134.5
1963	117.5	46.–	239.4	150.9	553.8
1964	22.9	32.4	145.4	71.6	272.3

Monthly average rainfall during summer months:
84.7 mm.

Regarding the rainfall factor, it should be stated that in order to make sure of the effectivity of the Mixed Hydroponics method, the plants were protected against rain; moreover, in respect of those crops which received the rainwater fallen during the summer, it was evident, that, in this region, wherever there is a sandy, high-drainage soil, cultivation cannot be carried out satisfactorily with such rain-water quantities at disposal. On the control fields kept at our Agricultural Station each year and treated in every respect in the same manner as the other fields except for having to depend exclusively on rainfall for their water requirements, the majority of the plants withered and crops were far smaller than those obtained after irrigation with sea-water.

In any case irrigation with a quantity of water corresponding to that of the rainfall mentioned in the above table is not sufficient on sand in the hot or temperate zone, since the volume needed lies between 6,000 and 12,000 m³ per hectare per crop (over four months), and rainfall at Orinon Beach only supplied 3,388 m³ over the same period, or half of the quantity required.

Mixed Hydroponics farming in our hot summer climate requires 4,000 m³ per hectare per crop over four months, so that at Orinon Beach we would have been faced with a deficit of 612 m³/ha if we had not been able to use sea-water without limitation, not only in order to provide the minimum requirements of the crops, but also to supply the water by means of irrigation at exactly the right moment, without having to depend on the vagaries of the rain, which in summer falls very irregularly indeed.

Actually, the rain, scarce as it was, presented us with serious difficulties, since, together with the salts of the sea-water, it tends to wash away the chemical fertilizer which has been applied to the fields and to deprive the root zone of its nutritious elements.

On the beach of Orinon, we began in 1959 to implement our system of Mixed Hydroponics Farming, which permits the use of sea-water or water of similar composition for irrigation.

Though actual farming by this method only began in April 1959, we had by then spent several years studying the beach, which combines a number of highly favourable natural conditions in addition to its remoteness in those days, which greatly facilitated our work. Once the necessary preliminary information had been collected, we were authorized by the Spanish Marine Authorities to establish the first Marine Agricultural Experimental Station.

DEFINITION OF THE "MIXED HYDROPONICS" METHOD

The name we gave to our system derives from the fact that it involves aspects and operations used in traditional farming as well as modern hydroponic techniques: hence the name "Mixed Hydroponics."

Our irrigation system is as follows: The sea-water is allowed to run into the irrigation channels and ditches surrounding the cultivated fields, as it is used in the infiltration system in classical agriculture. As a result of the excellent draining qualities of the soil, the water penetrates into the deeper layers and moistens the borders of the fields, without spreading laterally in this stage. Later it tends to rise to the surface by virtue of capillary force and evaporation, making its way through the pores of the sand, moistens the cultivated area from below, keeps the cultivated surface constantly at the required degree of humidity and supplies the roots of the plants with the water needed for their development. The cultivated area will not be salinated, since the salts tend to stay in the subsoil and are eliminated through the natural drainage zone. This may be understood if one considers that the concentration of salts – for instance, of sodium chloride, which because of its enormous concentration would be liable to cause the greatest damage – never occurs in this type of sea-water to the degree of saturation or anywhere near the saturation point.

As a matter of fact, the sea-water which we use for irrigation contains common salt (NaCl) in a concentration of 27.019 g/l, while a saturated aqueous solution at normal temperature would have a concentration of 360 g/l.

In order to explain the mechanism controlling the operation of irrigation, let us point out that every time a given amount of irrigation water sinks into the soil, it forces out an equivalent amount of water from the last irrigation supply, so that there never is a higher salt concentration in the zone of the channels than in the sea-water used for irrigation.

Between one irrigation and the next, evaporation of water into the atmosphere takes place in the channel zone, and the part of the sea-water which does not evaporate penetrates deeper into the soil, carrying along all the salts present in the soil, because, as has been shown, it is much farther from the saturation point than it would be after a complete irrigation cycle.

This theoretical-technical explication of the irrigation cycle in Mixed Hydroponics is, from the purely scientific viewpoint, yet complicated by the rules and laws of the "ion theory." It is in this field of "ion exchange" or "base exchange," in which we hold a patent for "Desalination of Sea-water by means of Ion Exchange" known as the Gomez-Patent. The ion Exchange theory provides an explanation for the electrolytic processes occurring in sea-water when it is used for irrigation and brought into contact with a suitable sandy soil.

Since sea-water occurs in the oceanic mass in a state of ionic dissociation, we may break down the average sea-water sample of Orinon, which has a total salt content of 34.73 g, as follows:

TABLE VII

Sea-water composition at Orinon

Cations		g/l	Anions		g/l
Sodium	Na^+	10.631	Chloride	Cl^-	16.388
Magnesium	Mg^{++}	0.966	Sulfate	SO_4^{--}	2.819
		0.328			1.304
Calcium	Ca^{++}	0.367			0.883
		0.041			0.479
Potassium	K^+	0.389	Carbonate	CO_3^{--}	0.063
		12.722			21.936

Total Cations	12.722 g/l	
Total Anions	21.936 ,,	
	34.659	
Other components	0.069	

The "other components," iodides and bromides, also occur in an ionized state, and account for the remainder of the examined sample.

As can easily be shown, the constituent ions of sea-water and of aqueous solutions in general can, according to the present state of our technical knowledge, be interchanged at will by using a multitude of natural and artificial substances such as clays, zeolites and synthetic resins. These substances, called "ion exchangers," have the property of replacing one type of ions by another in an aqueous solution. Thus a positive H^+ ion may be substituted by Na^+, Mg^{++}, Ca^{++} or another cation, and the hydroxyl ion OH^- for such anions as Cl^-, SO_4^{--}, CO_3^{--}, or other anions which are present.

In the case of irrigation of sandy soil with sea-water, which concerns us at present, it should be noted that the sand originating from the sea contains large quantities of calcium carbonate and acts as a so-called "calcium barrier" inhibiting a harmful ionic activity due to an excess of sea-water salts.

We cannot but conclude that, as the result of the different actions and physico-chemical processes involved, the highly saline oceanic water used for irrigation reaches the roots of the plants in the form of "fresh moisture." This is indicated by the many different crops harvested at our Experimental Station at Orinon over several consecutive years, some of which could be shown in the series of colour slides presented during this lecture.

As a final note, we would like to draw attention to the conspicuous absence of sodium chloride in the sand of beaches in the temperate and hot zone, even though they are covered by sea-water at high tide. This is shown by the fact that, when the tide runs out and the surface water begins to evaporate, there is no visible efflorescence of salt.

METHODOLOGY

Practical application of Mixed Hydroponics "in situ" requires
a series of operations. They are:

(1) An ecological study of the area, including the analysis of its soil and water and the collection of data on its geology, climatology, oceanography, natural resources, terrestrial and marine flora and materials for improving the soil.

(2) Conditioning of the soil according to the required standards.

(3) Marking out the plots for cultivation.

(4) Digging the plots to be cultivated to the depth required for Mixed Hydroponics cultivation.

(5) Balanced fertilizing of the plots, according to the crop to be raised on each plot, and covering them.

(6) Laying out and digging the irrigation ditches round the plots.

(7) Preliminary irrigation to consolidate the borders of the plots and to give the sandy soil the required consistency.

(8) Sowing and planting the plots in lines, furrows or whatever system is desired.

(9) Adequate care for the crops during the growing cycle.

The study of fertilizing requirements of the different crops grown at our Experimental Station at Orinon has led us to the conclusion that the best type of fertilizer to be recommended for Mixed Hydroponics is organic (manure) *and* the classical mineral fertilizers actually used. Fertilizing of sandy soil was carried out in an "incorporated" way, together with sowing and planting. This not only saved fertilizer, but also resulted in larger crops.

On seashore farms, Mixed Hydroponics permits the use of seaweed washed ashore by the tides to cover part of the fertilizer requirement, as has been investigated in our Plan for the Fertilization of the Spanish Sahara.

After the fields have been prepared, irrigation ditches are laid out and dug. Their width and depth depends on the crop to be grown.

Irrigation with saline well- or river-water or with sea-water, is the basic part of the whole process. It is carried out in three stages: preliminary irrigation for firming the borders of the fields; irrigation for tempering and seasoning the soil before sowing or planting; and ordinary irrigation.

(A) Preliminary irrigation, which firms the borders of the fields, is a protection against wind and permits the irrigation ditches to be made.

(B) Irrigation carried out immediately before sowing or planting, conditions and seasons the soil and provides the initial humidity needed for germination.

(C) Periodical regular irrigation with sea-water is carried out during the whole cultivation season and provides the plants with the water needed for their complete development. It is provided by the infiltration method by supplying the necessary quantity of water through the irrigation ditches or channels.

The reason why our method is called Hydroponics, to which we referred earlier, is that the sandy soil used is always kept slightly moist and the roots of the plants are at all times submerged in a bath of fresh water without being exposed to alternating water deficiencies or excesses, or in other words, are always kept at a constant humidity level.

There is a minimum water dosage (critical point) for each growing crop, whose irrigation volume depends on evaporation, air humidity and other factors. We found that this volume, as has been said before, is smaller than that required in classical agriculture. The most important aspect of Mixed Hydroponics is regular and frequent irrigation, because that is the way in which uniformity and economy in the basic water supply is assured.

Sowing and planting in Mixed Hydroponics is carried out much as in traditional agriculture, in lines, in heaps or by casting.

The different operations to be carried out in the course of the cultivation of the crops in addition to irrigation – light weeding, hoeing, top dressing, thinning, treatment against pests and disease and the like – are the same as those familiar to farmers, and as far as Mixed Hydroponics is concerned, are explained in the memorandums and information sheets issued by our Orinon Experimental Station.

The results are presented in Table VIII.

(The following general summary of the results with the main species cultivated by us, were accompanied by a number of slides during the lecture).

TABLE VIII

Principal Crops (grown with sea-water irrigation in the Marine Agricultural Station in Orinon)

Common name	Botanical designation	Variety sown	Harvest kg/ha [1]
Maize	Zea Mais	Double American hybrid	5,500
Maize	,,	Dwarf Canary	3,000
Beans	Phaseolus vulgaris	Garrafal Dwarf	375 [2]
Potatoes	Solanum tuberosum	Espanola	30,000
Lettuce	Lactuca sativa	Orinon	36,000 heads
Tomatoes	Lycopersicum esculentum	Espanola	30,000
Beets	Beta sativa	Amarylla-Lyon	31,500
Radish	Raphanus sativa	Espanola	15,000
Alfalfa	Medicago sativa	Espanola	10,000 [3]
Barley	Hordeum vulgare	Espanola	2,100 grain

[1] extrapolated from small plots
[2] grown together with maize
[3] kg hay per year.

Other plants grown in the normal manner were:

Food-plants: Peppers (*Capsicum annuum*), cabbage (*Brassica sativa*), onions (*Allium sativa*), carrots (*Daucus sativa*), groundnuts (*Arachis hypogaea*), sunflower (*Helianthus annuus*).

Green-Fodder: English ray-grass.

Flowers: Dahlia, carnation, gladiolus, zinnia, nasturtium.

Other plants: Three-coloured amaranth, Mexican agave.

The quality and appearance of the crops were excellent, with the weight and visible characteristics of the fruits harvested from fields irrigated with sea-water being superior to that of fruit from the corresponding control fields, which were irrigated with fresh water, as may be seen from Table IX and X.

TABLE IX

Chemical analysis of the Crop of 1960

POTATOES (Extract from Analytical Bulletin No. 953)

Irrigation	Average weight per tuber (g)	Moisture %	Mineral substances (ash) %	Nitrogenous matter %	Starches %
Sea-water	100	67.10	1.57	0.70	13.90
Fresh water	60	76.86	1.87	0.53	13.70

Visible characteristics: Tubers look sound and peels show no signs of disease.

TABLE X

Chemical analysis of the Crop of 1960

TOMATOES (Extract from Analytical Bulletin No. 909)

Irrigation	Average weight of fruit (g)	Moisture %	Mineral substance (ash) %	Nitrogenous matter %	Acids as H_2SO_4 %
Sea-water	62	92.—	0.60	0.41	0.08
Fresh water	45	91.08	0.59	0.41	0.07

Visible characteristics: Looks normal, pleasantly sweet taste.

HISTORICAL EVIDENCE

Spain is in a position to present the world with an unprecedented historical fact relating to the direct use of sea-water for irrigation; a fact which is of principal importance, since it offers an historical practical proof of the value of irrigation with saline water.

In the year 1719, Spanish barefoot monks of the Carmelite order founded the Monastery of the Desert of San Jose of the Island on a beautiful promontory known at the time as San Nicolas de Ugarte, at Sestao, capital of the province of Viscaya, on the left side of the estuary of Bilbao near the mouth of the river. In addition to the promontory, the monks owned a large stretch of land, of an area of more than 30 ha, in the lowland near the river. This land was known as "The Plantation." In this Plantation, which consisted of sandy soil deposited by the action of the sea, the Carmelites kept an unusual farm which they irrigated with sea-water, letting the sea flow into a lateral channel at high tide and from there into the water in channels of their original farm, which for more than a hundred years produced maize, wheat, grapes, apples and vegetables.

The crops obtained were considerable. We know that in times when the surrounding villages, which used conventional farming methods, had poor crops because of drought or other disasters, the monks came to their aid with the products grown on their salt-water farm. There is evidence to the effect that their maize crops amounted to 2,000 bushels (more than 80,000 kg), from an area which, according to our calculations, covered 15 to 20 ha. Further data on this important and successful case of marine agriculture are presented in the booklet: "Sestao, origen en el mundo del Riego directo de cultivos agricolas con agua di mar." Ayuntamiento de Sestao Centro Coordinador de Bibliotecas de Vizcaya, Sestao, January 1965.

This interesting locality was visited by world-famous personalities, including the famous Spanish novelist FELIX MARIA DE SAMANIEGO and CASPAR MELCHIOR DE JOVELLANOS, writer, lawyer and a great lover of agriculture, who was a minister of the 18th century to King Carlos IV. Also recorded is a visit by the learned German scientist CARL WILHELM VON HUMBOLDT, brother of the famous naturalist ALEXANDER VON HUMBOLDT.

It is a great honour for Spain that the chronicles and documents relating to the Monastery of the Desert of San Jose of the Isle and its religious and agricultural activities have been preserved to this day to bear witness before the whole world.

In view of all the contributions made to this international symposium and the additional information which I could present about the facts which we in Spain have established, there can, I believe, not be the slightest doubt about the importance for the whole of mankind of the direct use of sea-water or of saline water from terrestrian sources for the future development of the different communities of the earth. The enormous quantities of dissolved salts, their nutritious principles and the water contained in the oceans are an inexhaustible source of riches for present-day man and his future descendants, if we only know how to use them wisely.

SUMMARY

The world has been pondering on the use of sea-water since the period of the greek philosopher ARISTOTLE, without having until recently achieved positive advantageous results for its use by man.

A method of cultivation of economic plants by direct sea-water irrigation on sand is described in this paper and is called 'mixed hydroponics." The paper gives a short survey on

(1) past studies on the potability of sea-water and on the studies (experiments) of Mixed Hydroponics;

(2) The practical use of the "Mixed Hydroponics"-method at the experimental site on the Orinon beach (Santander province) discussed in detail;

(3) the results achieved with numerous economic plants on the sands of the Orinon beach by direct irrigation with sea-water of oceanic concentration.

RÉSUMÉ

Depuis le temps du philosophe grec ARISTOTE on a étudié dans le monde entier la possibilité d'exploiter l'eau de mer, mais ce n'est que récemment que l'on a obtenu dans ce domaine des résultats satisfaisants.

La méthode de culture que nous allons exposer a été nommée "hydroponie mixte." Dans ce rapport on va traiter les questions suivantes:

(1) Etudes faites sur la potabilité de l'eau de mer et études (expériences) d'Hydroponie Mixte;

(2) Description et discussion en détails de la méthode d'Hydroponie Mixte et son application pratique dans le centre experimental d'Orinon (province de Santander);

(3) Résultats satisfaisants des expériences faites dans les six années passées sur un nombre de plantes économiques par irrigation directe avec l'eau de mer.

RIASSUNTO

Fin dai tempi del filosofo ARISTOTELE, studiosi di tutto il mondo hanno considerato la possibilità di utilizzare l'acqua di mare, senza tuttavia ottenere, fino a poco tempo fa, risultati positivi in tal senso.

Il metodo di coltivazione che andremo ora ad esporre è stato da noi denominato "Idroponia Mista." La presente relazione tratterà brevemente i punti seguenti:

(1) studi condotti in passato sulla potabilità dell'acqua marina e studi (esperimenti) di Idroponia Mista;

(2) descrizione dettagliata e discussione del metodo di Idroponia Mista (irrigatione con acqua marina) e della sua utilizzazione pratica nel centro sperimentale di Orinon (provincia di Santander);

(3) presentazione dei risultati positivi ottenuti negli ultimi sei anni su un certo numero di piante economiche.

Author's address: Dr. Isidro Esteban-Gomez, Investigadora Quimica Industrial, Bilbao, Spain.

Forestry Trials with Highly Saline or Sea-water in Kuwait

by

Roger Firmin

1. Why Kuwait is interested in the utilization of highly saline and sea-water

We have scanty and costly sweet water resources but plenty of inexpensive and relatively large amounts of highly saline water-tables and aquifers that we would like to use for the reclamation of the country:

(a) by afforestation with mangrove trees in areas like Bubiyan Island and the Northern Bay Coast otherwise unsuitable for any other purpose (map 1);

(b) by afforestation on water-tables, with provisional watering by tube-wells, like the Failacha Island Project and Jahra windbreaks (map 2)

(c) by utilization, on a large scale, of saline aquifers, like at Sulaibeha and Abdaliya wells where salinity is reaching 4000–5000 ppm TDS;

(d) by utilization of 20 mgd [1] of treated sewage water of 2000 ppm TDS in areas having more than 5000 mm evaporation yearly (Piche), on soils usually characterized by a tendency to hardpan formation at variable depths. This will raise problems of secondary salinity which will become apparent in a few years (Map 4).

2. What have we to contribute

Now:

(a) Estimation of field and laboratory comparative experiments on the salinity tolerance of species of trees and shrubs in small experimental sample plots.

(b) Data of a first reconnaissance service of highly saline watertables at shallow depths, in relation with tree growth (tables IV and V and map 5 to 7).

(c) Estimation of the comparative methods of planting trees in saline conditions and of our small experiments in reducing evaporation of trees and shrubs planted in saline conditions.

1. One mgd = 1 million gallon per day.

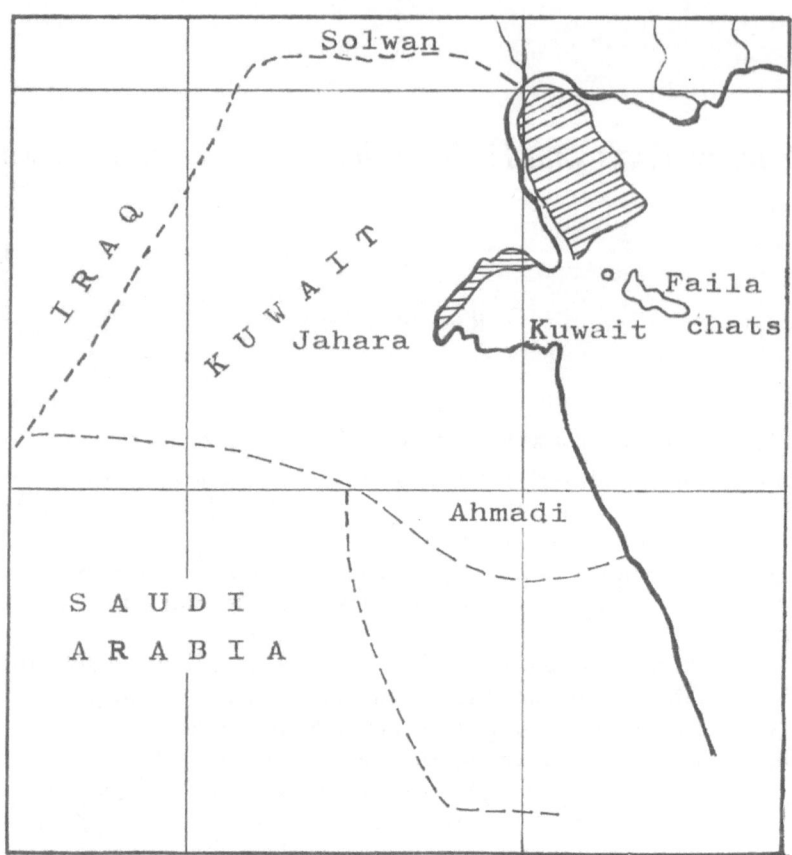

Map 1. Potential mangrove areas.

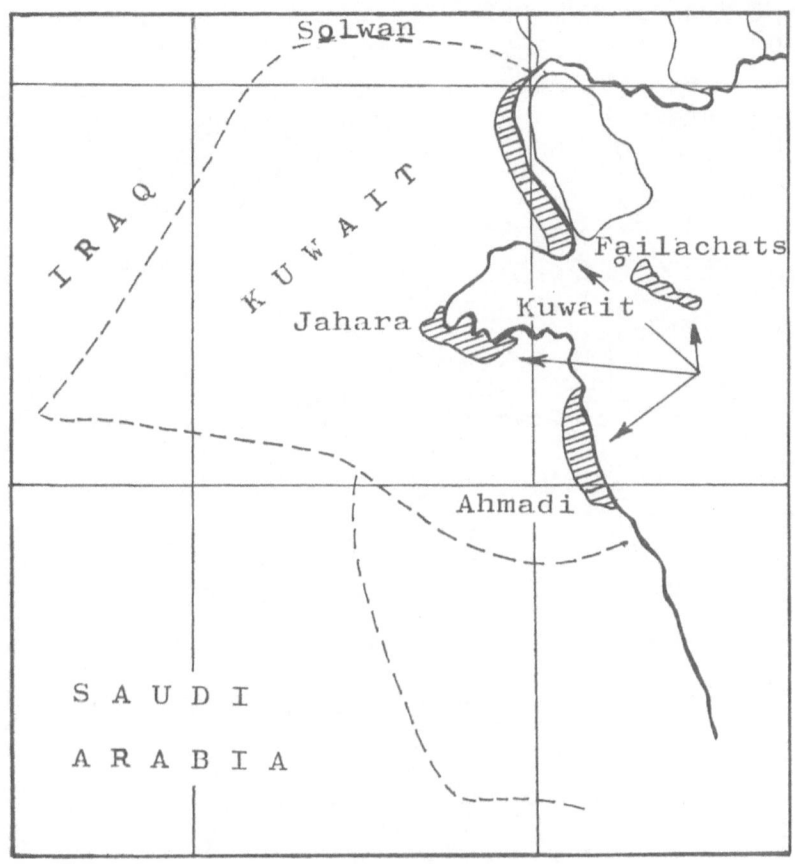

Map 2. Potential afforestation sites on groundwater-tables.

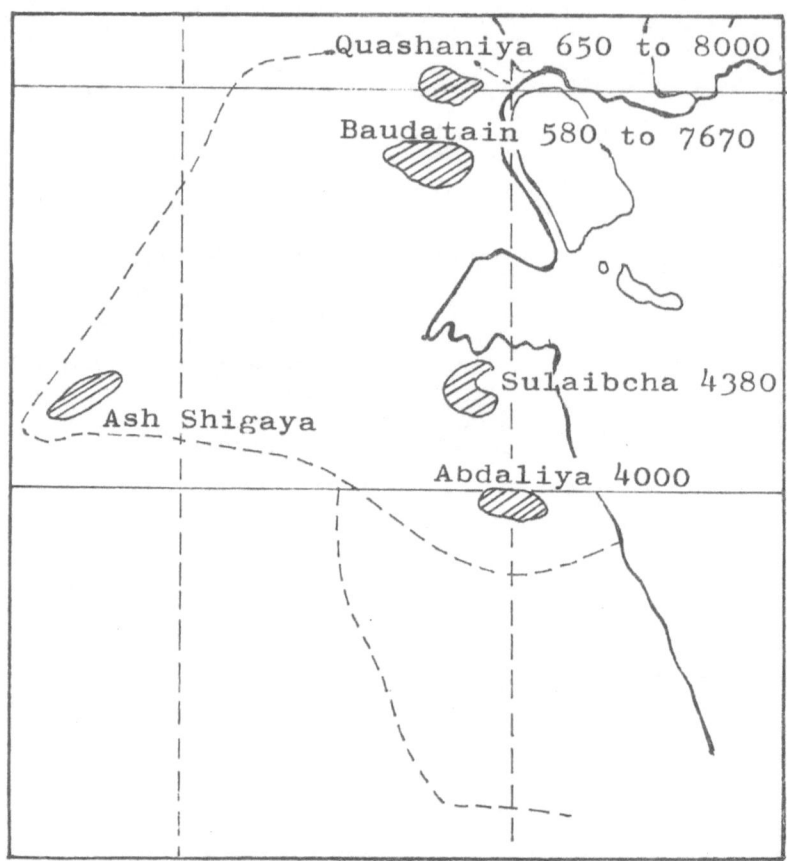

Map 3. Afforestation possibilities based on deep wells (aquifers).

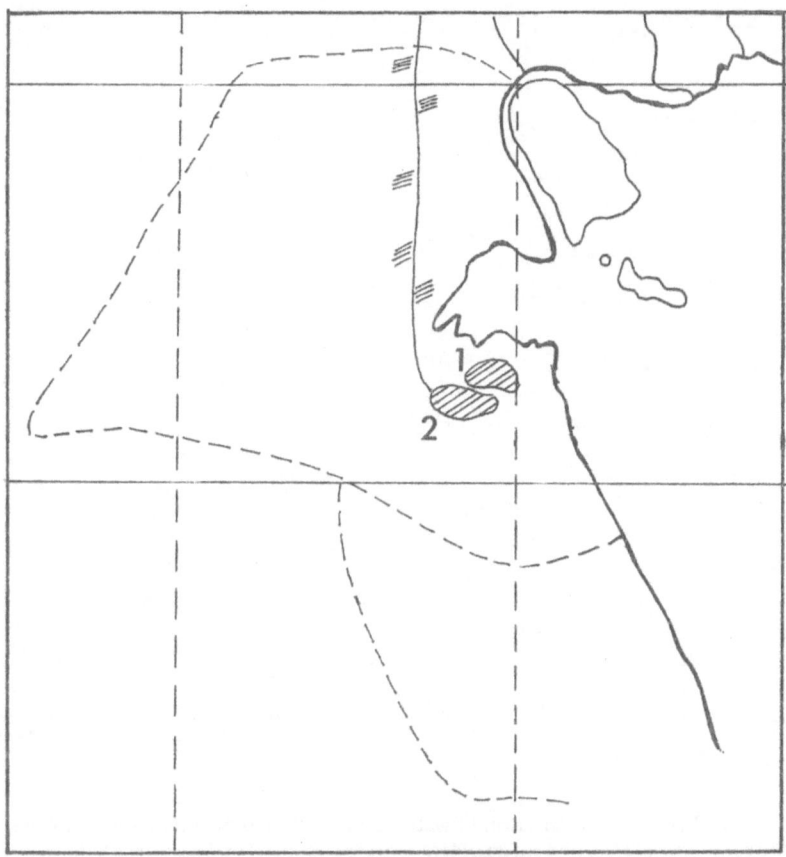

Map 4. Afforestation possibilities based on sewage and Shatt el Arab waters
(provisional siting).

1. Sewage Scheme: now 0.5 mgd 4000
                            20 mgd in 1968 of about 2000
2. Shatt el Arab Scheme: 80 mgd in 1970
                            reaching 1000 ppm

Map 5. Northern Kuwait. Location of water-tables collected by FAO Forestry in 1963/64 in the root-zone of trees and shrubs. Figures are indicating Total Dissolved Salts in ppm. Subiya and the very north are of low salinity, while Bubiyan is of sea-water salinity level and higher.

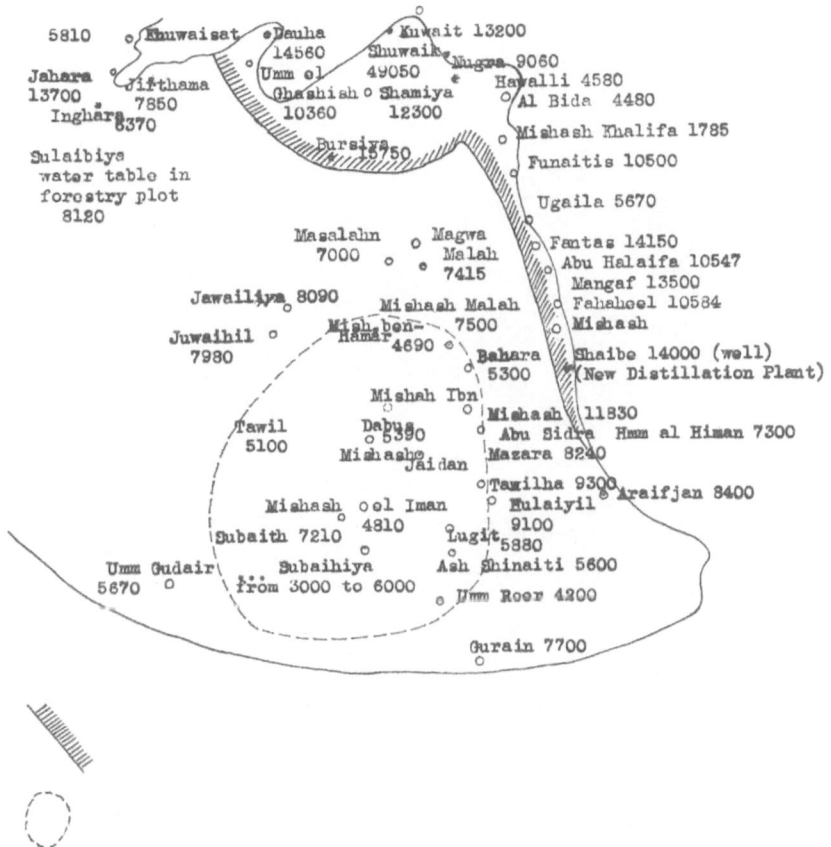

Map 6. Development Zone Jahra-Sheibe. Location of water-tables collected by FAO Forestry in Kuwait in 1963/64 near trees and shrubs. Figures are indicating Total Dissolved Salts in ppm (highest value per location only).

Note: Shallow water-tables, not deeper than 13 m maximum are recorded here in relation with the root zone of most trees although Prosopis can send roots as deep as 30 m.

Map 7. Neutral Zone. Location of water-table samplings near trees and shrubs. Figures are indicating Total Dissolved Salts in ppm, in samples collected by FAO Forestry Mission in 1964 and showing saline and highly saline waters at depths varying from 0.5–10 m deep.

(d) Results of Kuwaiti efforts in reducing the cost of the distillation of sea-water, saline water by electrodialyse, and other methods.

(e) Our problems and what we would like to know to improve our plantation.

In the future:

(a) Our plans of afforestation.

(b) Proposals for a Training Centre on Forestry salinity problems to be held eventually in Kuwait in 1966.

INTRODUCTION

Kuwait, situated on the northwestern shore of the Arabian coast, has an area of approximately 6,000 square miles (15,540 km²). Its physical resources are petroleum and natural gas. Second only to Venezuela as an oil exporter, Kuwait has reached the level of affluence, where economic improvement, in the form of a higher per capita income for her people, need not to be an overriding objective of national policy while the need for improvement of life of the increasing population (nearly half a million people for Kuwait State only, excluding the Neutral Zone) in the form of reducing somewhat the harshness of the climatological conditions by planting returns, is a problem receiving consideration. Most of the drinking water in Kuwait has to be distilled from the sea in a flashtype evaporator Plant, situated at Shuwaikh and producing now 6 mgd daily (8 mgd at the end of the year) to which 5.5 mgd are added from the ground fresh water field of Baudatain-Umm el Aish (reaching an average salinity of 625 ppm TDS). Umad Ahmadi has a distillation Plant supplying 1.5 mgd while the plant of Sheibe starting production soon this year will produce 3 mgd more. The water is sold at the price of 600 fils per 1,000 gallons, price covering production and maintenance costs. By the end of 1965 the daily production of distilled water, taken from the sea is reaching about 12.5 mgd in Kuwait State to which we may add 0.5 mgd produced in Ras-el Khafji and 0.2 mgd at Wafra in the Neutral Zone. On the other hand a relatively large amount of highly saline water is available at shallow depths, in the coastal areas and can be used "in situ" for afforestation or for reduced watering during the established period. Large quantities of saline waters of 4,000 to 5,000 ppm TDS are used daily for irrigation in Kuwait; in summer 1965, 16 mgd to 18 mgd are consumed for the Sulaibeha quality (4380 ppm) and up to 6.5 mgd from the Abdaliya quality (4,000 ppm TDS). This type of water is used for agricultural and forestry irrigation since 1951 in sandy soils, usually characterized by high pH (usually between 8 to 9) and a high calcium carbonate content which, under semi-desertic conditions, is generally concentrated, at rather shallow depths, in a layer of consolidated sand known locally as "gatch," resulting from the action of a high evaporation (more than 5,000 mm annually 80°C

METEOROLOGICAL DATA

Shuwaikh Lat. 29°20' N. Long. 47°57' E.

		Jan.	Feb.	Mar.	Apr.	May	June	July	Aug.	Sept.	Oct.	Nov.	Dec.	Average	Remarks
Ambient Air T°C 14/16 hl 1955/64 04/06h	Abs Max	26	33	39	42	48	48	49	49	46	42	37	31	40.8	Betw. 14–16h
	Avg Max	18	21	26	31	37	43	44	44	41	35	26	20	32.1	
	Mean	13	16	20	25	31	35	37	36	33	27	20	15	25.7	
	Avg Min	8	10	14	19	24	28	29	28	24	19	14	9	18.8	Betw. 04–06h
	Abs Min	−3	0	5	9	14	22	23	21	17	11	6	−1	10.3	
Dry-Bulb Temperature Mean 1961/64 Wet	Mean	13.5	16.7	20.1	24.9	31.1	36.0	37.2	37.1	33.0	27.5	20.1	14.9	26.1	In screen
	Mean	10.3	12.3	13.8	17.3	19.9	21.8	23.0	23.3	20.8	18.5	15.0	11.0	17.3	In screen
Sun Radiation 1958/64	Abs Max	57	65	75	78	79	79	79	79	76	72	66	62	72	Black Bulb
	Avg Max	49	54	59	66	71	74	74	74	70	65	55	49	63	Thermometer
Grass Minimum 1955/64	Abs Min	−4	−1	2	6	11	14	14	13	11	8	−1	−4	5.8	Temp just above
	Avg Min	7	9	12	17	21	23	25	24	21	16	12	8	16.3	ground
Mean sea temperature 55/64		16	17	20	23	28	30	32	33	31	27	23	17	24.8	Distil. Plant
Relative Humidity (%) 55/64	Avg Max	75	74	69	63	54	40	43	44	48	59	71	80	60	Hygrograph
	Mean	63	53	47	43	36	27	28	29	31	38	51	61	42.1	
	Avg Min	43	33	26	24	18	13	13	13	14	18	32	43	24.2	
Atmospheric pressure In mm 55/64	Abs Max	773	770	771	768	762	757	754	755	763	765	772	772	765	Mean sea level
	Mean	764	763	760	758	755	751	748	749	753	759	763	765	757	
	Abs Min	752	751	751	749	747	744	743	743	746	749	755	755	748	
Evaporation in mm 61/64	Piche	4.9	6.9	10.4	13.9	17.9	24.3	23.7	22.3	18.6	12.8	7.9	5.6	14.1 = 5146.5 yr Screen	
	Class A	3.1	4.5	6.3	8.9	13.1	18.6	18.0	16.7	12.9	8.7	4.9	3.1	9.9 = 3613.5 mm yr	
Prescott Evapotranspiration		2.8	4.1	5.7	8.1	11.9	16.9	16.4	15.2	11.7	7.9	4.5	2.8		"open"

Parameter	Jan	Feb	Mar	Apr	May	Jun	Jul	Aug	Sep	Oct	Nov	Dec	Year	Method/Note
Average hours of sunshine/day	7.7	8.4	8.3	8.2	10.1	10.7	10.3	10.8	10.3	10.0	8.1	6.9	9.2	Camp. Stokes rec.
55/64 Daylight % computed	7.34	7.02	8.38	8.71	9.50	9.44	9.63	9.20	8.33	8.02	7.22	7.20	8.3	Blaney Criddle
Mean Ground Temperature 5 cm (61/64)	15.2	18.4	22.4	27.7	33.0	37.2	38.8	38.8	35.4	29.7	21.9	16.5	2.8	Bent-Stem Therm.
10	15.5	18.8	22.7	26.2	32.4	36.2	37.9	38.3	35.2	30.0	23.2	17.3	27.8	Bent-Stem Therm.
20	16.0	19.3	23.1	27.6	32.5	36.0	37.4	37.8	35.3	30.3	24.0	18.3	28.2	Bent-Stem Therm.
30	17.7	19.6	23.1	27.2	32.2	35.6	36.8	37.3	35.3	31.1	25.5	19.9	28.4	Symons-Pattern th
60	19.9	20.5	23.1	26.5	30.6	33.8	35.3	36.4	34.9	31.6	22.3	22.4	28.1	Symons-Pattern th
120	23.1	22.2	23.4	25.4	28.1	30.9	32.4	34.3	33.6	31.9	29.2	25.2	28.3	Symons-Pattern th

Number of days visibility less than 2 miles

Year	Jan	Feb	Mar	Apr	May	Jun	Jul	Aug	Sep	Oct	Nov	Dec	Total
1955	0	1	2	1	1	0	1	0	0	0	0	0	6
56	2	7	1	9	6	9	2	2	1	1	1	3	47
57	3	3	3	2	2	2	2	0	1	1	1	1	22
58	2	0	5	2	2	6	1	1	2	1	1	2	25
59	2	1	0	1	2	9	8	6	3	1	2	1	35
60	2	2	4	5	6	6	5	2	1	0	2	1	38
61	2	3	6	2	9	10	14	4	6	2	2	0	60
62	1	5	2	5	4	4	3	1	2	0	4	1	32
63	3	4	9	7	7	11	8	1	1	1	0	1	53
64	2	1	3	3	3	7	9	7	1	0	1	1	37

Parameter	Jan	Feb	Mar	Apr	May	Jun	Jul	Aug	Sep	Oct	Nov	Dec	Year	Note
Total 55/64 daylight hr/yr	13,94	18,96	29,34	32,23	41,80	60,42	51,04	23,92	14,17	5,61	11,55	7,92	355	
(total 55/64 in days)	19	27	35	37	44	64	53	26	17	7	16	11		35,5 Year M

Rainfall (in mm)

Year	Jan	Feb	Mar	Apr	May	Jun	Jul	Aug	Sep	Oct	Nov	Dec	(yearly total)
1955	8.6	–	6.4	6.0	7.0	–	–	–	–	–	0.5	44.8	73.3
56	10.8	4.7	6.0	14.2	–	–	–	–	–	–	–	119.3	155.0
57	9.8	14.3	15.4	26.3	8.9	–	–	–	–	0.1	89.1	1.3	165.2
58	12.1	0.7	7.6	2.2	1.9	–	–	–	–	–	15.7	61.7	101.9
59	32.2	13.5	10.0	9.4	1.8	–	–	–	–	–	9.3	23.8	100.0

R. FIRMIN

													Mean
60	7.7	2.6	2.7	4.2	–	–	–	–	–	–	11.3	0.1	28.6
61	22.7	16.3	45.9	29.1	1.1	–	–	–	–	–	66.3	14.4	195.7
62	27.1	3.2	4.5	18.9	0.1	–	–	–	–	–	0.1	12.7	66.6
63	0.4	23.9	1.4	19.9	21.4	–	–	–	–	7.1	13.1	87.3	
64	12.2	2.2	2.2	–	–	–	–	–	–	1.1	8.6	26.3	
(Monthly avg 55/64)	14.36	8.14	10.2	13.0	4.2	–	–	–	–	0.01	20.0	30.0	100.00 MEAN
Daily avg 55/64	0.47	0.27	0.34	0.43	0.14	–	–	–	–	–	0.66	1.0	

Comparison between seasonal and annual rainfall at Shuwaikh (season Nov.–May)

											Mean
Year (calendar)	336.0	73.3	155.0	165.2	101.9	100.0	28.6	195.7	66.6	87.3	100.0
Season (Nov.–May)	193.5	81.0	194.3	115.1	144.3	52.0	126.5	134.6	80.5	25.1	104.7
season	1954	55	56	57	58	59	60	61	62	63	
	54.55	55.56 etc.									

Approx. rainfall from 1909 to 1954, recorded with less accuracy than from 1955 on:

	"	mm		"	mm		"	mm
1909	2.43	62.3	1926	6.36	161.6	1936	4.78	121.2
1910	9.79	248.2	1928	2.43	62.3	1937	3.70	94.0
1911	6.97	176.9	1929	7.49	190.4	1938	5.90	149.9
1912	2.67	67.6	1930	4.81	122.1	1939	3.54	89.9
1913	3.13	79.5	1931	3.86	98.1	1940	4.82	122.3
1922	0.95	24.1	1932	1.54	39.1	1952	5.23	132.7
1923	5.15	130.7	1933	4.99	126.8	1953	6.05	153.7
1924	4.32	109.7	1934	5.40	137.2	1954	13.23	335.3
1925	4.67	118.4	1935	7.11	180.5			

Approx avg yearly rainfall of 36 years being:

120.4 mm

between 1909 & 1964

in summer), acting on an annual rainfall of 120 mm (average of 36 years). The available moisture content of such soils is usually around 3% by weight in summer and the infiltration rate is of at least 20 mm an hour. The primary salinity is very variable due to parent material and topography.

DIRECT UTILIZATION OF SEA-WATER AND HIGHLY SALINE WATERS FOR AFFORESTATION IN KUWAIT

Due to the limitation of currents, the salinity of the Gulf is rather low for an island sea. The mean sea temperature for the period 1955/64 is ranging from 16°C in January to 33°C in August. The sea-water qualities along the Kuwait Gulf is variable as described in table I.

TABLE I

Total dissolved solids in ppm	42,020–65,000
pH Value	7.6–8.8
Total Alkalinity in ppm $CaCO_3$	130–165
Chlorides in ppm Cl	20,800–24,900
Sulphate in ppm SO_4	2,000–3,500
Phosphate in ppm PO_4	0.1–0.02
Total hardness in ppm $CaCO_3$	6,850–8,600
Permanent hardness in ppm $CaCO_3$	6,700–8,400
Temporary hardness in ppm $CaCO_3$	40–110
Bicarbonate as HCO_3	60 ppm
Sodium in ppm Na	13,050–13,920
Potassium in ppm K+	380–450
Calcium in ppm Ca++	400–530
Magnesium in ppm Mg++	1,490–1,700
Silica as SiO_2 in ppm	5–15 ppm
Bromine in ppm Br	65–85
Iodine in ppm I	0.03
Boron in ppm B+++	3.0
Manganese in ppm Mn++	0.001
Dissolved Iron as Fe+++	0.01
Zinc as Zn++ in ppm	0.005
Free caustic in ppm NaOH, Free chloride in ppm Cl_2 Free carbon dioxide in ppm CO_2 and Ammonia as NH_3 in ppm	} Nil
Suspended matters, max. in rough weather 55 ppm, min. in clam weather 5 ppm	
Evaporation rate	16 mm a day

The electro-conductivity in millimhos at 20°C, is variable along the shallow coastline about 100 miles long and ranges from 63 to 90 appr. (EC × 10³) with an average value of 70 corresponding to about 42 to 43,000 ppm TDS. Some shallow perched water-tables, as far as 1,000 m from the sea-shore, are reaching higher salinity than the sea itself sampled the same day like near Saidi, in the Island of Failacha or at Shuwaikh secondary school, a special case being Ain el Abd, Khor Muqta source, at the border of the Neutral Zone where the salinity is double that of the corresponding recorded concentration of the sea-water at Rasl Khafji but with a sodium: calcium ratio far better than the sea and explaining the existing surrounding vegetation of *Suaeda vermiculata* FORSK and *Halocnemon strobilaceum* (PALL.) M. BIEB. The outflow of 1,000 gpm is saturated nevertheless with hydrogen sulphide. Direct utilization of sea-water is under trial in Kuwait since 1963 but only under very small scale with introduction trials of mangroves, namely *Avicennia marina* (FORSK.) VIERH., *Bruguiera gymnorhiza* (L.) LAM. and *Rhizophora mucronata* POIR. at Bubiyan Island. The map "in fine" is giving the nearest stands of *Avicennia* in the Gulf; this species is ecologically the most suitable for the muddy flats of Bubiyan and of some parts of the northern coast of Kuwait Bay containing enough clay, clearly in the salinity range and at about the same latitude as on the Sinai shore where this species is still thriving. Reaching up to 6 m high, and living a century, the main value of this species is the tolerance to lopping under control for fodder production (sheep and cattle), its tannin content and its use by fishermen for curing of prawns, the firewood being of little value in an oil-country. The difficulty encountered for its practical propagation is that the fruits, available in August, contain viviparous seedlings, developing while attached to the branch. These must be collected at the stage when they are ready to drop, but long distance transport is difficult, as the fruits become rotten if not packed properly in plastic bags sent very rapidly after collection as viability does not exceed more than one week for such high moisture content of fruits.

Rhizophora mucronata POIR. is not indigenous in the Gulf area but introduced on the Iranian coast; its wood is well-known for centuries in Kuwait and was introduced from Zanzibar and the Ruzizi in Tanzania for roof-poles "dhows" (boats) and "shandals" (props); it does better on lighter soils than *Avicennia* soils and is therefore worth a trial. As for the most interesting *Bruguiera gymnorhiza* (L.) LAM., we have to go as far as the Yemen border to find remnants because the tree has been much exploited for its good timber. The heaviest soils of Bubiyan Island, practically at sea-level and partly recovered at high tide, are reaching up to 34% clay content and up to 57% silt. The soil becomes progressively more silty towards the western coast. The natural vegetation is extremely scanty, between the patches of pure salt, and is consisting, as at the extremely saline flats of Air el Abd (Neutral Zone), of *Halocnemon strobilaceum* (PALL.) M. BIEB. and *Suaeda vermiculata* FORSK. with other species able to tolerate high salinity as *Seidlitzia rosmarinus* BUNGE, *Zygophyllum coccineum* L. and its associate *Cistanche lutea* HOFFM. & LINK. The pH is generally higher than 8, $CaCO_3$ averaging as much as 34 to 38% for 940 meq/l/ Sat. Ext. of chloride and a very high sulfate content of 335 meq/l/Sat. Ext. There is, unfortunately, no practical utilization of this scanty natural vegetation and no possibilities exist for the utilization of the other indigenous species of Kuwait which might tolerate rather highly saline water-tables, namely: *Frankenia pulverulenta* L., *Cressa cretica* L., *Herniaria hemistemon* J. GAY, *Phragmites communis* TRIN., *Juncus acutus* L., *Tamarix passerinoides* DEF. Found in areas of lesser saline tolerance, probably under determined ionic ratios only, are the following indigenous species: *Alhagi maurorum* DC., *Bassia eriophora* (SCHRAD.) ASCHERS., *Limonium suffruticosa* (L.) O. KUNTZE and some species having a fodder value, best of them being *Nitraria retusa* (F.) ASCHERS. Where the surface layers are less saline during the growing season, the following plants are still able to thrive to a certain point on saline waters possibly characterized by more specific ionic ratios concentrated in certain layers of the soil,[1] allowing the growth of the root system: *Aizoon hispanicum* L., *Aeluropus lagopoides*, *Cynodon dactylon* (L.) PERS., *Cornulaca leucacantha* CHAE, *Fumaria parviflora* L., *Rumex vesicarius* L., *Astragalus tenuirugis* B., *Linaria chalepensis* (L.) MILL. and the subhalophytes *Launaea nudicaulis* (LESS.) HOOK., *Salsola baryosma* DANDY, *Sphenopus divaricatus* (G.) REICHB. and sometimes *Plantago coronopus* L. (the latter in winter and spring). It seems to appear clearly from table IV that healthy matured trees of *Prosopis juliflora*, *Zizyphus spina-christi* and to a certain degree *Eucalyptus camaldulensis* and *Tamarix passerinoides* (to mention only the most important species) are able to thrive on relatively highly saline waters, in a certain range of ionic ratio, not yet determined, especially the *Prosopis* with roots apparently able to tolerate salinities approaching the sea-water level. An important factor seems, however, to be the *permanent moisture* of the soil remaining always at rather high values at the roots' depth. These facts are long known as far as the date-palm is concerned although this species is not producing economically in highly saline waters.

The Na/Ca+Mg ratio of the sea-shore is generally between 3.5 and 3.7 in Kuwait and is better than some water-tables of Failacha and of the Neutral Zone where trees are found. For *Eucalyptus camaldulensis*, the ratio Mg/Ca is important because high concentrations of magnesium are more toxic for these trees than isosmotic values of other salts, but healthy mature *Eucalyptus* has nevertheless been found at a ratio as high as 6.4 in Failacha Island with a water-table at 1.50 m deep. This ratio is in general far lower for the other water-tables and varies between 3 and 4 along the shore for the sea-water. The highest HCO_3 values in meq/l are found in some wells at Fantas (12.9), Failacha Island (12 to 10 usually) and are always higher than the bi-carbonates of the sea-shore reaching no more than 2 meq/l. The ratio of Na/K is as high as 47 to 50 (in meq/l) at the seashore, the highest corresponding value near trees found up to now in Kuwait has been 36. Trace element in the sea are manganese at concentrations around 0.001 ppm, Zinc as Zn^{++} around 0.005 ppm, dissolved iron as Fe^{+++} is also around 0.01 ppm. Boron is found in the range of 3 ppm which is usually considered

1. See further details on table VI.

as toxic for plants. The direct use of such sea-waters requires probable modifications of the main ratios, mainly increase of potassium and calcium, phosphates and ammonium, as far as can be economically justified and depending on the soil solutions and the age of the trees.

RESEARCH

Small introduction trials with *Phoenix dactylifera, Atriplex nummularia* and *A. vesicaria* are apparently the most interesting species to test with *Prosopis juliflora*, under such concentrations of salts and ionic ratio as found in sea-water tables at depths less than one meter and usually characterized by:
  (a) a high Na/Ca ratio
  (b) a high Mg/Ca ratio
If the palm-tree in such cases is only considered as an ornamental rather than a date-production tree, *Atriplex* and *Prosopis* are useful fodder plants. Experiments with sea watering are under way with *Prosopis* trees established two and a half years ago. These small trials with sea-water irrigation were initiated, with and without addition of pure potassium phosphate or nitrate, in the same sample plot located at Jahra (2,000 m from the sea) in order to find out if we can use sea-water and up to which point trees can tolerate such salinity.

Similar studies are under experimentation since 1963 with lesser saline water and same fertilizers on *Eucalyptus camaldulensis*, at Sulaibeha, in soils showing the following salinity at establishment:

Depth in cm	pH	EC × 10³	CaCO₃ %	Cl Sat. extr. meq/l	SO₄ (Sat. extr.) meq/l
0–50	7.9	5.70	8.37	29.40	27.62
50–100	8.1	9.97	1.29	30.12	70.58
100–150	8.1	10.08	1.03	26.00	71.18

The water-table at 2 m depth is reaching the following values:

TDS	8120 ppm	Ca	41.84 meq/l
EC (10⁶)	11600 micr. ohm	Mg	2024
pH	7.3	Na	51.75
K	1.75 meq/l	Cl	70.60
CO₃	Nil	SO₄	43.97
HCO₃	1.60	B	unknown

The irrigation water applied in measured quantities is characterized by the following analysis, and is used on a large scale in Kuwait:

TDS	4380 ppm	Ca	21.98
EC (10^6)	6260 micr. ohm	Mg	14.85
pH	7.3	Na	24.85
K	meq/l	Cl	27.55
CO_2	Nil	SO_4	31.73
HCO_3	2.52	B	unknown

In Jahra, the sea-water is applied in known quantities, on a soil which had the following characteristics at establishment. This soil has been used in the past for local agriculture.

pH	Depth	EC $\times$ 10^3	Cl meq/l	SO_4	$CaCO_3$ %	Sand	Silt	Clay
8.2	150 mm	110	880	250	4.57	78.6	12.3	9.6
8.0	500	16.6	96	74	1.64	93.0	1.1	5.9
7.8	1,000	8.0	35.5	50.0	1.68	94.1	1.8	4.1
7.9	2,000	10.0	49.0	51.0	0.69	93.8	0.8	5.4

In Jahra, the water-table is also saline and at shallow depths of 2 m with the following characteristics:

pH:	8.2	EC $\times$ 10^6:	8,500	TDS:	5,950 ppm
Ca:	22.4 meq/l	Mg:	11.2 meq/l	Na:	51.5 meq/l
K:	0.10	CO_3:	0.40	HCO_3:	1.60
Cl:	54	SO_4:	29.06	Boron unknown quantity	

This type of water has been used for centuries in Jahra, in alluvial sandy loam, 3 m thick, overlaying coarser sands and gravels applying up to 30″ of water per month in summer and similar agricultural experiences are known at Mangaf, Fahaheel and Failacha.

Tables III–VI suggest a first estimation of the relative salinity tolerance of tree species, based on available world literature, communication, a few borings and some laboratory experiments carried out by some qualified officials on agricultural crops mainly in the range of medium salinity levels. Dr. Abu Fakr, Mr. Jami Banna and Mr. Subhi Attar are the main authorities having studied salinity problems in the State of Kuwait.

It should be put very clearly that the tables given "in fine" are only a first approach to the subject of relative salinity tolerance of tree species and are in bad need of proper systematic and repeated experiments by world specialists, under known conditions of all the interfering factors (i.e. bound water, osmotic pressure, specified ecological data etc.). In general, afforestation in saline conditions, as being tested in Kuwait with a number of species (each of them represented only in small quantity) is expected to give the best results in soils rich in calcium carbonates, with:

(1) organic manuring and potassium phosphate (in holes) and potash nitrate (water);

(2) complete mixing of the profile with aeration;

(3) reduction of the pH by sulphur addition;

(4) reduction of the soil evaporation by a plastic collar and other materials;

(5) migration of the salts outside the root-zone by appropriate ridging and deep planting in relation with the root-system of the species planted;

(6) irrigation at rather short intervals, by continuous nightly sprinkling, in order to avoid too much variation of the soil moisture content.

Evaporation is the most important factor to control by night irrigation under a collar of black plastic sheeting buried at 10 cm depth.

INDIRECT UTILIZATION OF SEA-WATER AND HIGHLY SALINE WATERS IN KUWAIT

Sea-water distillation began in 1951 at Mina-el Ahmadi, Oil Center, where the production of about 1.5 million gallons daily is now operated in nine Westinghouse triple-effect evaporators. In 1952, bids were invited to build the largest distillation system in the world, at Shuwaikh, the production of which is going to reach 8 mgd by the end of the year. The plant consisted in 1954 of 3 boilers of 80,000 lb/hr; 3 backpressure steam turbines 750,000 w, and 20 three-effect submerged-tube evaporators producing 2 mgd. In 1957, it became necessary to increase the capacity and four-stage flash evaporators – a new principle at the time – increased the capacity to 4 mgd. Again in 1959, a further 2 mil gal of distillation capacity was installed and comprised two nineteen-stage flash evaporators. In 1965, five additional 1-mgd thirty-stage flash evaporators are being erected; one being already under testing in August, a second starting in September bringing the production to 8 mgd at the end of the month and to 12 mgd in 1967.

Beside the Plants of Shuwaikh and Mina el Ahmadi, a Plant of 3 thirty-stage evaporators 1 mgd capacity each, will start before the end of the year at Sheibe Industrial Complex bringing the total daily production of Kuwait State to 12.5 mgd at the end of the year to which the production of the Plant erected at Ras el Khafji in 1963/64 and at Wafra in 1956/57 (oil-centers in the Neutral Zone) are adding respectively 0.5 mgd and 0.2 mgd.

At the time of writing only the production of Shuwaikh Plant is contemplated for forestry purposes in the form of sewage effluent mixed with Sulaibeha brackish water. Agricultural use of treated effluent is now of 0.5 mgd daily and will rise in the coming months to 1 mgd. This amount will increase suddenly in 1968 with the completion of the sewage scheme to 17 mgd with a progressive increase the next few years up to a maximum of 25 mgd. The present salinity ranges around 3,500 ppm TDS but is expected to decrease by 1968 to about 2,000 ppm TDS.

TABLE II

Costs of distilled water in kuwaiti fils per 1,000 gallons (1 KD equals 1 Sterling pound or thousand fils)

	Actual cost at 4,148,000 gallons per day unit (62/63)	Estimated cost in: 6 mgd Unit	1 mgd Unit (Plant "A")	2 mgd Unit (Plant "E")
Depreciation				
of Plant	331	229	850	114
	331	229	850	114
Operation				
Salaries – Evaporator staff	47	33	95	23
Boiler staff	27	19	41	15
Fuel, lubricants, chemicals	69	58	52	57
Electricity	34	29	2	12
	177	139	190	107
Maintenance				
Salaries – Evaporator	52	36	176	15
Boiler	7	5	10	4
Material	35	24	72	8
Workshop services	10	7	47	1
	104	72	305	28
Overheads				
General expenses	48	33	103	20
Technical and administration staff and control				
	48	33	103	20
	660 ($1.85)	473 ($1.32)	1,448 ($4.06)	265 ($0.75)

(Source: Ministry of Water and Electricity, Kuwait)

The cost of distilled water calculated for 1962/63 was 660 fils ($1.85) per 1,000 imperial gallons. But if the sweet water is distributed at 600 fils per 1,000 imperial gallons, the charges for haulage are about 900 fils per 1,000 imperial gallons, so that fresh water remains an expensive commodity in Kuwait. The Plant is combined with generators having an installed capacity of 162 megawatts charged at 3 fils for industries and 6 fils for domestic use.

If 6 million gallons a day were produced with the present equipment, excluding the use of the oldest units of the Plant, the price would drop to 473 fils/1,000 gal and in the newest type of unit, the cost is dropping to about 200 fils ($0.56) for a big modern installation. This type of Unit can thus produce at such a cost that, if this distilled water is mixed 50/50 with brackish water at 75 fils per 1,000 gallons (Sulaibeha quality), the resulting water would cost about 140 fils per thousand gallons and have a quality sufficient and economical for agriculture and forestry purposes.

Where electric power and distilled water are both required, the most economical arrangement is to use the exhaust from passout turbo-generators as the steam supply from condensation to produce distilled water. With the use of great amount of water for irrigation, power and water production could be tied together. If the policy is to continue, in Kuwait, with separate units, then the generation of power by gas turbines rather than by steam turbo alternators may give substantial savings. It remains clear, however, that at the present time, it is still cheaper for Kuwait to bring 120 mgd of sweet water by a pipeline of 170 km from the Shatt Arab (Iraq) to Kuwait as planified at an approximative cost of 30 fils ($0.08) per 1,000 gallons or 70 fils per 1,000 gallons depending whether the capital cost of the Project is included or not in the price estimation.

Distillation Plants are suitable for very high salinity while the electrodialyse process is preferred for lower salinity. At the time of writing, two plants exist in Kuwait; one located at Shuwaikh and treating 213,000 gal daily of Sulaibeha waters of 4,400 ppm TDS for agricultural purposes. The product water has 500 ppm TDS. The second Demineralisation Plant, based also on the electrodialysis process is located in the Neutral Zone at Wafra produces 50,000 gal daily at 500 ppm, the feed water being in this case of 7,200 ppm. This Plant has 4 stages while the one at Shuwaikh had five stages, each stage reducing the salinity by about 35%. Both Plants started production in May 1965 but a smaller Plant of 4,000 gal daily capacity has been in production in Northern Kuwait, at Umm el Aish, between 1957 and 1961, reducing the salinity from 4,000 to 500 ppm TDS. In the process covered by Ionics, the raw water passes through 4 cartridge filters operating in parallel. The feed water enters the stack section made of seven membrane stacks arranged to provide five stages of demineralization. It is possible to operate only 3 stages if the quality required is only 1,000 ppm TDS. Such a plant costs around 60,000 KD with a maintenance of 50 KD (3 stage Plt). Small mobile "batch-type" electrodialyse plants, producing 10,000 gal daily which reduce the salinity from 8,000 to 2,000 ppm TDS cost only 34,000 dollars. These are more interesting for afforestation on highly saline water-tables, for provisional watering during the establishment period of the trees, water being fed by tube-wells, to one unit Plant adapting easily to the variations of salinity. Very small units of 2,500 gal daily are available but they become relatively more expensive than 10,000 gal units.

Comparison between the feed and products waters used in Kuwait for agricultural purposes:

(Analysed by Dr. Temperle5)	Sulaibeha water	Electrodialysis Process effluent
Conductivity in Micromhos at 20°C	6,000	680
Total Dissolved Solids in ppm	4,350	495
pH value	8.2	7.8
Total causticity in ppm NaOH	Nil	–
Total alkalinity in ppm CaCO₃	125	40
Chlorides in ppm Cl'	1,105	80
Sulphates in ppm SO₄˙	1,560	200
Phosphates in ppm PO₄˙	Nil	Nil
Total hardness in ppm CaCO₃	2,050	140
Permanent hardness in ppm CaCO₃	2,000	–
Temporary hardness in ppm CaCO₃	50	–
Calcium in ppm Ca⁺⁺	470	25
Magnesium in ppm Mg⁺⁺	205	20
Free and Saline Ammonia in ppm NH₃	0.65	Nil
Nitrate in ppm NO₃'	4.0	0.5
Fluor in ppm F'	4.5	0.3
Zinc as Zn⁺⁺ in ppm and Boron in ppm B⁺⁺	Nil	Nil
Alluminium in ppm Al⁺⁺⁺ and Copper as Cu⁺⁺	Nil	Al: Nil Cu: 0.01
Total Iron in ppm Fe⁺⁺⁺	0.1	0.06
Silica in ppm as SiO₂	15	3

Our problems regarding the indirect utilization of sea- and highly saline waters are relatively simple and are related to the normal use of irrigation waters of low salinity. The approach to the problem of plant nutrition and fertilizer requirement in field experiments has to consider first the binary equilibria as defined by HOMES (1955). If the Na/K balance is considered for instance, the purpose is to obtain, by the effect of fertilizers combined with that of the soil and irrigation water, an optimum final balance in the soil after treatment. A method consisting of as many systematic treatments as there are nutrients in the cationic and the anionic groups of elements has to be followed, determining first the best ratios in fertilizers, then the anion/cation ratio and optimal application, adjusting the A/C value in order to obtain the best and cheapest complete formula. When we come to sea-water or highly saline waters, such a method can still be followed but it becomes complex, and the economy of adjusting a fertilizer's applicator to the flow of water becomes questionable, except may be for some trace-elements or collar additions of relatively small amounts of potassium phosphate and gypsum collected locally in certain cases. The detrimental role displayed by sodium, at such high concentrations, can nevertheless be held in check by a favourable combination of factors as proved by the analyses of soil solutions and water-tables found in the root-zones of trees which are theoretically not supposed to tolerate such conditions. As pointed out by scientists, the chemical analysis alone of a highly saline water cannot give an answer regarding its suitability, the same water being suitable under some conditions and unsatisfactory under others; under conditions of high salinity the plant nutrition concept should be completed by the concept of the balanced ionic environment (B.I.E.).

Drainage and organic manuring are important factors of success for such trials as well as a relatively low range of fluctuation moisture in the soil profile at root depths.

The determination of tables of maximum tolerance for specified ionic ratio with high electro-conductivity at known moisture levels, osmotic pressure and bound water values, is a problem which is new for trees and which needs, unfortunately, quite a number of years to give reliable results.

TABLE III

Provisional salt tolerance of trees and shrubs

Electrical Conductivity recorded (max.)
$EC \times 10^3$

50 or more for some spp.	*Avicennia marina, Suaeda vermiculata, Nitraria retusa.*
40	*Casuarina glauca, Conocarpus lancifolius, Phoenix dactylifera, Tamarix jordanis, Tamarix maris-mortui.*
35	*Atriplex nummularia, Atriplex vesicaria, Prosopis stephanianna, Prosopis tamarugo, Tamarix aravensis, Tamarix deserti, Tamarix mannifera, Tamarix orientalis, Tamarix pentandra, Tamarix Meyeri.*
30	*Acacia ligulata, Casuarina equisetifolia, Kochia indica, Prosopis juliflora, Tamarix aphylla, Zizyphus vulgaris.*
25	*Acacia sowdenii, Tamarix nilotica.*
18	*Acacia pendula, Acacia salicina, Casuarina stricta, Eucalyptus camaldulensis (Kuwait strain), Euc. sargentii, Euc. spathulata, Nerium oleander, Parkinsonia aculeata.*
16	*Acacia farnesiana, Callistemon lanceolatus, Casuarina cristata, Euc. camaldulensis var. obtusa, Euc. calcicultrix, Euc. Kondinensis, Euc. microtheca, Euc. coolabah, Prosopis chilensis (algarobilla), Prosopis juliflora var. velutina.*
14	*Acacia arabica, Albizzia chinensis, Casuarina lehmanii, Clerodendron inerme, Euc. pimpiniana, Haloxylon salicornicum, Sesbania grandiflora.*
12	*Acacia stenophylla, Bassia latifolia, Callitris glauca, Dodonaea viscosa, Euc. gomphocephala, Euc. kruseana, Melaleuca pauperifolia, Melia azedarach, Punica granatum, Thevetia neriifolia.*
10	*Albizzia lebbek, Butea monosperma, Euc. annulata, Euc. brachycorys, Euc. cornuta, Euc. melliodora, Euc. stricklandi, Ficus carica, Ficus religiosa, Hakea laurina, Lagenaria Patersoni, Ricinus communis (var. persicus), Salvadora oleoides, Thespesia populnea, Vitex agnus-castus.*
8.5	*Casuarina cunninghamiana, Caesalpinia gilliesii, Calligonum comosum, Dalbergia sissoo, Dodonaea attenuata, Euc. cladocalyx, Euc. forestiana, Euc. grossa, Euc. lansdowneana, Euc. largiflorens, Euc. le Soueffi, Euc. robusta, Euc. salubris, Inga dulcis, Terminalia arjuna.*
8	*Brachychiton gregorii, Euc. brockwayi, Euc. dundasi, Euc. intertexta, Euc. woodwardii, Ficus bengalensis, Myrtus communis, Prosopis spicigera, Schinus molle, Terminalia catappa.*
6	*Acacia deani, Agonis flexuosa, Balanites aegyptiaca, Cupressus arizonica, Euc. torquata, Grevillea robusta, Olea europea, Pritchardia filifera, Salix acmophylla, Tamarindus indica, Tecoma stans.*
5	*Cordia myxa, Cupressus sempervirens var. stricta, Elaeagnus angustifolia, Euc. astringens, Euc. campaspe, Euc. flocktoniae, Euc. lansdowneana, Euc. longicornis, Euc. patellaris, Euc. redunea, Euc. transcontinentalis, Lantana aculeata, Populus euphratica, Terminalia belerica.*

4.5	*Bombax malabaricum, Euc. citriodora, Populus Bolleana.*
3	*Acacia tortilis, Albizzia julibrizzin, Ficus sycomorus, Robinia pseudacacia, Salix alba.*
2.5	*Acacia cyanophylla, Acacia mellifera, A. raddiana, A. forestiana.*
2	*Euc. tereticornis, Hyphaene thebaica, Poinciana regia, Duranta plumieri, Populus oblegata.*
1	*Azalea, Bougainvillea, Populus thevestina* and *P. euramericana.*

TABLE IV

Electro Conductivity of Water-tables in Kuwait (EC × 10⁶)

6	As an expression of highly saline waters near which vegetation is thriving.
EC × 10	DEEP WELLS ARE NOT RECORDED HERE. Maximum depth 13.50 m.
180,000	Ain el Abd (Khor Muqta source); vegetation at 15 m of the spot.
90,000	Sea-shore at Ras el Quaid (Bubiyan-mangrove trials), *Suaeda fruct.*
76,000	Well at 15 m of sea-shore at Ras el Khafji, *Zygophyllum.*
75,000	Well 1 m depth at 1160 m of sea-shore at Saidi (Failacha), Alhagi and Shuwaikh secondary school (sea-shore).
67,000	Sea-inlet at Bubiyan (mangrove plot) with *Suaeda monoica.*

59,000	WITH TREES: Failacha, near Yacoub house; *Zizyphus* and *Phoenix.*
51,000	Quaraniya worst well, surrounded with *Zizyphus, Phoenix, Tamarix.*
46,000	Shuwaikh University Avenue with healthy *Prosopis juliflora* 6 years old.
37,000	Ain el Eid (Neutral Zone: camels (sometimes only).
27,000	Mishri (worst well used in mixture for tree irrigation).
22,500	Shuwaikh water table at 1.35 m under *Tamarix aphylla, Eucalyptus camaldulensis, Nerium oleander, Prosopis juliflora, Parkinsonia aculeata, Ricinus communis (persicus), Inga dulcis, Acacia farnesiana & Acacia arabica.*
20,000	At Dauha and Sheibe with *Zizyphus* and *Tamarix.*
18,000	Shadai, Shamiya and Abu Sidra wells; *Prosopis juliflora* – sheep –.
17,000	Mesilah and Quars el Sirra wells, no trees around.
16,000	Ain el Ar in Neutral Zone, near *Lycium.* Sometimes used for tea (rather bad).
15,000	Wells of Umm el Ghashish, Fahaheel, Roydat el Furat and Bahara (facing Bubiyan) with many *Zizyphus, Tamarix, Lycium* and *Prosopis.*
13,700	Worst well of Mangaf, Jahaliya wells, Rumaithiya with trees of various kind including *Dodonaea viscosa.* At Mangaf agriculture irrigation.
13,000	Jaleyha and Mishash hulayil (no trees, used for sheep).
12,500	Between Haswan and Tarfawi, facing Bubyan. Also Umm Safuq; a few *Zizyphus.*
12,000	Araifjan and Salmiyah old Jail; *Tamarix, Eucalyptus camaldulensis.*
11,500	Juwaihil, Sulaibeha forestry sample plot (2 m deep), Mina el Abdulah and Thamaniya; 115 experimental tree species, mainly *Eucalyptus.*
11,000	Gurain wells (sheep) and Salmyah (agriculture irrigation).
10,000	Masalahn, Nagwa Imghara, Subaih, Rachiwa and Sabriya; agriculture irrigation and trees of various species. Many *Zizyphus,* Two *Tamarix.*
9,500	Artawiya wells used for sheep and goats.
9,000	Imghara; Igdai and Chuwaichib wells used for many flocks of sheep.
8,500	Subaihiya (worst well for vegetable production), Tarfawi; many trees especially *Tamarix* windbreaks. Also Arhaya wells for sheep and Jahra – worst agricultural well (used since centuries) *alfalfa.*
8,000	Ash shinaiti wells, Khuwaisal, Mishash an Nuwama, Shadaf best well, and Ugailah, Atariz, Umm Ghudaiya, Umm Gudair, used regularly for flocks. At Mishash Ibn Dabus, a man was drinking (old bedouin).
7,000	Buhait, Nugra, Khatliya gardens, Mishash Ben Hamar, best agricultural well of Mishrif. Beside all the trees mentioned already; *Parkinsonia, Lantana, Clerodendron, Thespesia, Thevetia.*

6,500   Bida, Mangaf best well (agricultural irrigation) with many trees.
6,000   Bahara (Northern Kuwait bay coast) Mogha. Sheep and goats.
4,500   Best wells of Fantas, Fahaheel, Subaihiya and Jahra agricultural centers.

TABLE V

Utilization of highly saline water-tables according to ratios between calcium, magnesium and sodium

Na/Ca ratio	Sea-water:	Na/Ca + Mg ratio	
22.7	Khor Subiya sea water (mangrove)	4.7	worst recorded at Saidi, water-table at 1.10 m to 1.50 m, *Zizyphus*.
20.3	Bubiyan east coast (mangrove)		
18.3	Bubiyan south coast (mangrove).	4.0	also in Failacha, water table at 1.20 m near Mr. Yacoubis house Palm.
	Water-tables		
19	Quaraniyah agricultural well.	3.8	At Ain el Abd (Neutr. Zone)
16.4	Ras el Khafji, under *Prosopis*.		Ras el Khafji (") *Prosopis* Mishash Khalifa.
12.75	Saidi (Failacha), under *Zizyphus*.		
11.4	Failacha (Yacoub house), *Phoenix*.	3.6	Forestry sample plot at Antique site (Failacha) same at sea-water about at Bubiyan (3.75 mangrove) and Shuwaikh Port (3.6).
8.8	Umatina, *Euc. camald.*		
7.5	Ras Failacha, *Cenchrus* & *Panicum*.		
6.5	Shuwaikh, Merium, *Parkinsonia* & *Acacia farnesiana*, *A. arabica*.	3.3	Abu Halaifa well and Shuwaikh under *Prosopis*, *Euc. camaldulensis*, *Nerium, Tamarix aphylla, Acacia farnesiana*.
6.25	Ain el Abd; used for medicine.		
5.4	Mishash Khakifa, used for camels.		
5.25	Failacha south, Mina Abdullah with *Zizyphus sp. christi*.	3	other wells at Antique (Failacha). Xuaraniyah („) Mina Abdullah (*Lycium*), Mishrif.
5	Mesilah, Mishash Abu Sidra, Mishrif (worst well for tree irrigation at Mishrif). Used for sheep and goats.	2.75	Fantas, Funaitis, Mishash Sidra agricultural crops.
4	Worst well of Mangaf (*Calotropis*) and at Rumaithya; agriculture.	2.5	Abu Halaifa, Bursuya, agr. crops.
3.5	Worst well of Fantas (agriculture). Salmyah, and Hugaija (sheep). At Juwaihil, donkeys were drinking.	2.25	Ain el Eid, Umm Safaq, Hugaija, Bahara, Rumaithiya, Umatina, Used by people.
3.25	Ash Shinaiti, Mina, Mishash el Ashnan, with *Lycium*.	2	Ash Shinaiti, Atariz, Juwaihil Roydat el Furat, Mishash hulaiyil and Ras Failacha.
3	At Mishash hulaiyil, Mishash an Nuwama, Umm gudair (beduin well) with flocks.	1.75	Bahara (Neutr.), Fahaheel (agr.) Old Jail, Ain el Arq (tea).
2.75	Atariz, Bahara (near Bubyan) with flocks sheep, goats and camels.	1.50	Khuwaisat, Dauha, Igdai, Magwa, Arhaya, best well of Abu Halaifa Ugaila, Sabriya, Um Ghudaya, Umm Niga, Tarfawi, Artawiya, Maraga.
2.5	Ain el Eid, Ain el Arq, Artawiya, Dauha, old Jail, Jahra (*Tamarix* windbreak), Sheibe, Ugailah, Umm ghashish, Umm gudaiya, Roydat el Furat; and wells for flocks of sheep and goats.	1.25	Araifjan, Um Safaq, Subaith, Nagwa-Inghara, Thamamiya, Shadaf cattle well of Jahra.
2.25	Best agricultural well of Mangaf. Used for people between Haswan and Tarfawi (near Bubyan). Jahaliya.	1.0	Adaliya, Umm el Sabah, Mesilah, Imgha Imghara, Chuwaichib, Masalahi
2	Best agricultural well of Fantas Ig-	0.75	Jaaidan, Gurain, alah, ishash el Isjman, Mishash Ben Hamar, Rahaya, Tawil, Khatleya.

dai, Sabriya, Umm Niga (bed. well) Araifjan, Khuwaisat and Subaith. Arhaya, Thamamiya and Umm Safaq.

1.75 All used by people, for tea. Arfaya, Shadaf, Tarfawi, Nagra Inghar.

1.5 Chuwaichib, Umm Sabah (giving water to all beduins of SW Kuwait) Nugra (*Inga dulcis, Clerodendron*) Fahaheel, Jeleyha, Masalahn wells.

1.25 Khatleya, Mishash Ben Hamar, Tawil and forestry plot Sulaibeha.

best wells + for Na/Ca ratio-

1 Failacha (Jaber A.) *Populus euphr.* Gurain (sheep and beduins). Magwa (important center of gather). Malaha and the worst agricultural wells of Subaihiya vegetable production area.

0.75 Jahara best wells (used for cattle). El Bahara (N. Coast Bay), Mogha. Mishas Ibn Dabus (the old badu).

0.5 Jaaidan, Umm, Himan, Umm Reer, Subaid, Subaihiya.

0.50 BEST WELLS: of Jaaidan & Jahara, Mogha, Nugra, Bida, Umm reer, Mishash Ibn Dabus, Rachiwa.

0.25 Arfaiya, Hawalli, Sabriya, Subaib and Subaihiya.

TABLE VI

Pedological signification of halophytes growing near trees and shrubs found in Kuwait State in areas of highly saline water-tables, situated in the root-zone depth (Hydromorphous salines)

Halocnemum strobilaceum (PALL) M. BIEB. Tolerating the highest TDS ppm, but also found in non-saline soils. Generally medium carbonate. Tolerates high SO_4^{--}, Mg^{++}, bad drainage. Same for *Seidlitzia rosmarinus*. Both species are typical of solontchak soils.

Avicennia marina (FORSK.) VIERH. Always associated with high TDS but with low carbonates and very poor ratio between sodium and calcium.

Suaeda vermiculata FORSK. Typically associated with high chloride only, and low carbonates.

Juncus acutus L., *Phragmites communis* TRIN. Low carbonates due to good drainage. Accept the high TDS but also lower values especially Phragmites. Gives organic matters. High permanent moisture of the soil. Phragmites need more sweeter and permanent moisture, than *Juncus*.

Frankenia pulverulenta L. Is rather more specific as to the ratio and bound to an excess of CO_3 (free) up to 45% rather than to Cl.

Cressa cretica L. Is also found at quite high total salinity ranges with *Tamarix passerinoides* DEF. (the latter with low carbonates).

Zygophyllum coccineum L. Always associated with high carbonates, and thus in soils without free drainage (as *Cistanche lutea* HOFFMAN & LINK); is nevertheless of lesser ppm TDS high level. Special cases are usually represented by:

Aizoon hispanicum L. Found at high salinity levels but then generally corresponding to higher phosphates and $CaCO_3$ percentage, with little available moisture while *Plantago coronopus* which is extremely variable needs more moisture in winter.

Alhagi maurorum DC. Is also only found in soils rich in nitrates in case of rather high exceptional TDS.

Cynodon dactylon (Bermuda grass). Found usually in the so-called "soda-prairie" type (SIMONNEAU) with *Sphenopus divaricatus* (G.) REICHB. also indigenous to Kuwait. Salinity is usually higher between 1 to 2 m depth and is progressively reduced by action of carbon-dioxide released by the roots. Tolerates up to 7000 NaCl (dry soil).

Lepidium aucheri BOISS. and *Limonium suffruticosa* (L.) KUNTZE. Are on soil with a surface salinity dropping below 1% TDS during the growing season.

Author's address: Roger Firmin, FAO, P.O. Box 4130, Kuwait.

The Desert Garden of Eilat

by

ELISABETH BOYKO & HUGO BOYKO

ELISABETH BOYKO & HUGO BOYKO

INTRODUCTORY REMARKS

The introduction of plants and planting in an area which up to then has always been an almost absolute desert-area is naturally an exciting adventure. It required specific ecological knowledge and particular studies in order to achieve the aims, namely the establishment of a plantation, desirable and necessary for a township and harbour to be built in this barren area, without working on a hit and miss plan.

The plan for the then (1949) not yet existent harbour-town of Eilat necessitated also the creation of an adequate environment in order to make this most desolate place not only inhabitable but also attractive (E. BOYKO, 1952). The speedy growth of Eilat afterwards was to a considerable extent due to the garden, which helped its development to a green town with at present about 13,000 inhabitants, a population which is steadily on the increase (see photos Fig. 7 and 8).

In accordance with the principles mentioned in the lecture of HUGO BOYKO dealt with in all detail in the book "Salinity and Aridity – New Approaches to Old Problems" (H. BOYKO, 1966, p. 131–200), we chose the completely vegetationless gravelhills with their high permeability as the place for the garden, in order to prevent gradual salt accumulation.

Dr. Y. CARMON, the Coordinator of Agricultural Research and Education at that time, played a major role in the decisions on our planning and we also selected together the actual area for the garden.

Before dealing with the plants it seems necessary to present a clear picture of the most unfavourable environmental and technical conditions under which the whole scheme had to be carried out.

The extremely dry and hot climate with temperatures to be seen in the table below, gives rise to very high evaporation figures, and the high and dry winds add to the dessication.

Years have allegedly been experienced without any rainfall at all; we could, however, not find such an event in the records. Rainfall is erratic and an effective rainfall is a rare event. The yearly average fluctuates, according to the period taken, between 20 and 30 mm. Salt spray from the near Red Sea is an additional hardship for plant life in this region.

ENVIRONMENTAL FACTORS

The main factors causing absolute desert conditions on these gravel hills may be enumerated as follows:
(1) insignificant rainfall;
(2) the very low cloudiness-coefficient;
(3) extreme air dryness;
(4) strong winds blowing mainly in the same direction;
(5) very high temperatures and extreme evaporation, caused particularly
      by (2), (3) and (4);
(6) sand and dust storms.

The complex of these environmental factors is deadly inimical to plant life in general and leads to absolute desert conditions on these almost completely vegetationless gravel hills. This, however, is immediately and decisively changed when irrigation is applied, if only with saline water. But, on the other hand, the latter added another unfavourable factor apart from the high Total Salt Content (T.S.C.), namely its strong and frequent fluctuation, between 2000 and 6000 and sometimes up to 10,000 ppm.

In the following a few examples may be presented out of the great number of measurements carried out mainly by the Meteorological Service of Israel [1] and partly by ourselves. These few but characteristic data may further elucidate the climatic conditions caused by factors 1–6.

Rainfall   A graph taken from a paper by J. LORCH (1952) who accompanied us sometimes during our Negev expeditions in the years 1949–1951, summarizes the data of 8 foregoing years (see Fig. 1).

Since rainfall is erratic, it is difficult to take average figures. These average figures are always different and fluctuate by more than 50% according to the period taken as basis. They fluctuate between 20 and 31 mm as yearly average, showing that even the highest values are proof of extreme aridity.

Cloudiness   Eilath belongs to one of the regions with the lowest cloudiness-coefficient on the globe. This in addition to the low latitude (29°33′ N) raises the high influence of the IE-factor (Insolation-Exposure factor) on all living beings (plants, animals and man alike) and on the soil surface as well.

Humidity   In spite of the vicinity of the sea we are confronted here with extremely low figures of air moisture (see Fig. 2).

1. We should like to express here our sincerest thanks for their continuous cooperation.

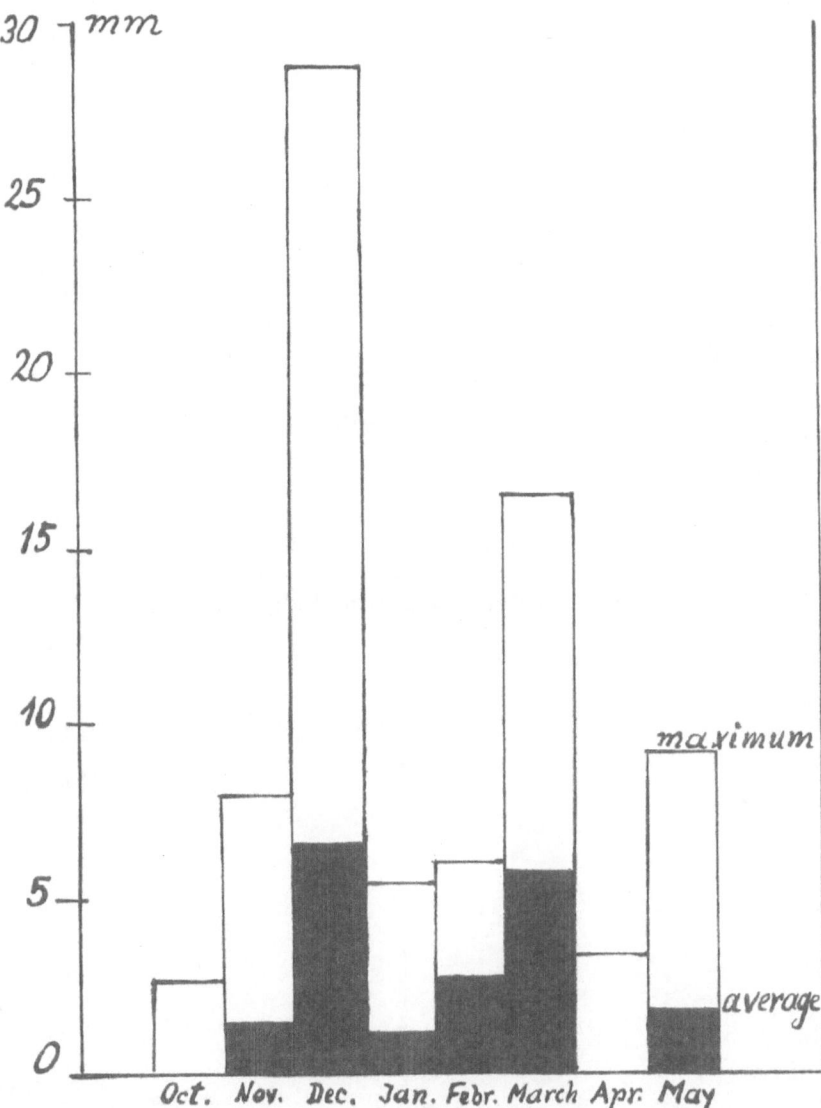

Fig. 1. Rainfall data of Eilat 1944–1951 (J. LORCH, 1952).
(1) The upper limit of each column indicates the highest monthly amount of rain;
(2) The black column indicates the average amount of rainfall;
(3) For all months the lowest monthly amount on record is 0.0 mm.

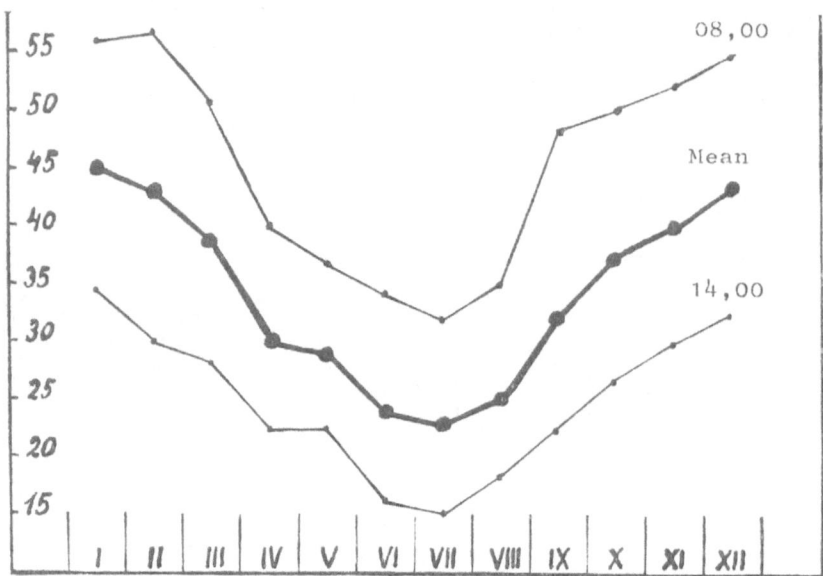

Fig. 2. Mean Relative Humidity in Eilat (J. LORCH, 1952).

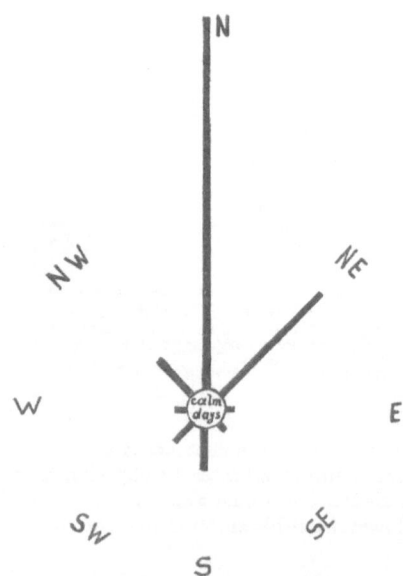

Fig. 3. Ratio of Wind conditions in Eilat.
(Diagram combining force, frequency and direction, based on the detailed data of the
Meteorological Service of Israel. For details see H. BOYKO, 1966, page 186).

Surface-Wind conditions   As to be seen from the diagram (Fig. 3), the by far strongest influence on plant growth is to be expected throughout the year from Northern and North-Eastern winds, both coming a long way over arid lands. On the other hand the much less frequent southern winds can sometimes blow with considerable force.

Temperatures and Evaporation   Fig. 4 shows the *mean* monthly values of Maximum and Minimum temperatures for a period of 8 years. Absolute yearly maxima are frequently reaching 45°C or even more. These high temperatures in combination with the factors (1)–(4) cause as their most significant effect on living beings, an extremely high evaporation.

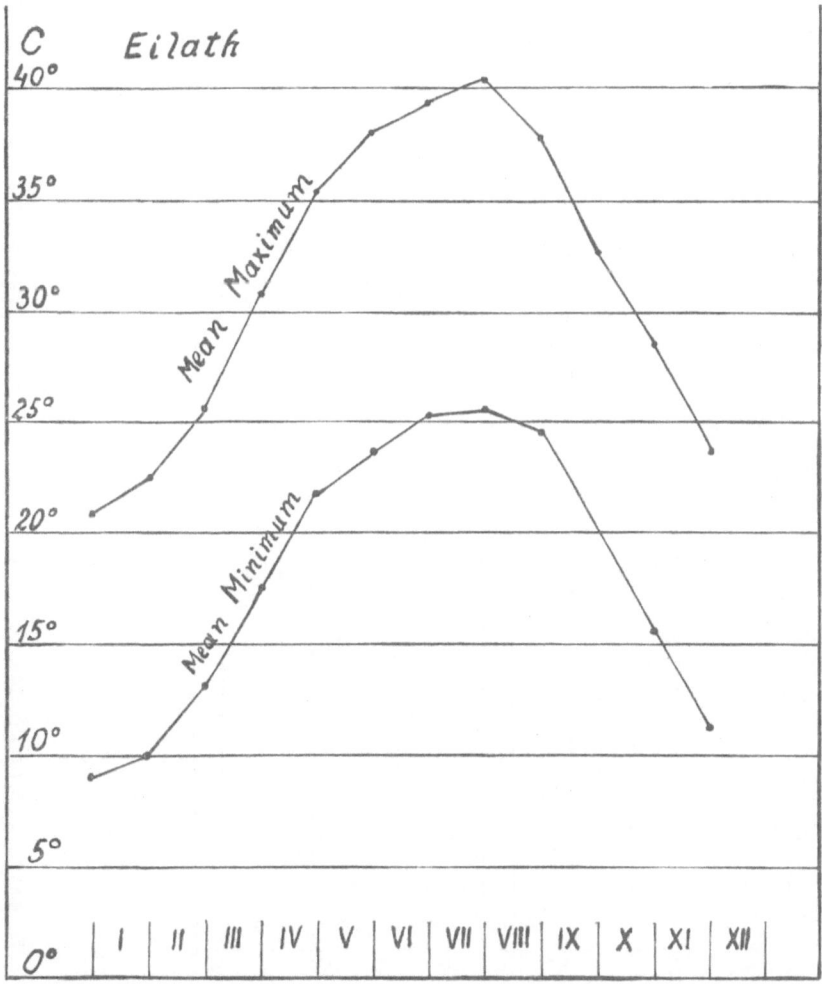

Fig. 4. Temperatures. Mean monthly Maxima and Minima (J. LORCH, 1952).

Evaporation figures recorded by two different methods in Eilat throughout the year 1964 on 362 days by the Meteorological Service are:

(a) recorded with a Piche Evaporimeter   4585 mm
(b) recorded with a U.S. Weather Bureau Class A Pan (protected
     by Chicken Hexagon Wire Netting 0.8 mm (= ¼ inch)   3573 mm

For comparison: The figure in Bet Dagon (mediterranean coastal plain),
for the measurement with the Piche Evaporimeter was for 366 days 1459 mm,
that is about a third (!) only of that measured in Eilat.

Fig. 5, taken from J. LORCH (1952), compares the evaporation figures of the

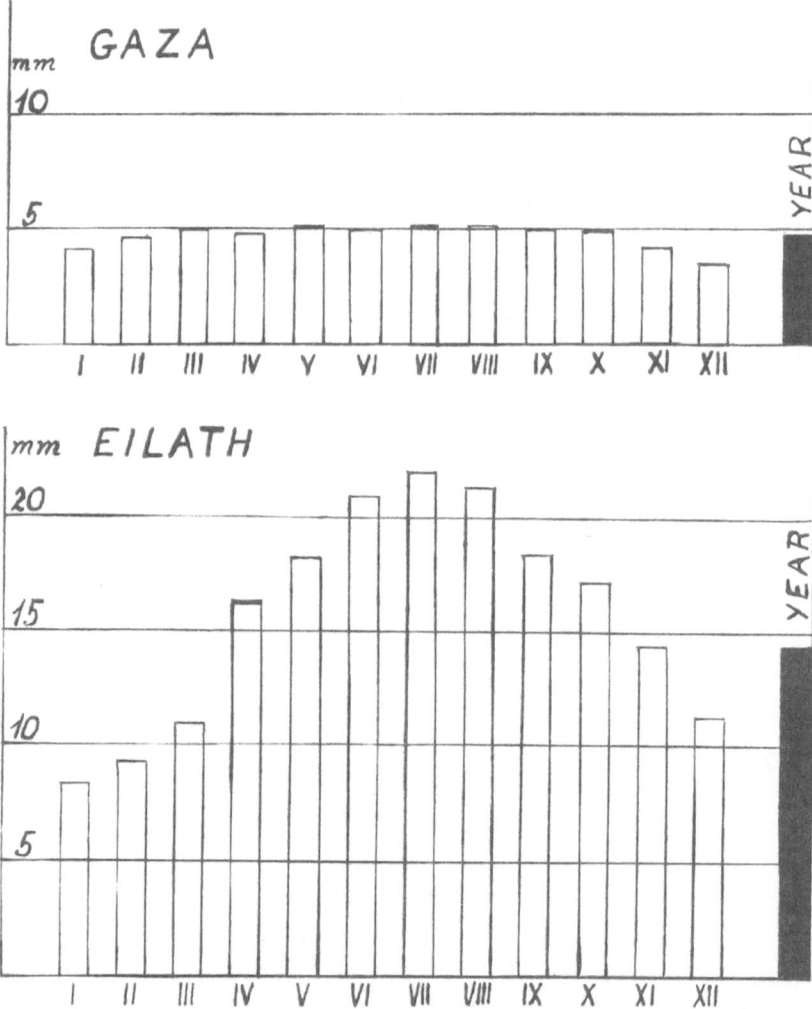

Fig. 5. Mean daily evaporation in Eilat and in Gaza (J. LORCH, 1952).

extremely arid place Eilat with the evaporation at the coastal town Gaza at the south-eastern corner of the Mediterranean Sea, having a semi-arid, south-mediterranean climate with high air humidity figures during the summer months.

Sand and Dust Storms Apart from the numerous Dust devils which one meets at almost every tour through Wadi Araba, there are from time to time also sand and dust storms of considerable strength and dust density. Instead of bringing figures which we do not have at our disposal, we can only report on our personal experiences. Such sand and dust storms were not infrequent and they sometimes reduce visibility to a few meters.

They too add to the unfavourable conditions for plant growing at this place.

The Soil

The soil is a coarse gravelly soil, consisting of mostly broken-up fragments of the rocks which form the mountains on the west side of the plantation. These rocks belong to various granitic, gneiss, Nubian Sandstone and also limestone formations.

The physical analysis of the soil shows:

Stones	61 %
Coarse sand	23 %
Fine sand	12.3%
Silt	2.7%
Clay	1.0%
	100 %

On some areas of the gravel slopes on which we planted, there was a gypsum-rock layer about 10 cm below the surface and up to 20 cm thick, which had to be broken with hoes before we could make planting holes. The plants on such slopes took far longer to get growing in a satisfactory way. In fact, we nearly despaired, the more so since the respective slopes were mainly those with an IE-factor (Insolation-Exposure Factor) of 20–30° South, adding the highest direct sun-radiation of all slopes. After a number of years, however, the trees and shrubs on these slopes caught up with those on the more favourable sites, and are now in no way inferior in size and development.

Owing to the quick percolation through this gravelly-sandy soil profile, which contains but a minimal percentage of fine particles of clay, there was no indication of salt accumulation whatsoever during all the years since the plants were put into the soil; the various plant species seem to select their specific nutrition from the offered high salt amounts in the irrigation water and all plants without any exception responded to it in a positive way.

The only exception of salt accumulation in the soil was an insignificant efflorescence in patches on the surface of the small irrigation plates ("Giessteller") around the plants, appearing after the moisture in the uppermost layer of a few millimeters had evaporated. These patches of salt efflorescence were only a fraction of a millimeter thick.

In full accordance with H. Boyko's principles for saline irrigation (1966 and summarized in 1967), the good permeability on these gravel hills explains the successful development even of those plants, which certainly cannot be called salt-tolerant. The raised salt-tolerance of many such plants, like *Morus alba*, (the Mulberry tree), or *Nerium Oleander, Ficus nitida, Vinca rosea, Acacia cyanophylla*, etc. in these gravelly-sandy soils is surprising, but also that of all other plants goes far beyond their salt-resistance in soils of a higher clay-content, i.e. in the so-called agricultural soils.

Soil temperatures

On places like these hills in a hot desert region, but near the Sea, attention has to be given to two factors, which are mostly of less significance in other places. Soil temperatures are the one and salt spray is the other.

We made measurements of soil profiles on northern and southern slopes and found to our surprise that:

(a) the maximum temperatures on the surface were not as excessive as we had assumed beforehand;

(b) the differences of the temperatures between northern and southern slopes were less significant than presumed;

(c) already in 20 cm depth even on the hottest days, the temperatures were so much lower than on the surface that no damaging of rootsystems had to be feared by this factor.

The reason for these favourable conditions of soil temperatures may be seen in the high content of gypsum crystals throughout the soil profiles, equilibrating the expected greater differences by their low heat conductivity and their hygroscopicity.

Salt Spray

This factor has been dealt with by the second author on several occasions, particularly in the book "Salinity and Aridity – New Approaches to Old Problems" (H. Boyko, 1966). Here it may suffice to stress again its influence on the soil. This problem complex has also been dealt with in detail and for the whole of Israel by S. Loewengart (1958) in his fundamental study on airborne salts.

The specific conditions in Eilat were first studied by the authors in the years 1949–1952 and included as one of the examples of the new principle of "Global Salt Circulation" in a report to UNESCO in 1958.

It could be shown that in Eilath the influence of the more salt bearing Southern winds on the composition of the soil is many times greater as that same influence by Northern winds although these latter, too, contribute to the Cl-content in the soil.

Once irrigation sets in, these influences are insignificant, since perco-lation in the area and natural drainage from it into the Red Sea quickly wash away the highly soluble chlorides.

Water

Water analyses were made right from the start and it was established that the water used from a well some 18 km north of the garden-site had a T.S.C. (Total Salt Content) of about 2200–3000 mg/l.

Soon however water was supplied by a pipe line from an oasis 42 km to the North, and this water showed a fluctuation from 2500–6000 and more mg/l T.S.C. Sometimes it even reached the high figure of 10,000 mg/l according to an oral information by Dr. WALTER STERN, the Head of the Hydrological Service of Israel.

The percentual composition of the irrigation water in Eilath, used from 1950–1956, is to be seen from Table IV:

TABLE IV

Composition of irrigation water in Eilat in percent

Ca	13.4%
Mg	7.4%
K	0.8%
Na	6.5%
CO_3	4.0%
SO_4	59.6%
Cl	8.1%
Fe ⎫	
Al ⎬ traces	0.2%
Si ⎭	
	100.0%

The amount of water supply was always far below the actual requirements necessary from a horticultural point of view. It was, however, sufficient to keep the plants alive and slowly growing in spite of the adverse conditions. Twice in the first year of planting the provisionally erected irrigation installation broke down for three weeks each in the hottest season; once in July 1950 and the other a few weeks later. All 200 plant species, however, survived this interruption of watering except two very shallow rooted ones: a *Kleinia* and a *Mesembryanthemum* species.

A partial explanation of this survival may be found in the water holding capacity and hygroscopicity of the gypsum layer situated in about 10–20 cm depth.

PLANTING OF THE DESERT GARDEN AND PLANT LIST

Plant Selection and Transport

Before we started to collect the specimens or to choose them in commercial nurseries for transplanting into the desert garden, we made a carefully selected list of a large number of those species, which we had reason to expect as success.

This list was made up on the basis of the following principles:

Priority was given to drought resistance and not to salt resistance, since we were convinced by our former ecological observations that the quick percolation in this gravelly soil would prevent salt accumulation. We also expected that the good aeration would help the plants to overcome the adverse effect of the high total salt content of the irrigation water.

We chose mainly such species which were known to us as drought resistant or the morphological features of which indicated this property, such as minimal leaf surface, xerophyllous leaves, a waxy layer on stems and leaves, succulence, etc. We did not refrain from chosing such plants, even when their natural habitat was near watercourses like *Nerium Oleander*, or in swamps with heavy soils, as for instance *Juncus maritimus*. We chose even a few species, about the survival of which we were rather sceptic, as for instance *Morus alba* (mulberry) which developed surprisingly well, or *Citrus aurantium* (orange tree) which survived but remained weak during the whole time and has never flowered.

Another consideration in the selection of our collection was the economic point of view. From the start we had in mind that species with an economic potential might constitute the basis for further development of this area, once it was established that they could stand the climatic hardships and saline irrigation. Plants for food, for fodder, and industrial raw materials were our main aims in this direction.

Last but not least we had to look for the basic necessity in this area, for shadow-giving trees and also for ornamental beauty. As is to be seen from the plant list, quite a number of the plant species we chose combined a number of these properties.

A rather difficult undertaking for men and plants alike was the transport through the partly roadless and mountainous desert region in the dry heat of the desert (Fig. 6), in october 1949.

Several days before the arrival of the first transport the place was already prepared and a number of empty former asphalt barrels were brought to the place of the garden, all filled with the desert underground water brought by command car from the well 18 km north of Eilat.

On the first of November 1949, the first saplings were in the soil and immediately watered. All plants survived the most critical time of transport and planting, and almost all that of establishment. Their further development is to be seen in general from the plant list for about 200 species and for some of them also from the measurements taken in their early stages in winter 1949/50 and for comparison from later years.

An earlier publication (E. BOYKO, 1952) can also be used to compare the development in the starting year with that 15–16 years later.

Almost 100% of the planted species, halophytes and non-halophytes alike, throve in this absolute desert through many years and the (for building and similar purposes) non eradicated ones are still growing there (see also Fig. 7 and 8).

Their very different ecological amplitudes prove that the number of species can be multiplied in experiments of this kind.

Rich sources of potentially successful plants are to be found in the standard works on the flora and vegetation of arid and semi-arid regions, particularly if one takes into account the geo-ecological law of plant distribution (H. BOYKO, 1947). A rich literature is to be found in the contributions to the UNESCO-volume on "Plant Ecology – Reviews of Research" (1955) and also in the other volumes of UNESCO's Series on Arid Zone Research.

A few particularly valuable works from South-West-Asia and North Africa may be mentioned here as examples: A. AUBREVILLE (1950); TH. MONOD (1957); R. NÈGRE (1959); P. OZENDA (1958); P. QUÉZEL (1965); CH. SAUVAGE (1961); V. AND G. TÄCKHOLM & M. DRAR (1941 ff.), for North Africa; L. BOULOAMOY (1930); V. DICKSON (1955); A. EIG (1931 and 1932); G. E. POST & J. E. DINSMORE (1932); J. THIÉBAUT (1936), for South- and West Asia; Most of these standard works of regional floras and vegetation contain also useful plantgeographical and ecological remarks.

Many plant species from non-desert regions should also be tried in similar experiments. A great number of the species presented in the following Plant-List are from humid regions, as for instance Morus alba L. from temperate Eurasia or Ficus nitida THUMB. from the humid Tropics. Thus the Desert Garden of Eilat, together with the experiments of plant growing with sea-water (H. AND E. BOYKO, 1959, 1966), may be seen as one of the first steps to a wide and most prospective new field of science of probably global impact: "Saline and Marine Agriculture" or in short "Salt Water Agriculture" (H. BOYKO, 1967), and to a productivizing of deserts in general.

PLANT LIST

Species planted in the desert garden of Eilat during 1949–1951

Species	Common name	Result [1]	Main economic potential (and/or other remark)
Abutilon denticulatum R.Br. (Malvaceae)	Indian mallow	+	fodder, orn. (= ornamental)
,, *muticum* SWEET.	,, ,,	+	seeds edible, fodder, orn.
Acacia cavenia BERT. (Mimosaceae)	Cavenia Acacia	++	perfume, wood, fodder
,, *cultriformis* CUNN. (,,) [3]		+	orn., h.[2]
,, *cyanophylla* LINDL. (,,)		+++	wood, fodder, shade, orn., h.
,, *farnesiana* WILLD. (,,)	Sweet Acacia	++	perfume, wood, fodder, h.
,, *Karroo* HAYNE (,,)	Karoo Acacia	++	hard wood, fodder, gum, tannin
,, *longifolia* WILLD. (,,)		+++	wood, fodder, shade, orn., h.
,, *melanoxylon* R.Br. (,,)	Blackwood	++	highly valuable wood (for veneers, furniture, boat-building, tools, pianofortes etc.
,, *raddiana* SAVI (,,)		+++	wood, fodder, shade (w.) (= wild in the region)
,, *sentis* VON MUELLER		+++	Wood, fodder, shade
,, *spirocarpa* HOECHST. (,,)		+++	wood, fibre (from the bark) fodder, orn. (w.) [2]
Aerva javanica JUSS. (Amaranthaceae)		+++	substitute for Kapok; veterinary medic., orn. (w.)
Agave americana L. (Amaryllidaceae)	century plant	+++	orn., beverage
,, *fourcroydes* LEMAIRE (,,)	Henequen Agave	+++	fibre (Henequen Sisal) for ropes, etc.; orn.
,, *sisalana* PERRINE (,,)	Sisal Agave	+++	fibre (Sisal Hemp) for ropes, hammocks, marine cordage, orn.
Aizoon canariense L. (Aizoaceae)		+++	native food in Sahara; orn.
Albizzia sp. (Mimosaceae)		+++	wood, shade, orn.
Aloe vera L. (Liliaceae)	Curacao Aloe	+++	medic., orn.

1. Development results are indicated as follows:
    +++ = excellent;   —— = died before establishment
     ++ = good   — = died during a break down of irrigation in July 1950.
      + = weak
2. (w.) = wild in the region and if irrigated, ornamental (= orn.). h. = honey or pollen plants.
3. The signs (,,) beneath the names of the species refer to the family name of the species mentioned immediately before.

Fig. 6. First plant transport, October 1949, through the mountainous desert of the Negev to Eilat.

Fig. 7. The Desert Garden of Eilath end of October, 1949, when planting started and the town of Eilat had just been founded.

Fig. 8. The town of Eilat 15 years after its foundation and after the establishment of the desert garden of Eilath, expanding its greeneries on the formerly vegetationless gravel hills (compare Fig. 7).

Species	Common name	Result [1]	Main economic potential (and/or other remark)
Anabasis articulata Moq. (Chenopodiaceae)	Jointed anabasis	+++	medic., orn. (w.)
Anastatica hierochuntica L. (Cruciferae)	Rose of Jericho	+++	(w.)
Aristida hirtiglumis Steud. (Gramineae)		+++	fodder, orn. (w.)
„ *plumosa* L. („)		+++	fodder, orn. (w.)
Artemisia absynthium L. (Compositae)	Absinthe	+++	essential oil, liqueur, medic., orn.
Asparagus acutifolius L. (Liliaceae)	Acute-leaved Asparagus	+	
„ *officinalis* L. („)		++	food, medic.
„ *Sprengeri* Regel (Liliaceae)		— —	orn.
Asphodelus fistulosus L. („)		+++	boiled bulbs eaten by Beduins, orn. (w.)
Astragalus spinosus Muschl. (Leguminosae)	Spiny Milk Vetch	+++	orn. (w.)
Atriplex halimus L. (Chenopodiaceae)	Saltbush	+++	fodder, orn.
„ *leucoclada* Boiss. („)	„	++	„ „
„ *nummularia* („)	„	++	„ „
Balanites aegyptiaca Del. (Zygophyllaceae)	Jericho Balsam	+++	
Bassia muricata Murr. (Chenopodiaceae)		+++	orn. (w.)
Blepharis edulis Pers. (Acanthaceae)		++	Seeds eaten in N. Africa
Caesalpinia pulcherrima Swartz. (Leguminosae)		+	medic., orn., h.
Calligonum comosum L'Her. (Polygonaceae)		++	fodder, orn.
Callistemon lanceolatus DC. (Myrthaceae)		++	wood, ess. oil, orn., h.
Calotropis procera R.Br. (Asclepiadaceae)	Auricula tree	+++	fibres for weaving (seeds) „ „ ropes (stems)
Capparis aegyptia Lam. (Capparidaceae)	Egyptian Caper	+++	orn. (w.)
„ *cartilaginea* Decne. („)	Fig Caper	++	orn. (w.)
Caralluma Aaronis Hart („)		++	orn. (w.)
Carissa edulis Vahl. (Apocynaceae)	Natal Plum	+++	fruit edible, orn.
Cassia obovata Collad. (Caesalpiniaceae)	Dog Senna	++	medic. (leaves and fruits); orn. (w.)

Species	Common name	Result [1]	Main economic potential (and/or other remark)
Cassia sturtii (,,)		++	orn.
Casuarina cunninghamiana MIG. (Casuarinaceae)		++	durable wood, shade
Caylusea canescens ST. HIL. (Resedaceae)		+++	orn. (w.)
Cerastium sp. (Caryophyllaceae)		+	orn.
Ceratonia siliqua L. (Caesalpiniaceae)	Carob tree	+++	fodder, seeds for plastics, shade
Cereus peruvianus HAW. (Cactaceae)		++	orn.
Chenolea arabica BOISS. (Chenopodiaceae)		+++	orn. (w.)
Citrullus colocynthis SCHRAD. (Cucurbitaceae)	Colocynth	+++	medic., oil (seeds)
,, *vulgaris* SCHRAD. (,,)	water melon	+++	food
Citrus medica L. var. *Limon* (BURM.) (Rutaceae)	Lemon	+	food
Cleome arabica L. (Capparidaceae)		+++	orn. (w.)
,, *droserifolia* DEL. (,,)		+++	orn. (w.)
Clerodendron fragrans VENT. (Verbenaceae)		+++	Climber, shade, orn.
Cordia myxa L. (Boraginaceae)		+++	wood, pharm., shade
Cucumis melo L. (Cucurbitaceae)	Sugar melon	+++	food
,, *prophetarum* L. (,,)	Globe cucumber	+++	medic.
Cupressus arizonica GREENE (Cupressaceae)		+	wood, shade, orn.
Cynara scolymus L. (Compositae)	Artichoke	— —	
Cynodon dactylon PERS. (Gramineae)	Bermuda Grass	++	fodder, lawn (var. *aegyptiaca*)
Daemia cordata R.BR. (Asclepiadaceae)		+++	orn., vine (w.)
Dalbergia sisso ROXB. (Leguminoseae)	Sissoo of India	+++	valuable wood, fodder
Danthonia Forskahlei TRIN. (Gramineae)		+++	Fodder
Diotis maritima SMITH (Compositae)	Sea Cud weed	++	medic., orn.
Dracaena Draco L. (Liliaceae)	Dragon Blood Tree	++	pharmac., orn.

Species	Common name	Result [1]	Main economic potential (and/or other remark)
Elaeagnus angustifolia L. (Elaeagnaceae)	Oleaster	++	oil, orn., h.
„ *pungens* THUMB. (,,)		++	orn., h.
Eragrostis bipinnata MUSCHL. (Gramineae)		+++	fibre for upholstery etc., orn. (w.)
Eriobotrya japonica LINDL. (Rosaceae)	Loquat	+	fruit edible
Erodium glaucophyllum L'HER. (Geraniaceae)	Heron's Bill	+++	fodder, orn. (w.)
„ *hirtum* WILLD. (,,)	„	+++	fodder, roots edible, orn. (w.)
Erythrina corallodendron L. (Leguminosae)	Coral Tree	— —	Cork-Wood, shade, orn., h.
„ *Crista-galli* L. (,,)	„	+	Cork-Wood, shade, orn., h.
Eucalyptus camaldulensis DENN. (Myrtaceae)		+++	wood, shade, h.
„ *gomphocephala* DC. (,,)	Tuart Eucalyptus	+++	valuable wood (also for ship-building), shade, h.
Euphorbia candelabrum TREM. (Euphorbiaceae)		+++	medic., orn.
„ *grandidens* HAW. (,,)		+++	orn.
„ *neriifolia* L. (,,)		+++	orn., medic.
„ *pulcherrima* WILLD. (,,)	Poinsetia	— —	orn.
„ *splendens* BOJER (,,)	Crown of Thorns	++	orn.
„ *triangularis* DESF. (,,)		+	orn.
Fagonia grandiflora BOISS. (Zygophyllaceae)		+++	orn. (w.)
Ficus bengalensis L. (Moraceae)	Bengal Fig	+	
„ *carica* L. (,,)	Fig Tree	++	fruit
„ *infectoria* ROXB. (,,)		+	wood
„ *macrophylla* DESF. (,,)		—	wood
„ *nitida* THUMB. (,,)		+++	wood, shade
„ *religiosa* L. (,,)	Peepul Tree	+	wood
„ *rubiginosa* DESF. (,,)		+	wood
„ *Sycomorus* L. (,,)	Sycomore	+++	fruit, wood, shade

Species	Common name	Result [1]	Main economic potential (and/or other remark)
Gazania splendens HORT. (Compositae)		+++	
Glaucium flavum BOISS. & HUET. (Papaveraceae)	Horned Poppy	++	oil, orn.
„ *grandiflorum* BOISS. & HUET („)		+++	orn.
Gomphocarpus fruticosus R.BR. (Asclepiadaceae)		+++	upholstering (seed hairs), (w.)
Grevillea robusta CUNN. (Proteaceae)	Silky Oak	+	hard wood, shade, h.
Gypsophila Rokejeka DEL. (Caryophyllaceae)		++	orn. (w.)
Haloxylon persicum BGE. (Chenopodiaceae)	Saxowl Tree	++	wood for charcoal, (carpentry), fodder
„ *salicornicum* BGE. („)		+++	fodder, orn. (w.)
Haplophyllum tuberculatum JUSS. (Rutaceae)		+++	fodder, orn. (w.)
Hibiscus esculentus L. (Malvaceae)	Okra	+++	food
Hyoscyamus desertorum EIG (Solanaceae)	Desert Henbane	+++	medic. (w.)
„ *muticus* L. („)		+++	medic.
„ *reticulatus* L. („)	Egyptian Henbane	++	medic.
Hyphaene Thebaica MART. (Palmaceae)	Dum Palm	+++	vegetable ivory, fibre plant orn., (w.)
Iris germanica L. (Iridadeae)	Iris	++	pharm., orn.
Juncus arabicus A. & B. (Juncaceae)		+++	fibre plant. (mats, cellulose, wickerwork)
Kleinia repens Haw. (Compositae)		— +++	orn. (died in the first summer during a break down of the irrigation work, succeeded at a second trial)
Kochia indica ROTH. (Chenopodiaceae)		+++	fodder
Lantana delicatissima HIRT (Verbenaceae)		++	orn.
„ *hybr. sp.* (red.) („)		++	orn.
Launea mucronata MUSCHL. (Compositae)		++	Fodder
Lavandula Stoechas L. (Labiatae)	French Lavender	++	ess. oil, orn., h.
Linaria floribunda BOISS. (Scrophulariaceae)		+++	orn. (w.)
Lycopersicum esculentum MILL. (Solanaceae)	Tomato	+++	food

Species	Common name	Result [1]	Main economic potential (and/or other remark)
Lythrum salicaria L. (Lythraceae)	Purple Loosestrive	+++	emergency vegetable, orn.
„ *tomentosum* VENT. („)		++	orn.
Melaleuca myrtifolia (Myrtaceae)		++	ess. oil, wood, orn.
Melia Azedarach L. (Meliaceae)	China berry tree	+	wood, pharm., shade, h.
Mesembryanthemum edule L. (Aizoaceae)		++	fruit edible, orn.
Morus alba L. (Moraceae)	Mulberry	++	wood, fruit, silkwormfodder, shade
Nerium oleander L. (Apocynaceae)	Oleander	+++	medic. orn.
Nitraria retusa ASCH. (Zygophyllaceae)		+++	fruits edible (w.) (for its salt accummulation in the vegetative parts and continuous shedding of leaves, eaten by animals, to be recommended for experiments in "Biological Desalination" (see HUGO BOYKO, 1966 b)
Ochradenus baccatus DEL. (Resedaceae)		+++	fodder, orn. (w.)
Oenothera Drummondi HOOK. (Oenotheraceae)	Coastal Evening Primrose	+++	sand binder, orn.
Olea europaea L. (Oleaceae)	Olive Tree	++	fruit, oil, wood, shade
Opuntia candelabriformis MART. (Cactaceae)		++	orn.
„ *ficus indica* MILL. („)	Indian Fig.	++	fruit, edible
„ „ *indica* (spineless variety) („)		+++	fodder, orn.
„ *vulgaris* MILL. („)		++	fruit, orn.
Panicum turgidum FORSK. (Gramineae)		+++	fodder, bird seed, seeds baked for bread in North Africa
Parkia sp. (Leguminosae)		+	seeds edible (?), orn.
Parkinsonia aculeata L. (Caesalpiniaceae)		+++	very successful even on the most wind exposed sites. wood, fodder, orn., h.
Pelargonium peltatum AIT. (Geraniaceae)		++	orn.
„ *zonale* WILLD. („)		— —	orn. (further trials recommended)
Pennisetum asperifolium KTH. (Gramineae)		++	fodder, orn.

Species	Common name	Result [1]	Main economic potential (and/or other remark)
„ *dichotomum* DEL. (,,)		+++	fodder, orn.
Phoenix dactylifera L. (Palmaceae)	Date Palm	+++	fruit, fibre, shade, orn.
Poinciana regia BOJ. (Caesalpiniaceae)	Peacock Flower	+++	wood, shade, orn.
Polypogon Monspelliensis DESF. (Gramineae)		+++	fodder, orn. (annual) (w.)
Portulaca afra JACQ. (Portulacaceae)		++	fodder, orn.
Prosopis juliflora DC. (Mimosaceae)	Mesquite	+	food (fruit) fodder wood, gum, h.
Psidium Guajava L. (Myrtaceae)	Guava	+	fruit, h.
Pulicaria undulata KOSTEL. (Compositae)	Desert Fleabane	++	orn. (w.)
Punica granatum L. (Punicaceae)	Pomegranate	+++	fruit, medic., tanning
Reichardia tingitana ROTH. (Compositae)		+++	food (w.) (used as salate) according to information by Beduin
Retama Raetam WEBB. (Leguminosae)	Roetam	+++	Cellulose, orn.
Romneya Coulteri HARV. (Papaveraceae)		+	orn., h.
Rosa Banksiae R. BR. (Rosaceae)		— —	orn. climber
Rumex vesicarius L. (Polygonaceae)		+++	Beduin food (cooked leaves)
Sabal palmetto LODD (Palmaceae)	Cabbage Palmetto	+++	fibre (wickerwork, mats, brushes), wood
Salsola foetida DEL. (Chenopodiaceae)		+++	orn. (w.)
„ *rosmarinus* SOLMS. LAUB. (,,)		+++	orn. (w.)
Salicornia fruticosa L. (,,)		+++	fodder, orn. (w.)
Sansevieria sp. (Liliaceae)		++	fibre
Santolina chamaecyparissus L. (Compositae)		+	orn.
Schinus molle L. (Anacardiaceae)	Pepper Tree	++	medic., gum, orn., h.
„ *terebinthifolius* RADDI (,,)		+++	medic., resin, orn., h.
Solanum Melongena L. (Solanaceae)	Eggplant	++	food
Spinacea oleracea L. (Chenopodiaceae)	Spinach	++	food

Species	Common name	Result[1]	Main economis potential (and/or other remark)
Statice pruinosa L. (Plumbaginaceae)		+++	orn.
„ *Thouini* Viv. (,,)	Egyptian Sea-Lavender	+++	good, orn.
Stenotaphrum secundatum Ktze. (Gramineae)	Augustine Grass	++	lawn
Suaeda fruticosa Forsk. (Chenopodiaceae)	Saltwort	+++	fodder, orn. (w.)
„ *Monoica* Forsk. (,,)		+++	fodder, orn. (w.)
Tamarix articulata Vahl. (Tamaricaceae)	Jointed Tamarisk	+++	medic., wood, shade, orn.
„ *maris mortui* Gutm. (,,)	Dead Sea Tamarisk	+++	wood, shade, orn.
„ *pseudo-pallasii* Gutm. (,,)		+++	wood, shade, orn.
„ probably n.sp. (,,)		+++	wood, shade, orn. (w.)
Tetragonia expansa Murr. (Aizoaceae)	New Zealand Spinach	+++	food
Thevetia nereifolia Juss. (Apocynaceae)	Yellow Oleander	+++	medic., orn., h.
Tribulus alatus Del. (Zygophyllaceae)	winged Caltrops	+++	orn. (w.)
Tricholaena Teneriffae Parl. (Gramineae)		+++	fodder, orn. (w.)
Vinca rosea L. (Apocynaceae)	Cape Periwinkle	+++	orn.
Vitex Agnus Castus L. (Verbenaceae)	Chaste Tree	+++	medic., wicker work, orn., h.
„ *pseudo-negundo* Hausskn. (,,)		+++	orn., h.
Washingtonia filifera Wendl. (Palmaceae)	Cañon Palm	+++	fibre, food, orn.
Yucca aloifolia L. (Liliaceae)	Spanish Dagger	++	fibre, orn.
Zilla spinosa Prantl. (Cruciferae)	Zilla	+++	Beduin food, (boiled leaves) orn. (w.)
Zygophyllum album L. (Zygophyllaceae)	white Bean Caper	+++	fodder, spice, orn.
„ *simplex* L. (,,)		+++	fodder, orn. (w.)
Zizyphus Spina-Christi Willd. (Rhamnaceae)	Christ-Thorn	+++	fodder, fruit. (w.)

About 20 non-determined species of Cactaceae have to be added, most of them have established themselves after planting and grew well in the following years.

Several of the species enumerated above and a collection of about 50 rare species from Argentine (Patagonia), Australia, Karroo, and Mexico, were unfortunately destroyed by Bulldozers in 1954, when this desert experiment garden was converted into a public garden in order to make room for a children's play ground. Irrigation, however, continued with brackish water and the remaining plants survive well in spite of insufficient irrigation.

Instead of the slides, shown during the lecture at the Symposium in Rome, a few examples of growth data may indicate the development. Many but not all measurements are taken from the best developed individuals, but all show the potential minimum, since this start shows the direction only of further development by selection, breeding, and better techniques. (Abbreviations: Meas. = Measurements; H = Height; W = Width; D = Diameter; DBH = Diameter Breast Height.)

Acacia raddiana SAVI: Tree, grown from seed in the small nursery (installed from the start as part of the desert garden); sown in winter 1949/50;

Meas.: 31. March 1960; H = 4 m; DBH = 11 cm.

Acacia cyanophylla LINDL.: Tree, transplanted in winter 1949/50 as 1 year old sapling.

Meas.: 21. March 1960; H = 5.50 m; DBH = 15 cm.

Capparis aegyptia LAM.: Shrub, grown from seed in the nursery of the desert garden; sown in winter 1949/50;

Meas.: H = up to 1.90 m; W = 6.10 × 4.90 m; branches near the basis up to 5 cm thick.

Casuarina cunninghamiana MIG.: Tree, transplanted in winter 1949/50 as 1–2 year old sapling.

Meas.: 21. March 1960; H = 6.50 m; DBH = 14 cm.

Cordia myxa L.: Tree, transplanted in winter 1949/50 as about 1 year old sapling.

Meas.: 21. March 1960; a) H = 3.80 m; DBH = 12 cm
　　　　　　　　　　　b) H = 3.50 m; DBH = 8 cm

Eucalyptus gomphocephala DC.: This tree was planted in a deep layer of pebbles on the sea-shore of Eilath as an about 1 year old sapling, in winter 1949/50.

Meas.: Oct. 1962; H = about 11 m, basis about 80 cm above the sea-shore and in 3.90 distance of it.

(At least parts of the root system must be in contact with sea-water and further investigations of ecological and physiological details are strongly recommended).

Salsola foetida DEL.; and

Arthrocnemum glaucum (DEL.) UNG. are growing wild near this remarkable tree. A little further away from the sea-shore some planted *Tamarix* sp. and *Nerium Oleander* L. are prospering.

Ficus Sycomorus L.: Tree, transplanted as about 1 year old sapling in winter 1949/50.

Meas.: 21 March 1960; H = 6 m; D (in 50 cm height) = 23 cm.

Juncus arabicus A. & B.: Perennial, planted as a small group of 2–3 clumps, each of them having a width of about 7 cm, in winter 1949/50.

Meas.: 12. Nov. 1964; H (fruiting stems) = up to 1.90 m; H (leaves) = up to 1.80 m; W (of the whole group) = 2.10 m × 1.20 m.

Remark: After transplanting the trimmed clumps the plant has never been cut.

Morus alba L.: (Mulberry) Tree, transplanted as about 1½ year old sapling in winter 1949/50.

Meas.: 21. March 1960; H = 4.50 m; DBH = 11 cm.

Olea europea L.: Tree, transplanted as 2 year old sapling in winter 1949/50.

Meas.: 12. Nov. 1964; H = 2.80 m; D (in 15 cm height) = 15 cm.

Parkinsonia aculeata L.: Tree, transplanted as 1–2 year old sapling, in winter 1949/50.

Meas.: 21. March 1960; H = 4.50 m; D (in 30 cm height) = 17 cm (on strongly wind exposed site).

Tamarix articulata VAHL.: Tree, planted as 25 cm long and about 1 cm thick cutting, in winter 1949/50.

Meas.: 21. March 1960; H = about 8.50 m; DBH = 23 cm. (one of the biggest trees in a large group).

Washingtonia filifera WENDL.: Palm tree, transplanted as about 40–50 cm high saplings, in winter 1949/50.

Meas.: 21. March 1960; a) H = 3.75 m; DBH = 40 cm
b) H = 4.50 m; DBH = 35 cm.

We take this opportunity to thank all coworkers, but particularly Mrs. EVA ZEUGER, JOSEF DROR and JOEL FREUDENBERG, who assisted in the difficult planting work during the first and most critical time with high pioneering spirit.

SALINE HORTICULTURE AND AGRICULTURE IN THE REGION OF THE DESERT GARDEN OF EILAT

Private gardens in the town of Eilat

Soon after the establishment of the desert garden the first settlers living in scattered barracks and a few houses followed this example and started with their own plantings utilizing the species shown as examples in the desert garden. Today, there is barely a house in this town of 12,000 inhabitants (1965) which is not proudly surrounded with the greenery of a few ornamental trees, shrubs and flowers or live fences. The most frequent species, now, to be found in these small gardens, are:

Acacia cyanophylla LINDL.

Agave americana L.

Carissa edulis VAHL.

Casuarina cunninghamiana MIG.

Clerodendron fragrans VENT. (sometimes as fence and properly cut in
                                                          Rococo-style)

Dalbergia sisso ROXB.

Eucalyptus camaldulensis DENN.

    ,,   *gomphocephala* DC.

Gazania splendens HORT.

Hibiscus rosa-sinensis L.[1]

Iris germanica L.

    ,,   sp. (Wedgewood)

Matthiola incana R. BR. (STOCKS) [1]

Mesembryanthemum sp. (red) [1]

Nerium oleander L.

Oenothera Drummondi HOOK.

Parkinsonia aculeata L.

Pennisetum purpureum (Elephant grass) (= *Pennicilaria*) (as fence) [1]

Phoenix dactylifera L.

Schinus molle L.

Tamarix articulata VAHL.

Thevetia nereifolia JUSS.

Tropaeolum majus L.[1]

Vinca rosea L.

Yucca aloifolia L.

Saline agriculture in Eilat and in Yotvata (Wadi Araba)

(a) The success with the cellulose plant *Juncus maritimus* in the desert garden of Eilat induced us to start in 1953 with a small field experiment on a loessy sand area near the oasis of Ein Ghadian, now Yotvata.

These experiments were carried out by a scientific team consisting of ELISABETH BOYKO, N. TADMOR and E. RAWITZ under the direction of HUGO BOYKO. The results were most satisfactory and led to a similarly satisfactory three years field experiment on sand near the Dead Sea in Neot Hakikar. There the irrigation water contained about 3000 ppm T.D.S. with 1000 ppm chlorine.

On the basis of these experiments a large scale plantation has successfully been started (1966) in the vast dune and flat sand area in Wadi Araba about 30 km north of Eilat. (T.D.S. of the irrigation water 4500 ppm with about 1200 ppm Cl content).

(b) Another highly informative field experiment with *Juncus* was carried

---

1. These species are more recent introductions.

out in 1955 against our ecological advice on 25 acres in the clayey almost impermeable highly saline and swampy depression, the "Sabha" of Eilat. As predicted by us, the saline irrigation on this heavy soil (the irrigation water had a T.D.S. of about 5000 ppm) soon led to high salt accumulation. As a matter of course, the leaves of the seedlings died, after the latter were transplanted from their freshwater beds into a brine of 120,000 ppm(!) T.D.S., standing after evaporation in a thick white salt layer. Without any reason (the area was then not thought of being of any further use) and inspite of the additional costs, the whole field was ploughed over, thus preventing any further observations. Nevertheless many plants, the roots of which were by chance not turned upwards and exposed to the burning sun, survived all this, developed fresh leaves and grew to short but strong plants.

In the Kibbuz Yotvata

After the planting of the desert garden of Eilat and encouraged by its success, a youth group (Nahal) started to settle 40 km north of it, in the Oasis Yotvata (Ein Ghadian). We distributed seeds of prospective fodder-plants and vegetables to them (as well as to other places for settlements in Wadi Araba) and in the course of a few years they developed to the well established "Kibbuz Yotvata." There the same saline water as in the desert garden of Eilath was used in the first years. Later a deeper drilling supplied them with less saline water, having only about 2000 ppm T.D.S. with about 600 ppm chlorine. Here, first in the oasis itself, mainly in the area where formerly an Eragrostidetum bipinnatae grew, and gradually also at its border, near the dunes beset with venerable old individuals of *Haloxylon persicum* BGE., good agricultural work is now carried out. The main crops, listed in April 1966, are:

Alfalfa, Rhodos grass, Mayze, Onions (for export), Tomatoes (for export), melons (for export), sugar beets and flowers.

Most of the flowers are also grown for export. Just flowering in April 1966 were:

various *Gladiolas* HORT. (excellent) (for export)
various *Centaurea* HORT.[1] (excellent) (for export)
various *Chrysanthemum* HORT. (excellent) (for export)
 Oenothera Drummondii HOOK. (for export) (?)
 Nerium Oleander L. (good)
 Tropaeolum majus L. (good)
 Parkinsonia aculeata L. (good)
 Avena sterilis L.[1] (flowering and fruiting) (grown as ornamental
               plant for cutting).

1. These flower species were new for the region.

In Timna

The successful gardening in Eilat with saline irrigation water on very permeable soil was followed up at various places in Wadi Araba. Thus, in 1956, another pioneer tried to make the desert bloom in the barren gravelly soil near the copper mines of Timna. There, the old mines of King Solomon have been revived with modern methods. 6000 m³ of saline desert water are used daily for this purpose. The enormous amounts of waste water flowing out after the mining process contain high total salt concentrations with sulphates as the main components. They are streaming into Wadi Araba and the new lake has sometimes the colour of copper vitriol. Its pH is fluctuating between 4 and 5. A refugee gardener from Germany, Mr. ARYEH EPHRONI, (his former name was Erich Freudenreich) decided to use this waste water. On the advice of professor H. HEIMANN he implemented our principles on this gravelly and very permeable soil as well as HEIMANN's principle of the Balance of the Ionic Environment and succeeded to create a park-like area of about 15 acres in the midst of the almost absolute desert of red rocks and gravel. In March 1960, we recorded the following list of plant species (ten species are new for the region and not to be found in the plant list of the desert garden in Eilath, they are marked by asterisks):

Species planted 1956–1959 in Timna (Plant list recorded on 21st March 1960)

	Result	Remark
Acacia cyanophylla LINDL.	+++	2½ years old, 3.50 m high
Acacia spp.	+++	
Arctotis grandis THUNB.*	++	
Casuarina cunninghamiana MIG.	++	
Ceratonia siliqua L.	+++	
Chrysanthemum HORT.*	++	
Erianthus Ravennae BEAUV.*	+++	(w.)
Eucalyptus gomphocephala DC.	+++	2½ years old, 6.50 m high
„ camaldulensis DENN.	+++	
Fagonia grandiflora BOISS.	+++	(w.)
Gaillardia pulchella FOUG.*	+++	
Gazania splendens HORT.	+++	
Gladiolas ¹	+++	
Iris germanica L.	+++	
„ sp. (Wedgewood) *	++	
Melia Azedarach L.	+++	
Mesembryanthemum sp. (red) *	+++	
Pelargonium zonale WILLD.	+++	
Pennisetum (Penicillaria) purpureum	++	
Portulaca grandiflora HOOK.*	+++	
Punica granatum L.	+++	
Thevetia nereifolia JUSS.	+++	
Tropaeolum majus L.*	+++	
Vitis vinifera L.*	++	grape vine

SUMMARY

A large variety of ornamental plants, mainly trees, shrubs and plants of potential economic value were introduced into an area which was in 1949 still an absolute desert. Yearly average rainfall there is about 20 mm only, and the place is one of the hottest and dryest regions of the globe.

The plan to lay a sound foundation for the harbour-town of Eilat (planned at that time, 1949) made it urgently desirable to create also an adequate environment in order to make this up to then most desolate desert-place not only inhabitable but attractive. It was due to a large part to the success of the garden which helped with the speedy development of this place into a green town of at present about 12,000 inhabitants, which number is steadily growing.

On the basis of combined ecological, horticultural, and economic reasonings about 200 plant-species were chosen. The main points of view in the selection of plants were the following:

    heat-drought-, and wind-resistance
    salt-resistance
    producing a maximum of shadow
    ornamental value
    economic potential, particularly as industrial raw-materials.

The first 2 points were of vital importance for the plants and were dictated by the very hot and dry climate accentuated frequently by strong desert-winds and by the saline water as the only source of irrigation. The T.S.C. (Total Salt Content) fluctuated from 2000 to 6000 mg/l and more.

According to the principles dealt with earlier by Dr. H. BOYKO (pp. 1–10) we chose the completely vegetationless gravelhills with their quick permeability as the place for the layout of the garden, in order to prevent gradual salt accumulation.

It could be shown that, after a difficult transport of the plants to the place and the start, 15 years later the place looks as green and adorned with as many parks and gardens as any city can ask for, for good normal living conditions. Gradually now this makes Eilat to an outstanding recreation place the whole year round and especially during the winter-months.

RÉSUMÉ

Une grande quantité de plantes ornementales, surtout des arbres, des arbustes et des plantes de valeur économique potentielle furent introduites dans une zone qui en 1949 était encore complètement désertique. Ici la pluviosité moyenne par an n'atteint que 20 mm environ et en outre il s'agit d'une des régions les plus sèches et les plus chaudes du globe.

Le plan de jeter les fondements les plus solides pour la ville-port d'Eilat (établi en 1949) fit naître le juste désir de créer aussi un milieu adéquat pour rendre ce lieu désertique et désolé non seulement habitable mais aussi agréable. Ce fut surtout grâce au succès du jardin que l'on réussit à transformer rapidement ce lieu dans une verdoyante ville qui compte à présent 12,000 habitants environ, et dont le nombre augmente constamment.

D'après un ensemble de considérations écologiques, économiques et concernant l'horticulture on choisit 200 espèces de plantes. Voici les principes qui sont à la base du choix:

résistance à la chaleur, à la sécheresse, au vent
résistance au sel
quantité d'ombre produite
valeur ornementale
potentiel économique, surtout comme matières brutes industrielles.

Les deux premiers points furent très importants pour les plantes et furent déterminés par le climat très chaud et sec accentué fréquemment par de forts vents désertiques et par le fait que la seule source d'irrigation est représentée par l'eau salée. Le T.S.C. (Total Salt Content) (contenu total de sel) variait de 2000 à 6000 mg par litre et au delà.

D'après les principes traités dans l'étude de M. Hugo Boyko on a choisi les collines graveleuses complètement dépourvues de végétation avec leur perméabilité rapide comme lieu pour le plan du jardin, dans l'intention d'éviter de graduelles accumulations de sel.

On prouvait par la suite, qu'après un transport difficile des plantes au lieu de départ, 15 ans plus tard, ce même lieu est verdoyant et agréable avec un nombre de parcs et de jardins suffisant pour toute ville qui veut jouir de conditions de vie normales. Peu à peu tout cela a fait d'Eilat une station de repos très connue pendant toute l'année et surtout pendant les mois d'hiver.

RIASSUNTO

In una zona che nel 1949 era ancora completamente desertica, sono state piantate una grande varietà di piante ornamentali, sopratutto alberi e cespugli, nonchè piante di potenziale valore economico. In tale località la piovosità media annua è di solo 20 mm circa ed è una delle regioni più calde e aride del globo.

Il progetto di gettare solide fondamenta per la città-porto di Eilat (elaborato nel 1949) rendeva urgente anche la creazione di un ambiente adeguato, tale da rendere non solo abitabile, ma anche attraente, quello che era un luogo deserto e desolato. Il successo del giardino ha contribuito largamente alla rapida trasformazione di questo luogo nella verde città che oggi conta circa 12.000 abitanti e il cui numero è in costante aumento.

In base ad una serie di considerazioni di ordine ecologico, agronomico ed economico, furono scelte 200 specie di piante. I principii fondamentali usati per la scelta delle piante sono stati i seguenti:

resistenza al calore, alla siccità e al vento;

resistenza al sale;

massima zona di ombra prodotta;

valore ornamentale;

potenziale economico, particolarmente come materie prime industriali.

I primi due punti erano di capitale importanza per le piante e derivavano dal clima assai caldo e secco: calore e siccità aumentati dai forti venti del deserto e dal fatto di potere disporre, come unica fonte di irrigazione, dell' acqua di mare. Il T.S.C. (contenuto totale in sale) variava da 2000 a 6000 mg/litro e oltre.

Secondo i principi discussi nella relazione del Dr. HUGO BOYKO, allo scopo di evitare una graduale accumulazione di sale, abbiamo scelto, come luogo dove situare il giardino, delle colline ghiaiose completamente prive di vegetazione, molto permeabili.

Dalle diapositive si può notare che, dopo un difficile trasporto delle piante nel luogo prescelto, quest'ultimo, dopo 15 anni, appare verde e adorno di tutti quei parchi e quei giardini necessari ad una città per garantire buone e normali condizioni di vita. Questo ha fatto di Eilat un luogo di riposo eccellente durante tutto l'anno e sopratutto nei mesi invernali.

REFERENCES

AUBREVILLE, A. 1950. Flore Forestière Soudano-Guinéenne Paris. pp. 523.

BOULOUMOY, L. 1930. Flore du Liban et de la Syrie Paris I: 431 (Text), II: 512 (plates).

BOYKO, ELISABETH. 1952. "The Building of a Desert Garden" J. Roy. hort. Soc. 76 (4).

BOYKO, HUGO. 1947. On the role of plants as quantitative climate indicators and the Geo-ecological Law of Distribution J. Ecol. 35: 138–157.

BOYKO, HUGO. 1954. A new plant-geographical subdivision of Israel (as an example for Southwest Asia) (2 fig.) Vegetatio, 5–6: 309–318.

BOYKO, HUGO. 1962. Main Principles of Direct Irrigation with Seawater without Desalination. Report to UNESCO, mimeographed. Negev Inst. f. Arid Zone Res., Bersheba p. 1–40.

BOYKO, HUGO. 1966. Basic ecological principles of plantgrowing by irrigation with highly saline or sea-water. In H. BOYKO [ed.] Salinity and Aridity – New Approaches to Old Problems (15 fig.) Mon. Biol. 16. Dr. W. Junk, The Hague, p. 131–200.

H. BOYKO [ed.]. 1966. Salinity and Aridity – New Approaches to Old Problems. Dr. W. Junk – Publishers, The Hague 38 fig. p. VIII + 408.

BOYKO, HUGO. 1967. Salt Water Agriculture. Scient. Amer., p. 89–95.

BOYKO, HUGO & ELISABETH BOYKO. 1959. Seawater irrigation – a new line of research on a bioclimatic Plant-Soil-Complex. Int. J. Bioclim. Biomet. 3 (II) Sec. B 1 p. 1–24.

BOYKO, HUGO & ELISABETH BOYKO. 1960. Principles and experiments regarding direct irrigation with highly saline and seawater without desalination. Trans. New York Acad. Sci., Suppl. to No. 8 Ser II 26: 1087–1102.

BOYKO, HUGO & ELISABETH BOYKO. 1966. Experiments of plantgrowing under irrigation with saline waters from 2000 mg/liter T.D.S. up to sea-water of oceanic concentration, without desalination p. 214–282. In H. BOYKO [ed.] Salinity and Aridity – New Approaches to Old Problems (15 fig.) Mon. Biol. 16. Dr. W. Junk, The Hague.

DICKSON, VIOLET. 1955. The Wild Flowers of Kuweit and Bahrein. London p. 144.

EIG, A. 1931 & 1932. Les éléments et les groupes phytogéographiques auxiliaires dans la flore palestinienne. *Beih. Fedde's Rep. spec. nov. Berlin-Dahlem*. 62 part I: Texte p. 1–201, part II: Tableaux analytiques, (120 Tableaux).

LOEWENGART, J. 1958. Geochemistry of waters in Northern and Central Israel and the origin of their salts. *Bull. Res. Counc. Israel*, Jerusalem, 7G(4): 176–205.

LORCH, J. 1952. Climatological Data for the Negev. Meteor. Service of Israel. *Meteor. Notes* 4: 1–12.

Meteorological Service of Israel: Monthly and Annual Reports, Beth Dagon.

MONOD, THÉODORE. 1957. Les grands divisions chlorologiques de l'Afrique. Londres. p. 147, 2 pl.

NÈGRE, R. 1959. Recherches phytogéographiques sur l'étage de végétation mediterranéen aride (sous-étage chaud) en Maroc occidental. Travaux de l'Inst. Scient. Chérifien, Rabat. Serie Botanique, No. 13.

OZENDA, PAUL. 1958. Flore du Sahara Septentrional et Central. Cent. nat. Rech. sc. Paris p. 486.

POST, G. E. & J. E. DINSMORE. 1932. Flora of Syria, Palestine and Sinai. Beyrouth. I: 658, II: 928.

QUÉZEL, PIERRE. 1965. La Végétation du Sahara du Tchad à la Mauritanie. Stuttgart, G. Fischer Verlag.

ROSENAN, N. 1951. Monthly course of mean daily evaporation 1940–1949, Meteorol. Service of Israel.

Salinity and Aridity – New Approaches to Old Problems 1966 (ed. H. BOYKO). Dr. W. Junk Publishers, pp. VIII + 408, (38 figs.), Den Haag.

SAUVAGE, CHARLES. 1961. Recherches Géobotaniques sur les Subéraies Marocaines. Tanger. I: IX + 462, XIV + 252, 3: cartes.

TÄCKHOLM, VIVI and GUNNAR & MOHAMMED DRAR. Flora of Egypt. Cairo. (from 1941 (Vol. I) on).

THIÉBAUT, J. 1936. Flore Libano-Syrienne. Institut Français d'Archéologie Orientale Cairo. 1: 175, 2:372 1940.

UNESCO. Series on Arid Zone Research.

Authors' address: Dr. Elisabeth Boyko, Dr. Hugo Boyko, World Academy of Art and Science, 1 Ruppin Street, Rehovoth, Israel.

Irrigation with Saline Water in Puglia

by

NICOLA FICCO

This paper refers to experiments which were carried out on the Adriatic coast of Puglia, on sandy-calciferous soil.

The water used for irrigation originates from the carstic fold and contains a total variable salinity ranging from a minimum of 1-2‰ in October-November to a maximum of 8-8.5‰ in July and August.

EXPERIMENTAL PROGRAMME

Experimentation was based on the following treatments:

T_1 (– short intervals and small volumes)
T_2 (– long intervals and large volumes)
C_o (– organic fertilizers)
C_m (– mineral fertilizers)
C_{om} (– mixed fertilizers)

METHODS OF IRRIGATION

The irrigation methods adopted were flooding and lateral infiltration from furrows. Concerning the sizing of plots and quantities of water, the principle was maintained that the small plot size and the corresponding small quantity of water ensures the least hydric consumption in every single watering. It was tried, therefore, to employ for every short interval watering the lowest possible specific volume by elevating the superficial modulus and building small sized compartments.

CROPS EXPERIMENTED WITH AND PRODUCTION RESULTS

The main crops experimented with may be summarized as follows:

Tomatoes (local smooth round variety) were irrigated by infiltration from furrows at intervals of 8 and 16 days, and mean specific watering volumes were between 237–268 m³/ha for the 8 days interval and 511–565 m³/ha for the 16 days interval. The average yield was 14.7–20.1 t/ha for interval 8 and 12.0–12.3 t/ha for interval 16. The best yields were obtained on plots at interval 8 and with organic and mineral fertilizers.

Forage maize (var. *Wisconsin* 641/AA) was irrigated by flooding at intervals of 15 and 20 days; specific watering volumes ranged from between 354–427 m³/ha for the 15 days interval and 593–594 m³/ha for the 20 days interval. The average green fodder yield was 42.6–46 t/ha for interval 15 and 21.5–26.7 t/ha for interval 20. The best results were obtained on plots irrigated at interval 15 and with organic and mineral fertilizers.

Sweet sorghum (var. *Hegary*) was irrigated by flooding at intervals of 15 and 30 days; mean specific watering volumes ranged from between 266–312 m³/ha for interval 15 and 536–625 m³/ha for interval 30. The average green fodder yield was 36–41.8 t/ha for interval 15 and 18.5–29.1 t/ha for interval 30. The best results were obtained on plots irrigated at interval 15 and with organic and mineral fertilizers.

Vigna sinensis was treated with two auxiliary irrigations whose mean specific watering volumes were 246–268 m³/ha. The average green fodder yield was 19 t/ha.

Sunflower (var. *Russian Giant*) was irrigated by infiltration from furrows at intervals of 24 days; specific watering volumes ranged between 400–469 m³/ha and organic and mineral fertilizers were used. The average seed yield was 1.6 t/ha.

EFFECTS ON SOIL

The survey of this important aspect of the experiments was carried out on the correlation between rainfall and salinity of soil.

Daily precipitations were, therefore, recorded in the field and annual results were as follows:

            1955 . . . . 378.8 mm
            1956 . . . . 452.6 mm
            1957 . . . . 541.9 mm

Samples of water were taken periodically in order to determine total salinity and chlorine-ion. Sometimes samples were taken at 6.00, 12.00 and 18.00 hours of the same day so that possible salinity variations during the day could be controlled. From the analysis it has been noted that the salt content increases from 6 to 18 hrs. This is mainly the result of pumping which is done throughout the zone during the spring/summer period.

Prior to the start of each experimental irrigation season and after the harvest, soil samples were taken from each group of three plots which underwent the same experimental process in order to determine the total salt content, chlorine-ion and pH. The results of the analyses showed an increase in the saline residue of the soil, in consequence of irrigation by saline waters, and a pH trend towards alkalinity.

On the basis of elements resulting from periodical analyses performed on irrigation waters, and of hydric consumption recorded for each crop during the irrigation season, it was possible to follow up the balance of salt in the soil, i.e. to determine the effective quantity of salt contributed by irrigation and that left in the soil. This balance could be gauged owing to the considerable number of analyses performed: these, during the experimental period, were 109 for the irrigation water and 78 for the soil.

The following table only shows data of the soil analyses which were taken at the beginning and at the end of the survey.

March 15th, 1955 Results of analysis at beginning of experimentation			January 10th, 1958 Results of analysis at end of experimentation		
Total salt content ‰	Chlorine-ion ‰	pH	Total salt content ‰	Chlorine-ion ‰	pH
0.376	0.035	8.00	0.450	0.022	8.06
0.350	0.053	8.08	0.450	0.018	8.03
0.492	0.014	8.15	0.340	0.013	7.98
0.675	0.062	7.96	0.520	0.035	8.02
0.606	0.014	7.76	0.470	0.035	8.01
0.529	0.044	8.16	0.490	0.035	8.00
1.146	0.053	8.01	0.410	0.018	7.82
0.695	0.071	8.00	0.400	0.018	7.96
0.425	0.040	8.08	0.410	0.018	8.28
1.470	0.260	7.42	0.390	0.031	8.01
1.450	0.260	7.70	0.590	0.134	8.25
0.660	trace	7.90	0.440	0.022	8.52
1.050	0.220	7.90	0.520	0.071	8.30
0.580	0.100	8.15	0.490	0.040	8.45
3.210	0.700	7.80	0.460	0.035	8.08
0.690	0.050	7.90	0.450	0.075	8.50
1.800	0.560	7.85	0.540	0.129	8.65
1.000	0.530	8.00	0.690	0.022	8.42
3.300	1.100	7.87	0.410	0.018	8.50
2.430	0.700	7.87	0.390	0.018	8.00
1.770	0.400	8.00	0.490	0.040	8.12
0.670	0.044	7.40	0.440	0.053	8.31
0.530	0.035	7.50	0.410	0.035	8.10
0.710	0.062	7.70	0.360	0.013	8.28
0.600	0.053	7.70	—	—	—
0.540	0.035	7.70	—	—	—
0.838	0.062	7.88	0.490	0.031	8.02

It can be observed from these data that there was both a diminution of total salt content and of chlorine-ion. This variation, calculated as an average, is as follows:

	Total salt content	Chlorine-ion	pH
March 19th, 1955	1.058	0.208	7.58
January 10th, 1958	0.765	0.039	8.18
as percentages:	−28%	−81.3%	+7.9%

The comparison of data of the analysis performed in 1955, before the commencement of the experiment, and of those relating to the analysis of soil samples taken some time after the 1958 crops were harvested, shows clearly how, contrary to the usual agricultural techniques followed in that area, after four years of uninterrupted soil irrigation there has been no verification of any increase in the total salt content and, therefore, of chlorine-ion.

The table shows that there was even a considerable decrease in both total salt content and chlorine-ion. This demonstrates that, within the period of a solar year, salt supplied to the soil by irrigation disperses underground without causing even the slightest damage to vegetation. The slight increase recorded for pH, if we compare the 1955 data with those of 1958, is not significant since it is expressed in such a measurement as not to cause the soil reaction to pass from alkaline to acid.

CONCLUSIONS

Data drawn from our complete field experiments agree with PANTANELLI's general principle, according to which in utilizing saline water it is necessary to operate at brief intervals with small volumes of water.

However, we wanted to go more deeply into our investigation and arrive at an effective evaluation of both watering volume and interval in respect of each different crop. This was desirable in view of our aim to achieve real elements on which to base and elaborate irrigational transformation plans which would then permit us to impart concrete instructions on irrigation techniques to the farmers concerned.

Speaking of small volumes and brief intervals makes little sense as it depends on what is meant by a small volume; this could refer equally to 250 or 350 m³/ha. But this difference of 100 m³/ha creates, in our case, an increase in salt of about 0.8 t/ha for every watering; hence the necessity of employing those irrigation methods which use the smallest volume of water that can be determined with fair approximation.

Our conclusions may be summed up as follows:

(1) Elaboration of farm crop plans with introduction of forage crops (forage maize, sweet sorghum, forage beet, *Vigna sinensis*) and industrial crops (sugarbeet, sunflower).

(2) Application of suitable irrigation techniques (small plots for use of classical irrigation methods; flooding and lateral infiltration from furrows,

for which it is comparatively simple to determine watering volumes) bearing in mind the principle of small watering volumes at brief intervals.

(3) The irrigated soil to rest every two years in order to avoid deterioration of its mechanical and chemical properties; this practice is only prudent if we consider that in the course of a four-year irrigation treatment of the experimental plot no increase of salt content was detected.

(4) If maximum salt content of water during the irrigation season should reach 5–6 g/l, instead of adopting complete rest of the soil (see para. 3) crops may be cultivated whose growing cycle extends from the end of summer to December, for instance: cabbage, lettuce and chicory which have the advantage of requiring only few irrigations (two or three) at transplantation time and then continue to grow without further irrigation.

(5) Taking into consideration the variability of the water's salt content which applies to the whole Puglia coast, and bearing in mind the particular sensitivity of the seeds or saplings to salt, it is advisable to anticipate the seeding and transplantation time of spring and summer crops.

(6) Supply the irrigated soil with abundant organic fertilizers (where manure is not available) composed of sea algae (*Posidonia oceanica*), which are to be found in large quantities all along the coast during the autumn-winter period, treated, desalted and mixed with vegetal residue compost. This biological corrective, which is a substitute for the manure lacking in that zone, is very good, cheap and effective for the activation of soil treated with saline waters.

The above-mentioned directions should, of course, be considered valid only for those soils whose permeable conditions, and the presence of limestone, ensure normal drainage of the washed salts from the arable layer of the soil on cracked rock.

SUMMARY

The author refers to experiments which aim at determining the resistance of some vegetable crops being irrigated by saline waters and the most suitable methods of irrigation technique in order to utilize the above mentioned waters.

The survey serves the purpose of ascertaining how the yield of the various crops varies in accordance with the length of intervals, specific watering value and fertilizing (with mineral fertilizers or manure, or a combination of the two), considered together.

The results obtained with the following different methods have been compared:

- irrigation with large volumes and long intervals;
- irrigation with smaller volumes and shorter intervals;
- and, in both cases, using on some plots only mineral fertilizers, and on others only manure, and in others, combinations of the two.

Experiments have been made on the following crops: tomato, maize (fowl and forage), beet (sugar and forage) sweet sorghum, sunflower, cabbage, soya, alfalfa, pepper and egg-plant.

The subterranean water used in these experiments has a salt content at the beginning of spring of about 4 g/l, which gradually increases up to August to about 8.5 g/l.

Summing up, from the observation made and yield results obtained, it appears that for the crops tested it is best to adopt:
– brief intervals of watering with small volumes which can only be achieved with certain methods, such as flooding and furrow;
– mineral and manure fertilizers.

RÉSUMÉ

L'auteur se rapporte aux épreuves faites dans le but de déterminer la résistance de certaines cultures sous irrigation avec des eaux saumâtres et les techniques d'irrigation consentant la meilleure exploitation de ces eaux.

Les recherches prouvent que les récoltes des divers produits subissent des modifications qui sont en fonction de la durée des intervalles et de la valeur spécifique de l'irrigation et de la fumaison (engrais organiques ou mineraux ou mixtes).

On a comparé les résultats obtenus avec les méthodes suivantes:

Irrigation avec une grande quantité d'eau, par longs intervalles.

Irrigation avec des quantités d'eau inférieures et par intervalles plus courts: dansles deux cas pour certaines parcelles on n'a employé que des engrais mineraux, pour d'autres que du fumier et pour d'autres encore un engrais mixte.

On a fait des épreuves sur les suivantes cultures: tomates, maïs fourrager et non fourrager, betteraves à sucre et fourragères, sorgho, tournesol, choux, soja, alfa, poivrons, aubergines.

L'eau souterraine employée au cours de ces épreuves comporte au début du printemps un contenu en sel de 4 g/l qui augmente graduellement jusqu'à août et atteint près de 8.5 g/l.

Conclusion: d'après les observations faites et les récoltes obtenues on conclut qu'il convient d'adopter pour les cultures sous-mentionnées: irrigations par intervalles courts et avec une petite quantité d'eau, pouvant être effectuées seulement d'après des méthodes particulières telles que la submersion ou l'infiltration latérale; engrais mineraux et organiques.

RIASSUNTO

L'autore si riferisce alle prove condotte allo scopo di determinare la resistenza di alcune colture irrigate con acque salmastre e le tecniche d'irrigazione più adatte per usufruire di queste acque.

La ricerca prova che i raccolti dei diversi prodotti, variano congiuntamente in funzione della durata degli intervalli e del valore specifico dell' irrigazione e delle concimazioni (concimi organici o minerali o entrambi).

Sono stati messi a confronto i risultati ottenuti con i seguenti metodi:

Irrigazione con grandi volumi e turni lunghi; irrigazione con volumi minori ed intervalli più corti; in ambedue i casi usando: in alcune parcelle soltanto concimi minerali, in altre solo letame, ed in altre ancora una combinazione di entrambi.

Le prove sono state fatte sulle seguenti colture: Pomodori, mais da foraggio e da granella, bietole da foraggio e da zucchero, sorgo, girasole, cavolo, soja, alfalfa, peperoni e melanzane.

L'acqua sotterranea usata in queste prove presenta all'inizio della primavera un contenuto salino di 4 g/l, che aumenta gradualmente fino ad agosto, arrivando a circa 8.5 g/l.

Conclusione: dalle osservazioni fatte e dal raccolto ottenuto risulta che per le colture sperimentate sarebbe meglio adottare: irrigazioni a brevi intervalli e con piccoli volumi, che possono essere effettuate soltanto con determinati metodi, quali la sommersione e l'infiltrazione laterale; concimi minerali ed organici.

Author's address: Dr. Nicola Ficco, Ente per lo sviluppo dell'Irrigazione e trasformazione fondiaria in Puglia e Lucania, Bari, Italy.

Irrigation with Salty Water in Tunisia

by

J. W. van Hoorn, Ch. Ollat, R. Combremont & G. Novikoff

In Tunisia and the adjacent North African and Middle East countries, the water used for irrigation contains larger or smaller quantities of soluble salts, while the groundwater of the irrigated areas may also be more or less saline. In these circumstances, irrigation has often resulted in the deterioration or even the total sterilization of the soil.

In Tunisia, the use of brackish water has been the subject of a research project conducted since 1935 by YANKOVICH and NOVIKOFF at the Botanical Service. Field studies undertaken by the soil experts DESSUS and SABATHE, covering the period between 1940 and 1955, have also resulted in the determination of limit values for the sensitivity of crops to salt. Recently, this research has been resumed by the Pedological Service, the Agricultural Service, the Department for the Rehabilitation of the Medjerdah Lowlands, and the Special Fund Project (Research Centre for the Use of Saline Water in Irrigation).

IRRIGATION AS PRACTICED IN TUNISIA

The oldest irrigation systems are found in Southern Tunisia, in the oases of the Gabès District and of Chott Djérid (see Fig. 1). Irrigation water is more or less brackish, with a salt content of about 1 to 3 g/l, while the soil often contains gypsum. According to observations made at the oases, the most important question is that of providing good drainage. Though local varieties (such as Alig, for instance), are less sensitive, it seems that lowering the water-table from 0.80 m to 1.30 m should result in doubling the crop of a variety like Deglat en Nour.

In the olive groves of Oued Mellah, near the oasis of Gabès, which are irrigated with water of a salinity of 4 g/l, of which 200 mm per year are given in addition to a rainfall of about 160 mm, drainage has also proved to be important. A rise of the water-table from 1.50 m to 0.80 m results in a

Fig. 1. Map of Tunis.

serious reduction of the crop (from 60 kg per tree to 20 kg per tree). Equally important is the choice of a ligh-textured, permeable soil for the plantations.

On a study plot in the oasis of Gabès, different market and fodder crops have been grown over a period of eight years, using water of a salinity of 3 g/l on a sandy-gypsous soil. At Zarzis, water of a salinity of 6 g/l has been used on a sandy-chalky soil. Table I shows the average crops obtained on the two plots:

TABLE I

Crop	Gabes	Zarzis
Lucerne, green fodder	115 t/ha	110 t/ha
Sorghum, green fodder		45 t/ha
Barley	3.1 t/ha	2.1 t/ha
Sorghum, grain		2.4 t/ha
Carrots		27 t/ha
Tomatoes	21 t/ha	
Turnips	30 t/ha	
Asparagus	4.6 t/ha	2 t/ha
Artichokes, selected	31,000 heads/ha	
Artichokes, unselected		10,000 heads/ha

The figures given in this table refer to crops harvested from small plots. Cultivation on a larger scale would be less productive. However, it has been established that certain fodder and market crops can be grown with satisfactory results while using saline water.

In the same region of Zarzis, where the farmers irrigate their crops, it has been found necessary to interrupt the irrigation from time to time because of the rise of the water-table.

In the Kasserine region, an area of about 300 ha has been under irrigation for the last 60 years with water containing 2.8 g/l; crops include grain (2.0–2.5 t/ha without nitrogen fertilizer), maize (3.5 t/ha for a strain known as hybrid maize 462), lucerne and carrots (25 t/ha). Though the soil texture is on the heavy side, no dangerous degree of soil salinity has ever been found, probably because the land is well drained by a neighbouring wadi. Since 1960, 850 ha of orchards have been planted in the same area (apricots, peaches, plums, apples and pears); these are being irrigated with water containing 1.6 g/l.

Orchards – about 1000 ha – have since 1955 also been planted in the Kairouan region. They are being irrigated with water containing 2.8 g/l. In this area, the water-table is low.

Starting in 1958, the Tunisian Government has undertaken the amelioration of four areas in the Sfax region which are irrigated with water containing 4–5 g/l. The soil is mostly sandy and originally not or slightly salty, with calcareous and gypseous outcrops in depth. The surface of the areas concerned varies from 40 to 100 ha. Since in three of them, Melloulèche, Ahzeg and Nakta, the water-table was at a height of between 3 and 6 m even before cultivation, drains were dug at distances of 50 m. At Melloulèche they are 0.80 m deep, at the two other sites 1.50 m. In the fourth area, Henchir El Hicha, the water-table lies at considerable depth (more than 30 m). Where drains of 1.50 m have been used, the water-table has not risen to more than about 1.20 m; in the lowest part of the Melloulèche area, it has risen to 0.80 m, with the result that the soil has become very salty and even sterile. After the drainage system was extended in depth to 1.50 m, considerable improvement was noted and cultivation could be resumed. This confirms already mentioned observations made in other parts of Tunisia, which tend to show the importance of drainage in order to prevent the water-table from rising permanently above 1–1.20 m.

Crop rotation on these fields is: oat vetch or ray grass (winter), cotton (summer), beans, grain sorghum, wheat, fodder sorghum. Outside the crop rotation, 15% of the area is covered with lucerne and 15% with fescue grass. These are left for three years, and then followed by crop rotation. Experimentally, vegetable crops such as tomatoes, melons, asparagus, onions, carrots and artichokes have also been grown. At Henchir El Hicha, the olive groves have developed satisfactorily. The apricot and almond trees, though, consistently had their leaves burned by the salt. Table II shows the average crops obtained from these plots.

TABLE II

Crops on Irrigated Areas at Sfax

Crop	Harvest in t/ha
Wheat	2.8–4.0
Maize	4.5–5.0
Grain sorghum	2.8
	1.7 after grain
Fodder sorghum	60–100 green
Lucerne (hay)	6 (1st year)
	18 (2nd year)
Cotton	2.0–2.5
	4.5 after lucerne
Beans	7–10 green
	plus 3– 4 dry
Asparagus, short	2
Tomatoes	28
Melons	12

After building dams (see fig. 2) and constructing the primary irrigation channel system, the Tunisian Government began in 1958 to rehabilitate the Medjerda Lowlands, which are to be put under irrigated crops. The dam on the Mellègue is capable of impounding 300 million m³, while the average annual inflow is of the order of 150 million m³. The main irrigation canal starting at the El Aroussia Dam at the entrance of the Lowlands has a capacity of 14 m³/sec., which is sufficient for the irrigation of about 20,000 ha in summer and twice as much in winter. Annual rainfall in the Lowlands is of the order of 400 mm per year, with the rains falling between the end of September and the end of April. The salt content of the irrigation water varies between 1 g/l in winter and 3 g/l in summer, half of which consists of sodium chloride; the Na/Ca ratio is 2. The Oued Ellil Dam is mainly designed for supplying fresh water (0.2–0.4 g/l) to the city of Tunis. The

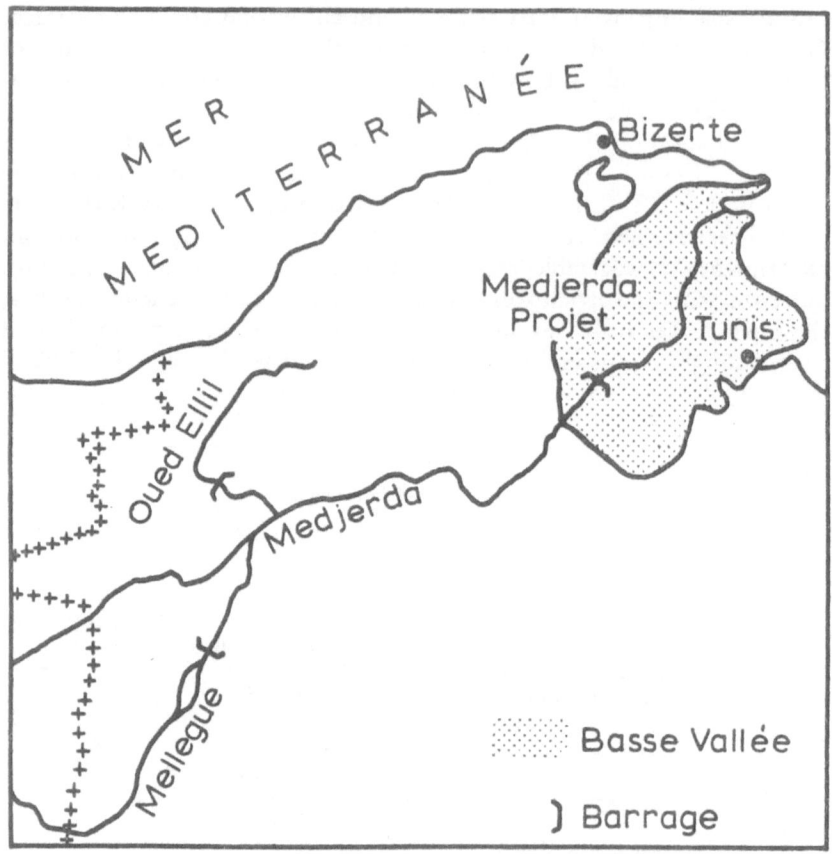

Fig. 2. Map showing the Medjerda Project.

Government envisages the construction of dams on other affluents of the Medjerda in the same region as the Oued Ellil Dam, where the river water is fresh, due to ample rainfall, which is of the order of 1000 mm per year. This would on the one hand make it possible to increase the irrigated area, and on the other hand improve the quality of the irrigation water by reducing its summer salinity to 1.5–2 g/l.

Except for the hilly area bordering the valley, which is covered with orchards, vineyards and market gardens, the Lowlands soil is mostly medium to heavy, with a calcium carbonate content of about 30%. After land reform is carried out, these low soils, which traditionally were used for grain because of the low rainfall and high groundwater-table, will be drained, parcelled and cultivated under irrigation. Their agricultural use will be determined by means of pedological and hydrological research and will be based on highly productive fodder crops for cattle raising and on vegetable growing.

The first problem to be solved in recovering these low-lying lands will be that of adequate drainage. The soil of the plains is generally not very saline in the upper layers (with an EC_e between 1 and 3) and highly saline at 1 m and lower (EC_e between 8 and 12). Since the water-table is highly saline, often to the degree of 10–15 g/l, and fluctuates between 0–1 m in winter and 1–2 m in summer, one should expect the soil to become salinated rapidly if the existing balance between the rainfall in the winter and the capillary rise in the summer is upset by the addition of brackish irrigation water, unless care is taken to provide sufficient drainage.

Starting in 1959, a beginning has been made with the draining of these lands by means of tile drains at a depth of about 1.50 m and at intervals of between 40 and 80 m according to the permeability of the soil. This drainage has proved sufficient to prevent any permanent rise of the water-table or permanent salination of the soil. In 1961, the Department for the Rehabilitation of the Medjerda Lowlands started a research project on an experimental plot at Bejaoua, on a heavy-textured soil containing about 50% of clay ($< 2 \mu$) with the double aim of investigating the effect of drainage on the water-table and the development of soil salinity, and of determining the water requirements of different crops. The plot is divided into four identical sections; on each section, a crop rotation system requiring a different quantity of water is practiced. All sections are equipped with drains at 30 and 60 m distance and at a depth of 1.60 m. Two years of experimentation (1962 and 1963) already allow certain conclusions to be drawn:

Intensive cultivation with continuous irrigation, such as lucerne, results in a slight increase of soil salinity.

Semi-intensive cultivation, involving a biennial crop which requires interruption of the irrigation during part of the spring and summer months, results in a fairly stable salinity condition.

Extensive cultivation, in which the land lies fallow in summer and an unirrigated crop is grown the next winter, results in a slight improvement in soil salinity.

On no occasion during the two years of the experiment has any really serious situation developed, notwithstanding the fine soil texture and the high water salinity (up to 3.8 g/l). This is due to the deep and well-spaced drain system and to the soil being leached by the winter rains.

Water consumption is of the order of 12,000 m³ for lucerne, 8,000 m³ for tomatoes, and 5,000 m³ for artichokes and maize. The evapo-transpiration potential of lucerne, measured in lysimetric boxes, rises from a minimum of about 1 mm per day in winter to a maximum of about 12 mm per day in summer (monthly average for July). The average calculated for the three summer months is 9.5 mm per day, or 50% more than the value obtained with the Penman formula and 0.8 mm more than the evaporation of a class A pan.

The water-table rose several times to the surface level after irrigation or heavy rain, as may be expected in view of the low volume percentage of large pores in a soil of this type, but the rises rarely last long and have no adverse effect on the crop.

During the period covered by the experiment, there were no indications of major importance requiring a drain distance of 30 rather than of 60 m.

PROJECT OF THE RESEARCH CENTRE FOR THE USE OF SALINE WATER IN IRRIGATION

With a view to the future extension of the irrigated area, the Tunisian Government, with the assistance of the United Nations Special Fund and of UNESCO as Executive Agency, at the end of 1962 established a project designed for the study of the use of brackish water in irrigation. In addition to a headquarters building in Tunis, which contains chemical, physical and physiological laboratories, and a Documentation Centre, the project consists of three experimental stations, each representing a characteristic region of Tunisia:

(1) The Cherfech Experimental Station in the Medjerda Lowlands, where the soil has a medium texture with about 30% clay, the water-table is high and saline, and the irrigation water moderately saline with 2–3 g/l in summer;

(2) The Ksar Rhéris Experimental Station in Central Tunisia (annual rainfall averaging 150 mm), with light, sandy soil having a chalky layer at a depth of about 70 cm, a low water-table (below 25 m), and irrigation water containing 4.0 g/l;

(3) The Tozeur Experimental Station, where the soil is light, gypseous and saline, and the irrigation water has a low salinity of 2.0 g/l.

At the Cherfech Experimental Station, which has an area of 12 ha, the experimental programme consists of the following experiments:

(a) An experiment with four different kinds of irrigation water, presenting a salinity scale ranging from 0.5 g/l to 4.0 g/l, and four different crops. In 1965, the first year of the experiment, the crops chosen were: fodder sorghum, sunflower followed by lucerne, maize, and artichokes. This experiment will make it possible to study soil development as well as crop reaction to water of different salinity, which in turn will provide a basis for a better estimate of the value of possible improvements of the irrigation water of the Medjerda as the result of the construction of further dams.

(b) Two experiments involving two different frequencies and three different dosages of irrigation. The frequency difference is obtained by spacing out irrigation at intervals of n and 2n days, n varying according to the crop and its growth stage. Volumes of irrigation water provided are chosen in such a manner that the first is less than evapo-transpiration, the second equal to it, and the third higher than it and therefore also produces a leaching effect. In order to measure evapo-transpiration for each crop, a meteorological station has been set up, equipped with a set of lysimetric boxes on which the crops involved in the two experiments are planted. One experiment started in 1964 with fodder sorghum followed by maize, the other in 1965 with a lucerne crop.

(c) A study of the water and salt balance (see below).

The problems of Central Tunisia differ in part from those of the Medjerda Lowlands. Since the water-table is low and the area to be rehabilitated not extensive – which even under the given circumstances might cause the water-table to rise – there is no drainage problem, nor is it possible to conduct a study of the water balance similar to the one being carried out at Cherfech. The question of improving the irrigation water by mixing it with fresh water is also not of primary importance: as a matter of fact, there is no fresh water available in sufficient quantity. On the other hand, the land in this region often shows slopes of a highly varied pitch, which presents problems in the cultivation and irrigation of fields. The programme of the Ksar Khériss Experimental Station, which consists of 30 ha – 20 ha of annual crops and 10 ha of orchards – includes:

(a) Two experiments on irrigation frequency and dosage, involving in 1965 (the first year of the experiment) fodder sorghum and maize. Since the lots, with an area of 200 m² and a length of 17 m, are practically free of slope, they are irrigated on the basin principle, which makes it possible to select the irrigation dosage in proportion to the evapo-transpiration.

(b) The same experiments are repeated on larger plots, with an area of 800 m² and a length of 68 m, so as to study also the distribution of the water and the development of the soil within the length of the border or the furrow.

(c) The establishment of the relation between the slope and length of a field and its irrigation requirements (maximum and balance stream).

The Tozeur Station, with an area of 17 ha, has been planted to date palms for the last thirty years. In the past, the trees have suffered from under-irrigation and lack of drainage. As a result, they have developed rather unevenly. In order to lay the foundation for a satisfactory analysis of future experiments, the project has been started off with two "blank" years; that is to say, irrigating the whole station equally and measuring the crop of each tree. As a first result, this will permit conclusions as to the improved production of poorly kept and under-irrigated adult palm trees after two years of normal cultivation. As from 1966, the following experiments will be undertaken:

(a) An experiment with two irrigation dosages with and without undergrowth, designed to show whether the introduction of undergrowth crops leads to changes in water requirements – a question of some importance for the future rehabilitation of the oases of the Djérid region.

(b) A study of the water and salt balance on a plot with drains spaced at 40 m and laid at a depth of 1.60 m.

As this outline shows, the plan is to start the project with the main stress on the reaction of soils and plants to saline water. Fertility tests involving the introduction of manure and fertilizers and experiments on the resistance of different strains to salt have been postponed to a later stage of the project.

STUDY OF THE WATER AND SALT BALANCE AT CHERFECH

The purpose of this research project is to study the water and salt balance, as well as the relation between the height of the water-table and the output of the drains. To this end, a plot of 4 ha was provided with drains at a distance of 40 m and at a depth of 1.50 m. (see Fig. 3). Though wider spacing

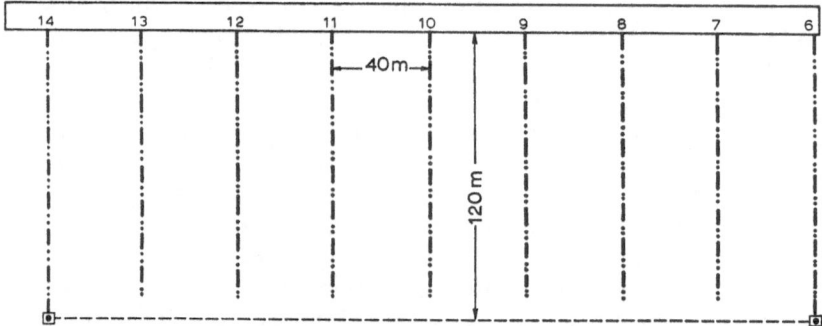

Fig. 3. Drain system at the Cherfech Experimental Station.

might well have been sufficient to secure adequate water disposal and control the water-table – soil permeability measurements having indicated the need for a drain distance of 60–80 m in this region – spacing at 40 m was chosen in order to provide for a sufficient number of repeats. On the one hand, irrigation water intake and drainage water output are measured in order to determine the water balance; on the other hand, measurements are made of the salt contents of the water and samples are taken at the beginning of each month from different layers of the soil profile, in order to determine the salt balance and keep track of the development of the soil. A network of piezometers has been installed for measuring groundwater fluctuation following irrigation and rain periods, and these serve at the same time for the taking of groundwater samples in order to observe the development of its salinity.

The experiment was started in the spring of 1964. The plot was planted with maize for grains, and irrigated from the beginning of May until the end of August. By then, it had received 450 mm irrigation water and 100 mm rain water. After the maize crop – 5.5 t/ha – had been harvested, vetch – barley was sown in late October. This crop was given a dose of 110 mm of irrigation water immediately after sowing; after that, the rainfall was heavy enough to make irrigation unnecessary. At the end of March, a crop of 25 t/ha of greenfodder was harvested.

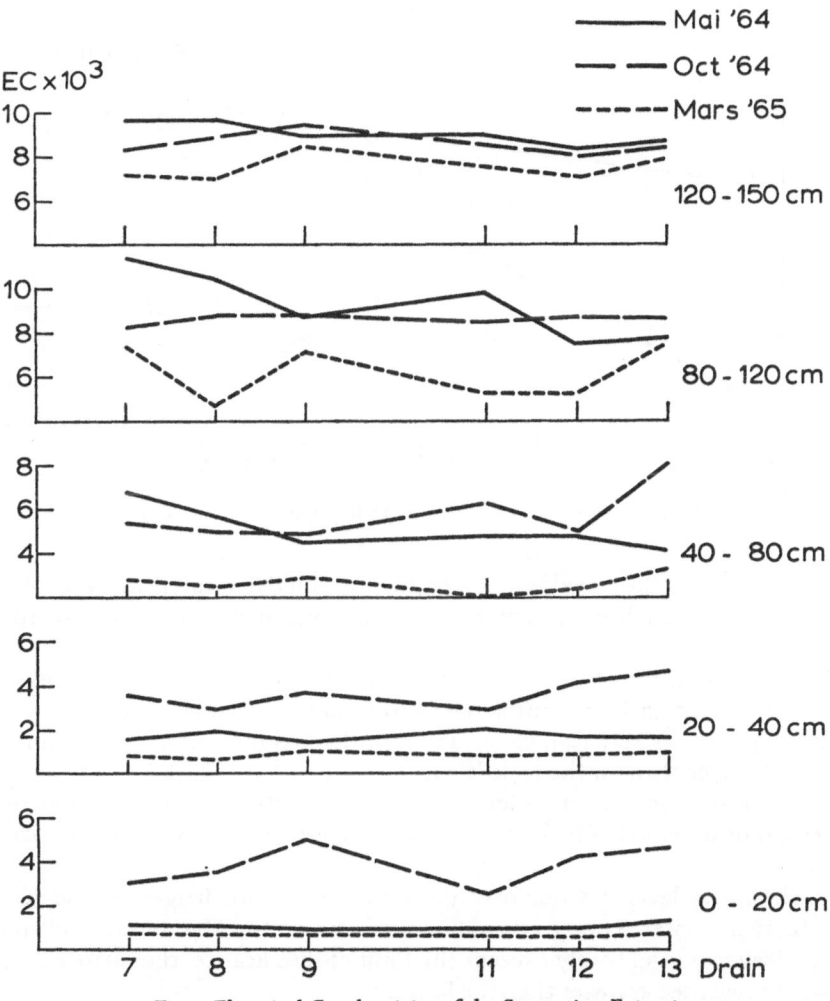

Fig. 4. Electrical Conductivity of the Saturation Extract.

Fig. 4 shows the development of the conductivity of the saturation extract at three sampling dates, as well as the differences between the drains. Though the soil texture of the whole plot is fairly homogeneous, there is a gradient of infiltration speed, which decreases towards Drain 13; this is confirmed by a higher increase in conductivity in October.

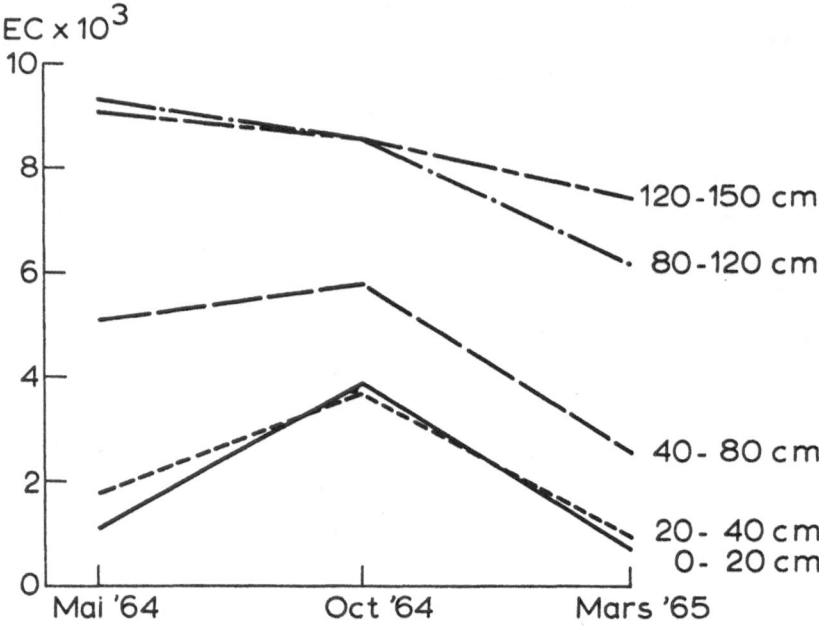

Fig. 5. Electrical Conductivity of the Saturation Extract. Mean of six drains.

Fig. 5 shows the development of the conductivity as an average of 6 drains. The conclusions drawn from this overall development are the following:

The salt content of the 0–20 and 20–40 layers increased through irrigation and decreased again in winter. The end level in March 1965 lies below the starting level in May 1964. Quantitatively, the difference is less marked for the 0–20 layer than for the 20–40 layer.

The 40–80 layer became slightly more saline with irrigation and clearly less saline in winter. The level in March 1965 was clearly lower than in May 1964.

The 80–120 level did not become more saline with irrigation and lost salinity in winter. Though the difference between the March 1965 level and the May 1964 level barely exceeds the limit of significance, the former may still be regarded as lower than the latter.

In the 120–150 level there have been no developments. Though the number of repeats was insufficient for providing evidence of development in this layer, it seems quite likely that it has, in fact, also become less saline. This could have been shown by an irrigation dose and frequency test with a larger number of repeats.

The following figures show the same development. Fig. 6 shows the development of salinity calculated as kg salt per 100 t of soil, and as total

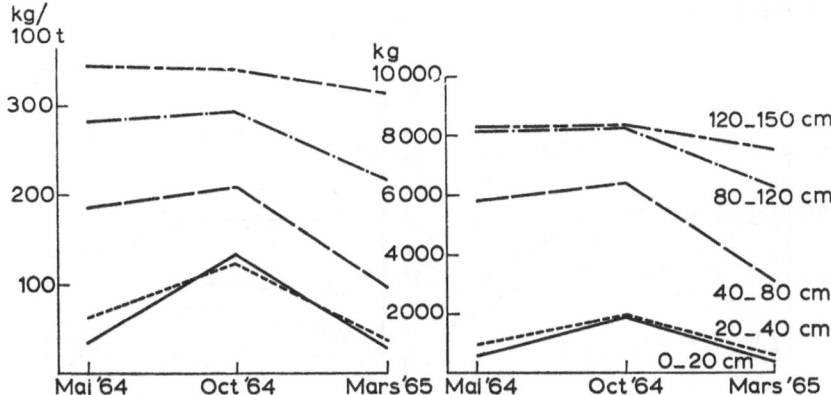

Fig. 6. Salinity calculated as kg salt per 100 t of soil (at left) and as total weight of salt contained in different soil levels (at right).

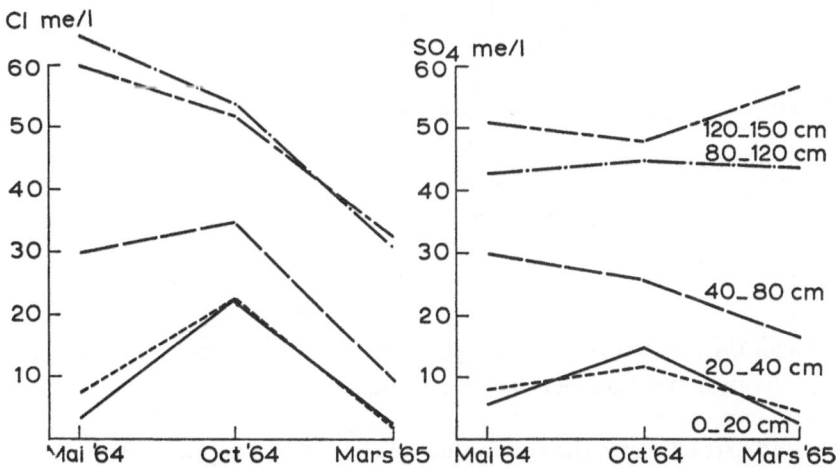

Fig. 7. Development of chloride and sulfate contents in different soil levels.

weight of salt contained in different soil levels. Fig. 7 deals with the development of the anions, chloride and sulfate; figures 8 and 9 relate to the kations, calcium, magnesium, sodium and potassium.

Tables III and IV summarize the results of the salt and water balances. The table on salt balance is divided into two parts. The first contains the calculated quantities of salt introduced by irrigation and the calculated quantities of salts removed by drainage, as well as the difference between the two. Over the whole of the two periods, the quantity of salts introduced by irrigation was lower than that removed by drainage. The

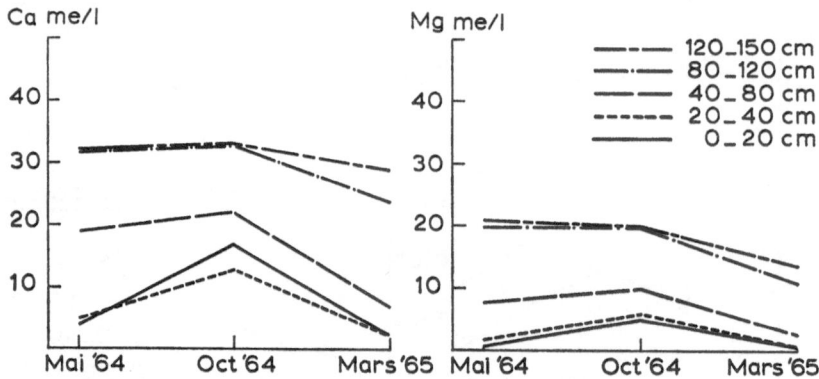

Fig. 8. Development of calcium and magnesium in different soil levels.

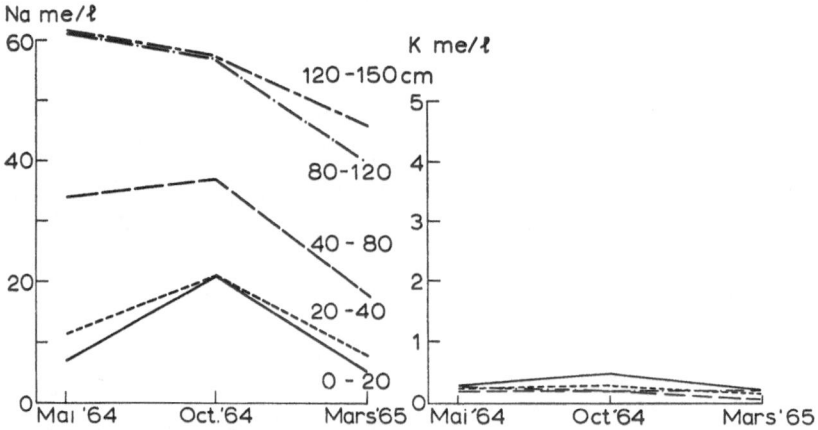

Fig. 9. Development of sodium and potassium in different soil levels.

second part tabulates the salt content of the soil to a depth of 1.50 m (drain depth) at different sampling dates. It is evident that during the first (summer) period the salt content of the soil increased; during the second period, on the other hand, it decreased, and over the whole year it decreased as well. The difference between the results of the two parts of the table, and particularly the fact that the salt content of the soil increased between May and October 1964 even though the quantity of salts removed by drainage was larger than that introduced by irrigation, may be attributed to two causes:

(a) Differences in the composition of the soil must be taken into account.

(b) Since the flow lines of the drainage water pass through levels lower than 1.50 m, it is possible that the higher soil levels are salinated in summer by salts introduced with the irrigation water and by capillary ascension,

TABLE III

Salt balance

Water

Period	Salts introduced with irrigation water in kg	Salts removed with drainage water in kg	Difference − salts removed + salts added
1. 5.64 to 1.10.64	5,381	6,661	−1,280
1.10.64 to 1. 3.65	1,050	6,683	−5,663
1. 5.64 to 1. 3.65	6,431	13,344	−6,913

Soil

Period	Salts at 1. 5.64 or 1.10.64	Salts at 1.10.64 or 1. 3.65	Difference − removed + added
1. 5.64 to 1.10.64	23,865	27,076	+3,211
1.10.64 to 1. 3.65	27,076	18,205	−8,871
1. 3.64 to 1. 3.65	23,865	18,205	−5,660

while the drainage water, during its passage through lower, highly saline layers, leaches those layers; in other words, that the quantity of salt in the drainage output is not related to the soil layers down to drain depth, but to lower strata.

The water balance includes the quantities of irrigation, rain and drain water. Drain water quantity amounts to about 25% of the total irrigation and rainwater input.

The readings from the piezometer network provide information on the fluctuations of the water-table – which rises to about 1 m at the time of irrigation and then falls again rapidly to the drain level – and on relations

TABLE IV

Water balance

Period	Crop	Irrigation mm	Rain mm	Drainage mm
1. 5.64 to 1.10.64	Maize	452	107	113
1.10.64 to 1. 3.65	Vetch barley	110	368	127
1. 5.64 to 1. 3.65		562	475	240

between the height of the water-table and the output of the drains during the evacuation period, and the relation in time between the lowering of the water-table and the drain output. This relation makes it possible to calculate the permeability and the depth of the permeable subsoil below drain level, while the time required for the water-table to drop makes it possible to calculate the storage coefficient of the soil above drain level. The results obtained show that the subsoil of the plot is fairly homogeneous, and that the permeable layer reaches a depth of 3.50 m, has a permeability of 2.5 m/day, and is followed by a layer with low permeability. The storage coefficient is 0.04, meaning that a drop of the water-table of 10 cm corresponds to a drain output of 4 mm – a reasonable figure for this type of soil.

On the subject of the leaching period, the following remarks should be made:

On the one hand, one can leach permanently, by adding a leaching dose to each irrigation dose and thus increase summer irrigation requirement. The other possibility is to confine oneself to certain parts of the year. Taking the practical case of the Cherfech crops: if the maize fields are to be wetted properly after sowing, one needs ample first irrigations, well beyond what the crop itself requires; in other words, one must leach. In view of the climatic conditions of the Lowlands, it is also advantageous to soak the soil well at the beginning of winter cultivation, so that the leaching action of the winter rains shall be more effective. In 1964, leaching first took place in spring, during the first irrigations, then in autumn after the first heavy rainfall, and finally in winter, when there was a period of abundant rain. This raises the thought that unless the soil salinity develops in the course of the season in a manner which endangers the crop, one might confine oneself to applying leaching doses twice a year: in the spring after planting the summer crop, and again in autumn after planting the winter crop.

Another question directly connected with sowing is that of germination in an environment irrigated with saline water, for this affects the future development of the plant and its tolerance to salt. In order to throw light on this question, a study of germination percentages in different concentrations of soluble salts (4, 8, 12 and 18 mmhos) has been undertaken for lucerne, vetch and bersim. The results (see fig. 10) show that the germination percentage is highest for Gabès lucerne, with vetch and bersim following. Moreover, when one and the same crop is watered with solutions of different concentrations, one notes a clear delay in germination. For instance, (see fig. 11), when vetch was watered with a solution of 4 mmhos, 80% of the grains germinated after 53 hours; when a solution of 8 mmhos was used, it took 58 hours, and with a solution of 12 mmhos, 74 hours. Such graphs allow one to determine the critical conductivity for a given crop which must not be exceeded if one wants to obtain a germination percentage of more than 50, as well as the concentration of the soil solution above which germination would be delayed too much. With irri-

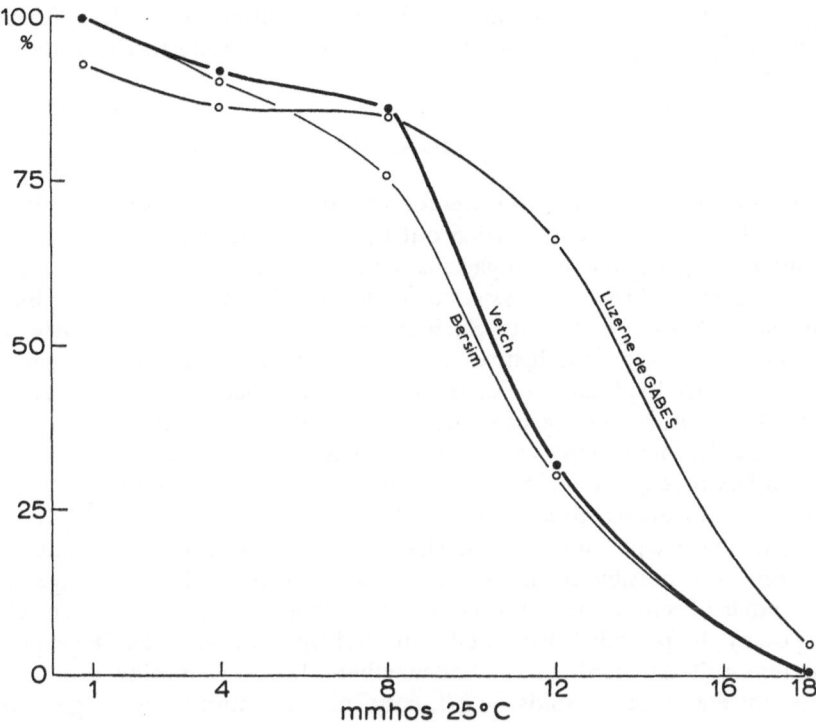

Fig. 10. Percentage of germination of different crops, related to conductivity of the solution in mmhos at 25°C.

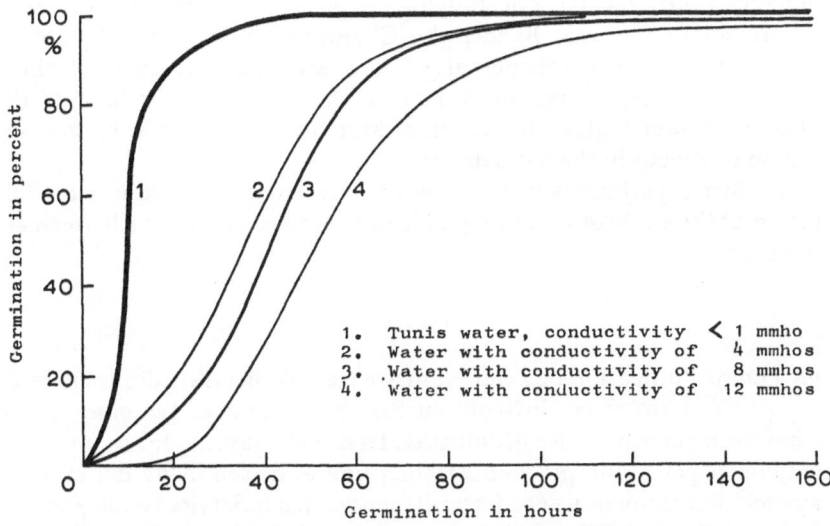

Fig. 11. Percentage of germination of vetch in relation to water with increasing conductivity and chlorides dominating the sulfates.

gation water of 5 mmhos and a sandy soil, conditions at the Ksar Rhéris Station encourage a concentration of soil salinity liable to affect germination.

SUMMARY

Since 1935 the use of brackish water for irrigation purposes in Tunis has been the subject of researches carried out by YANKOVICH and NOVIKOFF at the Botany Institute. The investigations made in the field from 1940 to 1955 by the pedologists DESSUS and SABATHE made it possible as well to fix the limit of plants' resistance to salts. Such investigations have been recently resumed at the Pedology Institute, the Agricultural Service, the Medjerda Lower Valley Exploitation Department and the Special Fund Project (Research Station for the Use of Saline Water for Irrigation purposes).

The oldest irrigations were performed in the oases district, in South Tunisia, where water holds a quantity of salt varying from 1 to 3 g/l and soil is gypseous to a greater or lower extent. On the ground of such investigations we are now in a position to affirm that drainage is extremely important and can make it possible to double crops. In the district where an irrigation system has been recently set up – in which water holds 4–5 g/l – we received a good yield, provided that an efficient drainage was available. The exploitation of the Medjerda Lower Valley, where the soil is of middle to heavy structure and water holds 2–3 g/l, confirms once more how largely the success of plantgrowing under irrigation depends on drainage.

In view of the development of irrigated districts, the Government of Tunis, thanks to the Special Fund granted by the United Nations, sketched out a plan with the object of studying the possibility to use brackish water for irrigation purposes. Besides the Chemistry, Physics and Physiology Laboratories three experimental stations are available, each of them representing a typical region of Tunisia: the station of Cherfech, in the Medjerda Lower Valley; the station of Ksar Gheriss in the Centre and the station of Tozeur in the oases district.

The first experiments in this line of research were made in 1964. The results of the study of water and salt balance at the station of Cherfech are reported.

RÉSUMÉ

En Tunisie l'utilisation de l'eau saumâtre a été l'objet d'études menées de 1935 par YANKOVICH et NOVIKOFF au Service Botanique. Les observations faites sur le terrain par les pédologues, DESSUS et SABATHE, de 1940 à 1955 ont également permis de préciser certaines limites de sensibilité des cultures aux sels. Récemment les études on été reprises par le Service Pédologique, le Service Agricole, l'Office de la Mise en Valeur de la Basse Vallée de la Medjerdah et le Projet du Fonds Spécial (Centre de Recherches pour l'Utilisation de l'Eau Salée en Irrigation).

Les irrigations les plus anciennes ont été pratiquées dans la région des oasis dans le Sud de la Tunisie, où les eaux contiennent entre 1 et 3 g de sel par litre et où les sols sont plus ou moins gypseux. D'après les observations le drainage est d'une grande importance et peut permettre de doubler les récoltes. Dans des périmètres récemment mis en irrigation, où l'eau contient 4 à 5 g per litre, on a pu obtenir de bons rendements, à condition de disposer d'un drainage suffisant. La Mise en Valeur de la Basse Vallée de la Medjerdah, où les sols sont d'une texture moyenne à lourde et où l'eau contient 2 à 3 g par litre, a confirmé de nouveau l'importance du drainage pour la réussite des cultures irriguées.

En vue de l'extension future des périmètres irrigués, le Gouvernement Tunisien, avec l'aide du Fonds Spécial des Nations Unies, a créé un projet ayant pour but d'étudier la mise au point de l'irrigation avec des eaux saumâtres. Outre les laboratoires de Chimie, de Physique et de Physiologie ce projet dispose de trois stations expérimentales, chacune réprésentant une région typique de la Tunisie: la station de Cherfech dans la Basse Vallée de la Medjerdah, la station de Ksar Ghériss dans le Centre et la station de Tozeur dans la région des oasis.

En 1964, la première année de l'expérimentation, un début a été fait avec l'étude du bilan hydrique et du bilan des sels à la station de Cherfech, dont les résultats ont été exposés.

RIASSUNTO

In Tunisia l'utilizzazione dell'acqua salmastra è stata oggetto di studi condotti dal 1935 da YANKIVICH e NOVIKOFF al Servizio Botanico. Anche le osservazioni fatte sul terreno dai pedologi DESSUS e SABATHE, dal 1940 al 1955, hanno permesso di precisare certi limiti di sensibilità delle colture con acque salse. Gli studi sono stati ripresi recentemente dal Servizio Pedologico, dal Servizio Agricolo, dall'Ente per la valorizzazione della Bassa Vallata del Medjerdah e dal Progetto di Fondi Speciali (Centro di Ricerche per l'Utilizzazione dell'Acqua Salata per Irrigazione).

Le prime irrigazioni sono state praticate nella regione delle oasi nel Sud della Tunisia, dove le acque contengono tra 1 e 3 Grammi di sale per litro e dove i terreni sono più o meno gessosi. In base alle osservazioni il drenaggio è di una grande importanza, tanto da consentire di raddoppiare il raccolto. Nelle zone recentemente messe ad irrigazione, dove l'acqua contiene 4–5 grammi per litro, si sono potuti ottenere buoni rendimenti a condizione di disporre di un drenaggio sufficiente. La valorizzazione della Bassa Vallata del Medjerdah, dove la struttura dei terreni è media o grossolana e dove l'acqua contiene da 2 a 3 grammi per litro, si è avuta nuova conferma della importanza del drenaggio per la riuscita delle colture irrigate.

In vista della futura estensione delle zone irrigate, il Governo Tunisino, con l'aiuto del Fondo Speciale delle Nazioni Unite, ha attuato un progetto avente lo scopo di studiare la messa a punto dell'irrigazione con le acque salmastre. Oltre i Laboratori di Chimica, di Fisica e di Fisiologia, questo progetto dispone di tre stazioni sperimentali, rappresentanti ciascuna una regione tipica della Tunisia: La stazione di Cherfech nella Bassa Vallata di Medjerdah, la stazione di Ksar Ghériss nel Centro e la stazione di Tozeur nella regione delle Oasi.

Nel 1964, primo anno di sperimentazione, si è cominciato con lo studio del bilancio idrico e salino alla stazione di Cherfech, di cui saranno esposti i risultati.

REFERENCES

BUREAU, P., J. P. COINTEPAS, P. ROEDERER & J. GILBERT. Tolérance à l'eau salée de quelques cultures pratiquées en Tunisie (Tolerance to saline water of some crops grown in Tunisia).

COMBREMONT, R. Les cellules de mise en valeur de la region the Sfax. (Rehabilitation units in the Sfax region). Ministry of Agriculture, Republic of Tunisia.

VAN 'T LEVEN, J. A. & M. A. HADDAD. 1962, 1963. Report on Research at the Experimental Station of Bejaoua, 1962 and 1963. Department for the Rehabilitation of the Medjerdah Lowlands, Republic of Tunisia.

Authors' addresses: Centre de Recherches sur l'Utilisation de l'Eau Salée en Irrigation, Route de la Soucra, B.P. 10, Ariana, Tunisia.

Irrigation with Salt Water and Drainage in Tunisia[1]

by

JEAN-PAUL COINTEPAS

Use of brackish water in agriculture is not a new phenomenon in Tunisia as the Oases of Southern Tunisia have been producing dates and vegetables since time immemorial. Yet, having recently perfected powerful means of water harnessing such as dams, deep drillings etc., the surfaces to be irrigated have greatly increased; it has been undoubtedly established that most available waters in Tunisia contained more or less salt.

In Tunisia this phenomenon can be explained by important developments of more or less salty or gypseous outcrops, therefore it is unusual to find water suitable for irrigation containing less than 1 g/l; only some well-watered sandstone mountains in northern Tunisia are not submitted to this rule.

The setting up of many delimited regions on the whole of the Tunisian territory has enable us to determine, according to climatic regions, the criteria of irrigation by salty water; among those criteria we will retain only two which are valid for all irrigated zones:

(1) selection of cultures which can bear the salt of the soil.
(2) fight against excess of water which risks soaking the soil.

In this present report we will insist on this second characteristic and as example, we will single out one definite region where the non-observance of this rule has been the cause of quite serious disappointments.

STUDY OF SALTINESS AND WATER STRUCTURE OF AN IRRIGATED REGION (MELLOULECHE)

The region under study is situated 50 km north of the town of Sfax along the sea; the average rainfall measured at Sfax is 220 mm, average temperature is 19°C; the rainy season goes from September to April and is at its peak during November, January and March.

1. Note taken from a series of studies made on behalf of the Tunisian Pedological Service, to be published by the Pedological Service of the O.R.S.T.O.M.

The irrigated zone covers 110 ha; topographically it is a smooth slope with a declivity of 6‰ and it is made of two types of soil:

(1) in the half western part the soil is composed of layers of alluvial light sand, 50 to 60 cm thick, on top of a hard calcareous crust 30 cm thick, partly broken up and varying from a nodulous crusting of 75% $CaCO_3$ to calcareous-clay spots of 50% $CaCO_3$.

(2) in the lower half of the eastern part, the soil is made of alluvial light sand or argillaceous sand laying at a depth of 1 m covering a thin soft calcareous-gypseous crust made of 25–35% of $CaCO_3$ to 13–30% of gypsum; it seems that this zone used to be a filled-up old swamp ending in a small salty and gypseous depression which was originally excluded from the region under study. (Conductivity 8–9 mmhos per 1 m) Practically this whole region is not salty (containing less than 2 mmhos/cm).

The irrigation water originated from an artesian boring yielding 80 l/sec., salinity of this water amounts to 4 g/l and its ionic balance is particularly bad (S.A.R. = 15).[1] The fictitious yielding is 0.80 l/sec/ha, i.e. an average contribution of 6.9 mm/per day while the potential evapo-transpiration is estimated at 5.5 per day.

When this area was irrigated in 1958, the groundwater-table was varying between 6.50 m in its north-western part (the upper one) and 3.50 m in the southern part at the limit of the lower stretch: its salinity was very high varying between 7 and 11 g/l.

Only after four years the upward thrust of this sheet of water was felt: its effects were demonstrated by barren patches covering the area of 20 ha of the lower stretch, where plants of fescue grass – known for their salt-resisting qualities – failed regularly to grow and an olive grove in the downstream region was giving very poor results.

The Pedological Service started a systematic study of this water sheet and was able to prove an upsurge of 3 m in the northern part of the region, and from 2–3 m between this area and the sea-shore.

A network of piezometres has been able to show that the irrigation watering of the upper part was the cause of a state of temporary choking for a duration of 5 to 6 days: still this state of choking was not harmful because the underground water remained generally at a depth of 2–3 m, while in the lower part of the region the underground water was at a depth of 1 m.

Each irrigation made the water level raise 30 cm and it returned to its original level not before 8 or 10 days; it has also been established that every time the water sheet reached a depth of 1.30 m the soil salinity was at its peak; when the water reached 1 m of depth, the soil was practically barren. (Conductivities of 80 mmhos/cm have been traced on the soil surface).

---

1. S.A.R.: Sodium absorbtion ratio $= \dfrac{Na}{\sqrt{\frac{1}{2}(Ca + Mg)}}$ (me/l).

The upsurge of the underground water sheet was evident, particularly between January and March – the region's rainy season, which is also sowing time – thus explaining the failure of cultures in certain plots.

A drainage network made of earthenware pipes has been set in 1958 but was situated on the crust's level, was therefore functioning only 2 or 3 days after irrigation and in the lower part only.

This led us to lower the network to a depth of 1.70 m which – considering the soil permeability and the drain intervals (50 m) – would have maintained the underground water sheet between 1.30 m and 1.50 m.

Two years after carrying out the works, the network was still not functioning in the upper part of the region, but in the lower part drains were running 8 to 10 days after each irrigation; following a rainy winter – between January and February 1964 – most of the drainages functioned uninterruptedly.

The ensuing results were spectacular: the halophilic vegetation has almost totally disappeared with the exception of spurry; the fescue grass and lucerne sprouted again without, however, reaching the output of well drained plots. As a matter of fact their real output has never been estimated because they were used for grazing grounds and not mowed.

Parallel to the study of the underground water, a study of the soil has given an outline of the salinity evolution; the following table schematizes the results which have been obtained:

Depth of the soil	Soil salinity (Conductivity in mmhos/cm)								
	Upper part of the region		Lower part of the region			Downstream of the region		
	Before irrigat.	Before drain.	After drain.	Before drain.	After drain.	Before irrigat.	Before drain.	After drain.
0– 20 cm	1.5	6–10	6–8	10–75	13–20	7	20–80	40
20– 30 cm	1–2	7–10	5–8	11–14	8–14	6–7	9–10	20
30– 60 cm	1–2	9– 9	4–6	8	9–13	6	6– 7	10–12
60–100 cm				6	9–13	6	6– 7	7

An exact test has even been undertaken by which known doses of irrigation water were increased from 7000–10,000 m³/ha.

Salinity in cotton cultivation varied between 4–7 mmhos/cm with a slight maximum density at crust level, and this can be explained by a defect in permeability in this section, or by thick-size roots unable to sink further to exploit waters to be found at this level.

Analysis of the results so obtained shows that soil salinity has not considerably changed after drainage; the lower plots were not desalted which leads us to believe that the soil was insufficiently irrigated.

Nevertheless, in spite of a steady salinity, vegetation has come back to life again; the dusty structure of the surface, frequently appearing in barren patches, has slowly disappeared since the network functions again.

Underground water had therefore a double harmful role:
  (1) to asphyxiate plants weakened by the presence of salt.
  (2) continuous supply through capillary debit which by evaporation at the soil surface produces high salinity, generating at the same time a dusty structure on top and thus failure in sowings.

When underground water drops to a 1.50 m level as it happens now, the soil salinity maintains itself at an acceptable level and its equilibrium depends on methods of irrigation: frequent irrigations (700 m³/ha every 8–10 days) are likely to be the optimum to maintain a weak salinity.

Taking into consideration the above mentioned basic principles a bi-annual intensive rotation has been established; we are stating here a succession of cultures, as well as medium and maximum yield obtained:
  beans mid-September up to mid-April   0.4 t/ha, dry (max. 1.1 t/ha)
  cotton mid-April mid-Oct.   1.5 t/ha (max. 2.8 t/ha)
  common wheat mid-Oct. up to mid-May   2 t/ha (max. 3.5 t/ha)
  sorghum for fodder mid-May up to Sept.   8–10 t/ha of hay;
  Temporary meadows may supply during three years the needs of 100 milk-cows
  fescue grass and lucerne   8–10 t/ha of hay
  oat-vetch   5 t/ha of hay.

Agricultural outputs justify irrigation costs on condition that mineral and in particular organic fertilizers are used.

Experience acquired in the Mellouleche region shows that the water structure constitutes the great danger for culture irrigated by salty waters; many examples gathered in different parts of Tunisia prove that it is a general phenomenon.

OTHER EXAMPLES OF WATER STRUCTURE IN IRRIGATED ZONES

In comparison with the Mellouleche region, we will analyse another area taken in southern Tunisia: the region of Henchir el Heicha 100 km south of Sfax. This region is situated on sandy-argillaceous calcareous soil (12%), becoming argillaceous-sandy with a high percentage of chalk (25%) at 50 cm depth.

At a depth of about 140 cm we reach a gypseous layer delicately crystallized (from 10–12% of gypsum) being the result of an old underground layer locally called "terch"; this region is under cultivation since 1958; although of higher salinity (5.5 g/l) than the water used at Mellouleche, the irrigation water has a better equilibrium (S.A.R. = 8); for seven years not a single underground water sheet has appeared. Its salinity varies between 4 and 6 mmhos/cm, outputs are a bit better than those of Mellouleche (lucerne 30 t/ha of hay, cotton more than 2 t/ha).

While studying an olive grove at Bou Ficha – between Tunis and Sousse – Mr. HAMZA (1961) has stressed the importance of water structure symptoms which on an olive grove are different than those of salinity alone. Salinity superior to 5 mmhos/cm nips the leaves; at a higher percentage (0–10 mmhos/cm) the crown is not well developed, the foliage thin with yellow extremities. Effects of water structure is felt by producing bright green foliage – underdeveloped – and tuberculosis on the trunk; as soon as the underground water reaches 1.30 m these symptoms appear: upon reaching 1 m and less, the olive tree withers even if salinity is inferior to 5 mmhos/cm.

Similar observations were made by J. P. COINTEPAS (1963) in southern Tunisia near Gabes on an olive grove irrigated by water with a salinity contents of 4.0 g/l; the soil is an alluvium-soil made of argillaceous-sand calcareous-gypseous. Salinity between 7 and 8 mmhos/cm is more or less uniform in the soil. Production varies between 25 and 30 kg of olives per tree in the best plots and drops to 11 or 12 kg per tree when underground water reaches 1.40 m and less.

Miss H. LARGUECHE (1960) reaches the same conclusions after studying different varieties of *Eucalyptus*: for the standard type to be found in Tunis region (*E. camaldulensis, E. occidentalis*) irregular water structure at less than 50 cm causes withering of the trees; this phenomenon is accompanied by sulphuric formations around the roots.

Besides, oases of southern Tunisia are here to confirm our thesis; most of them are situated on terraces or along Sebkhas in order to exploit sources bursting out from cracks or reliefs; in such a way the different stages of alterations in the water structure can be observed. G. NOVIKOFF (1960), plant sociologist of the Pedological Service, mentions a certain number of observations made in the Oasis of Djerid.

The "Oasis" palm grove near Tozeur is situated on a slope along a small depression; in the upper part the underground water is found at more that 2 m of depth; in the middle part the water sheet is at 1.60 m of depth and in the lower part its level is at 1.25 m; the draining network reaches 1.50 m and the ditches are at an interval of 100 m along the slope. The "Oasis" soil is made of rough sand, and is gypseous in its depth. The yields of the palm grove of the Deglat en Nour variety goes from 3 to 1 as the underground water gets nearer to the surface; the roots become superficial and "gley" (root-sickness) and sulphur spots are noticed in the ground section.

Another industrial palm grove is situated near Kebili, the surface of which is on gypseous sandy-alluvial soil and turns into alluvial-sandy or alluvial-argillaceous in its depth; its water level stabilizes around 1.40 m or 1.50 m. In general the roots take advantage of the whole decontaminated layer; outputs are higher, however a difference is noted in the outputs from 5 to 8% between the rows along the ditches and the rows which are situated in the middle of the drain intervals which are spaced in this plantation at a distance of 60 m.

During systematic pedological data collecting done by this Service such observations have been renewed in all oases; the Deglat en Nour palm variety which is a top export type is less sensitive to salinity than to hydromorphy; all ascent of underground water causes a reduction of the root system. Certain groundwaters have a more harmful ionic equilibrium (in proportion of SO_4 and Ca compared to Cl and Na).

Besides, the underground water motions are the cause of a gypseous delicately crystallized formation, highly consistant, called "terch." This formation constitutes a mechanical obstacle to the penetration of roots. Those different factors constitute a serious obstacle in output increase; tunisian farmers dispose of several varieties of palms whose resistance increases when hydromorphic conditions are getting tougher such as Allig, Bouhattam etc.; unfortunately their crops sell nearly only on the local markets and are therefore of limited interest.

CONCLUSIONS

Policy of irrigation by salty water dictates therefore a high degree of supervision of the underground water; before setting up an area to be irrigated, it is essential to study the hydrological conditions of the region (depth, maximum yield of the water sheet). When underground water is non-existent or at great depth, one must verify if such underground water will not come into existence in a less permeable layer: checkings of such permeability in depth are therefore necessary.

As a matter of fact the underground water has a harmful role:

(1) provoking asphyxia in plants which – in soils with similar salinity – are more sensitive than plants in more drained soils.

(2) by creating – using evaporation on the soil surface – a higher salinity percentage which destroys vegetation with superficial rooting and by substituting halophilic vegetation such as spurry, *Suaeda, Halocnemum, Arthonemum.*

It is therefore necessary to drain waters deeply. In examples mentioned in this report we have found that the underground water depth limits are set between 1.30 m and 1.70 m according to the vegetable species and nature of soils.

Here again we see that a simple evaluation of the soil composition is not enough to determine the necessary depth of the draining; in the case of the Kebili Oasis we notice an adequate lowering of underground water to a 1.40 m level in a soil made of sandy-alluvial composition into an alluvial sandy-type; in the Tozeur palm grove, although water depth is 1.60 m and its composition of a coarse sand, the water structure of the underground water is felt at 70 cm or 80 cm; beside soil texture its composition seems to play a great role.

Problems of irrigation by salty water seem to reach a high degree of complexity by factors involved: physical and chemical composition of the soil which has a direct influence on the water circulation, beside salinity of underground water and seasonal variations of its level or during irrigation itself; as a consequence each particular case has to be thoroughly studied before the region is irrigated.

In order to determinate the different parameters of the evaluation of a definite underground water sheet and its soil salinity, the United Nations Special Fund in cooperation with the Tunisian Government, has created the "Centre for Researches for the Use of Salty Water for Irrigation" which program has been submitted in a separate communication.

SUMMARY

Among mesures to be taken when irrigating with salt water, the most important one is the supervision of the underground water sheet. In four years an upthrust of 3 m has been noted in a region of central Tunisia and as a consequence 20% of the region's surface was barren.

A draining of 1.70 m in depth has provoked the renewed growth of the vegetation; the soil salinity was not perceptibly modified.

Examples taken from different regions in Tunisia – and in particular in the southern oases – prove that a deep draining is indispensable to guarantee a fair vegetable production; as a matter of fact, proximity of the underground water sheet, provokes at the same time asphyxia of plants and due to capillary suction, a strong salinity in surface.

Depth of the draining to be executed varies according to each vegetable species and also depends upon the physical and chemical characteristics of the soil; deep draining must accurately be determined in each particular case.

RÉSUMÉ

Parmi les précautions à prendre quand on irrigue à l'eau salée, la plus importante est la surveillance de la nappe phréatique. Dans un périmètre du Centre de la Tunisie on a constaté en 4 ans une remontée de 3 m de la nappe phréatique, ce qui a rendu à peu près stérile 20% de la surface du périmètre. Un drainage à 1.70 m de profondeur a permis un nouveau démarrage de la végétation. La salure du sol n'a pas été sensiblement modifiée.

D'autres exemples choisis en divers points de la Tunisie et notamment dans les oasis du Sud montrent qu'un drainage profond est indispensable pour assurer une production végétale convenable. La proximité de la nappe provoque en effet à la fois une asphyxie des plantes et une forte salure de surface dûe à la remontée capillaire. La profondeur du drainage varie pour chaque espèce végétale. Elle varie aussi en fonction des caractéristiques physiques et chimiques du sol. Elle doit être déterminée soigneusement dans chaque cas particulier.

REFERENCES

COINTEPAS, J. P. 1963. Observations made on a southern Tunisia palm grove. Pedological Service - Tunis ES 49.

COINTEPAS, J. P. 1963. Observations made on the draining of the Mellalouche region. Pedological Service - Tunis No. 236.

DIMANCHE, P. 1962. Pedological Study made on the Nefta Oasis. Pedological Service - Tunis No. 240.

HAMZA, M. 1961. Contribution to a Study made on asphyxia and salinity tolerance of olive trees. Pedological Service - Tunis ES 35.

LARGUECHE, H. MLLE. 1960. Roots behaviour of some forest essences in salty soils. Pedological Service - Tunis ES-32.

NOVIKOFF, G. 1960. Studies of the soil of Nefsaouas and Djerid palm groves. Pedological Service - Tunis ES 26.

SABATHE, R. 1952. Provisory Pedological studies of the Mellouleche region. Pedological Service - Tunis 402 E.

U.S. Department of Agriculture. Diagnosis and improvements of saline and alkali soils agriculture, Handbook No. 60.

Author's address: Jean-Paul Cointepas, Service Pedologigue de Tunisie, Office de la Recherche Scientifique et Technique Outre Mer (ORSTOM), 8 Boulevard Didon, Carthage, Tunisia.

A New Way of Culture to Decrease Salinity

by

MANUEL MENDIZABAL & GUILLERMO VERDEJO

This brief communication has the purpose to inform about a new method of cultivation named "Arenado" in which sand is used in an unusual way of agriculture in the South East of Spain. We think it may be of general interest because:

(1) It allows the use of saline water with a high percentage of salt.

(2) It saves water of irrigation.

(3) It increases the temperature of the soil.

(4) It brings about a quick desalination of the ground, so that after three years the soil which at the beginning was unproductive and salty becomes quite normal.

(5) It achieves an early and plentiful harvest that gets better prices in the market.

In this short paper we are leaving out the historical origin of the system and are proceeding immediately with the explanation of the method itself.

The land used for this way of cultivation is generally prepared with two crossway plowings. After that it is levelled carefully and then covered with a layer of good manure of 3–4 cm (60–65 t/ha). Over that, a layer of very coarse sand about 10–12 cm, is distributed which amounts to about 1000 m³/ha. The most convenient size of the sand grains is 3–5 mm; this is about the size of a rice grain. Later on, if it is necessary, ridges are made as to be seen in Fig. 1, and windbreaks are constructed of canes (*Arundo donax*) in order to prevent the plants from damage by strong winds.

The plots are separated by concrete walls (specially if they are at different levels) and the canals for irrigation are similarly built of concrete in order to keep the water without direct contamination with the soil. This system (see Fig. 1 and 2) is applied especially for green vegetables (peppers, cucumbers, peas, tomatoes, green beans, egg-plants, etc.). Generally, it is possible to grow two kinds of plants at the same time: some years ago, we have started to use this method of cultivation also for citrus trees and table grapes with very satisfactory success (see fig. 3).

The success of this kind of cultivation has been proved during the last 20 years; each year the "Arenado" spreads to more than several hundred of additional hectars of cultivation in East Andalusia and in the Mediterranean Zone corresponding to the coast of Almeria, Granada and Malaga (see map).

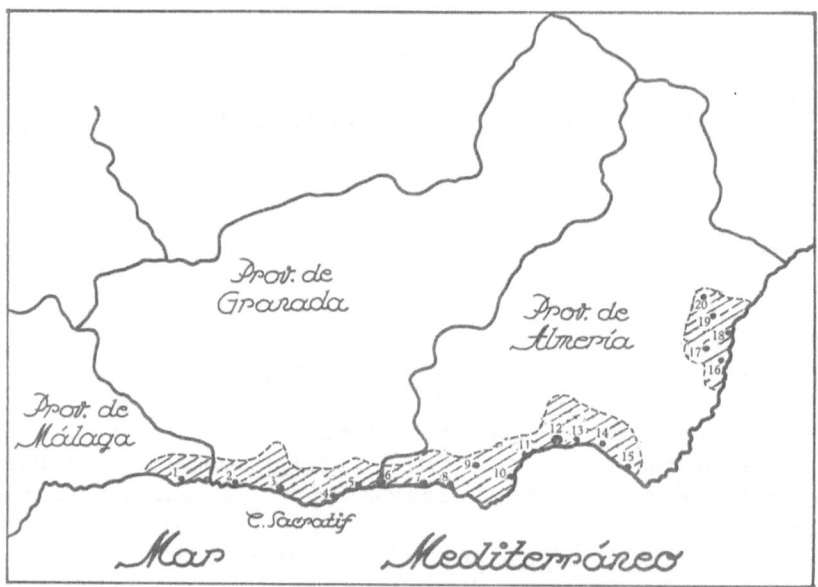

Map 1. The distribution of "Arenado" cultivation in the Spanish littoral of mediterranean Andalusia.

1 Nerja	11 Aguadulce
2 Almuñecar	12 Almería
3 Motril	13 La Cañada
4 Catell de Ferro	14 Alquián
5 Rábita	15 C. de Gata
6 Guainos	16 Mojacar
7 Adra	17 Turre
8 Balerma	18 Garrucha
9 El Ejido	19 Vera
10 Roquetas	20 Cuevas del Almanzora

In fact, to explain the basic principle of this method we must remember that in arid zones, the sun increases strongly the amount of evaporation, whereby in clayey soil the salt increases.

The underground water rises by capillarity and results in the same process of evaporation so that the salt is accumulated in the ground year after year. But if we use a manure compost and the sand layer, as described

Fig. 1. Table grapes grown with the new method of "Arenado".
Fig. 2. A green bean plantation with wind-breaks. Note the concrete walls and canals in order to keep the water free of soil contamination.

Fig. 3. For tomato cultivation it is necessary to hold up the plants on canes with ligatures.

above, capillarity is broken down and the water from rain, dew, or irrigation penetrates the soil dissolving the salt, and deposits it far below the roots; at the same time the plants are placed in a ground rich in air and with a higher temperature.

After 2–3 years, the sand must be removed (very carefully so as not to mix it with the soil); the area is ploughed, the compost manure and fertilizers are restored and the ground is again covered with clean sand. The old sand can be used for 8–10 years if it is kept clean of earth.

This is a short report on one of the most important works we are doing in the Instituto of Aclimatación in Almeria.

During the communication a set of slides were shown about this kind of cultivation and at the same time all the facts and points about it were explained. We would be glad to give further information to Centers of Investigation or any one interested on this matter.

It should be mentioned that the "Arenado" culture can anticipate the normal date of harvest by 15–20 days. The costs of this kind of cultivation are 40–50% higher than that of normal production, but the profits are very high.

Generally, the work is done by the owner or in sharing a plot, whereby the members of each family can dedicate their work to a piece of land of about 1 ha, and earn enough money for a good standard of living.

In many cases the peasant who begins the "Arenado" as a labour-worker, becomes himself the owner of a plot within 4–5 years. Thus this method, leading to a higher standard of living, is therefore of high interest also from a social point of view.

Authors' address: Instituto de Aclimatación, Generalisimo 116, Almeria, Spain.

PART III: SALINE SOILS AND BIOTOPES

Vegetation under Saline Conditions

by

V. J. CHAPMAN

The response of plants to salinity, either as single species or collectively as a community, has excited the interest of botanists, be they ecologists or plant physiologists, for many years. A fuller knowledge of the causes underlying the response has become increasingly important in the past two decades because of the need to bring saline soils into cultivation in order to increase the world's agricultural output.

Cultivation of saline soils can conveniently be divided into two independent aspects:

(a) The incorporation and utilisation of maritime saline soils, generally by the process of building stop banks and allowing natural rainfall and leaching to remove the surplus salt.

(b) The reclamation of inland saline soils by the use of irrigation water.

In the humid tropical regions of the world there are vast acreages of land at the mouths of the great river deltas which are more than ready for reclamation. These deltaic areas need intensive scientific study followed by the expenditure of sufficient money to build the necessary stop banks and sluices. The incorporation of this land, however, will go some way towards alleviating the food problem of East Asia. There are similar vast areas of inland saline soil the reclamation of which we are particularly concerned with in this Symposium. The agricultural use of these soils will serve sectors of the world's population, which at present are not adequately supplied with food.

Natural vegetation can be studied in relation to soil salinity, and when, as a result, communities or individual species can be related to specific groundwater conditions or salinity ranges then clearly they can be regarded as useful indicator species.

Thus, for instance, *Haloxylon persicum* in sand deserts indicates usable underground water in 15–30 m depth (BOYKO, 1951, 1966) or, as examples for heavier soils, the appearance of *Aster subulatus* or *Erigeron canadensis* on "inned" salt marshes indicates that the ground is ready for sowing by glycophytes.

Wild halophytic species can also be studied separately with a view to understanding their response to salinity. If saline land is to be reclaimed, detailed
study of the response of crop plants to increased soil salinity is of major
importance if maximum utilisation of saline soils is to take place, and such
a study must take cognizance of different races and strains, a feature which
has been ignored in the past.

There is no difficulty of course in distinguishing between extreme halophytic conditions and those characteristic of glycophytes, but there is a real
problem at the lower limits of salinization. GENKEL's (1954) definition of
halophytes as plants growing in saline habitats with a capacity to become
readily adapted, during ontogeny, to high salt concentrations does not
assist us, nor does his definition of glycophytes as plants growing in non-
saline habitats and with little capacity for adaptation to saline conditions
assist, because we must need define the point where one passes from glyco-
phytic to halophytic conditions. STOCKER (1928) proposed that 0.5% NaCl in
the soil water should mark the boundary between the two conditions, and
as subsequent work has not demonstrated that any other value is of greater
significance, the present writer (1960) incorporated this value in a modifi-
cation of IVERSEN's (1936) schema and further suggested its application to
black alkali soils as well as to white alkali soils. Plants growing on a saline
soil may also be grouped according to specific features: thus GENKEL and
SHAKOV (1946) have proposed that solontchak floras comprise four groups
of plants:

(a) Salt accumulating or Euhalophytes
(b) Salt excreting or Crynohalophytes
(c) Salt impermeable or Glycohalophytes
(d) Salt localized in special structures.

Whilst such a classification may have its value, it is evident from more
recent work that the nature of the saline ions, e.g. Na^+, Mg^{++}, Cl^- or SO_4^-,
is of perhaps even greater significance. This was clearly demonstrated by
VAN EIJK (1939) and over a series of years by STROGONOV (1962, transl. 1964) and
emphasized by the present writer in 1960. In fact cogent argument can be
adduced for insisting that future progress in this field will be most significant if it is directed to studies of the response of halophytes and glycophytes to the different ions that occur in excess in saline soils.

PLANT STRUCTURE AND SALINITY

Plants typical of saline soils exhibit certain morphological features that give
them a characteristic physiognomy. The most pronounced feature is
succulence, but the leaves and stems are often glaucous, water storage
hairs (*Atriplex*) may occur, leaf reduction is common and in some genera
(*Spartina, Statice, Aegialitis, Aegiceras, Avicennia*) salt secreting glands are abundant on the leaves. In the past most of the emphasis has centred around the

development of succulence in relation to salinity and the other facets have been largely ignored. Induction of succulence (= water storage tissue development) was experimentally produced by G. W. CHAPMAN (1931) and POKROVSKAYA (1954, 1957) later induced it in *Statice gmelini* and *Atriplex tatrica*.

In 1939 VAN EIJK demonstrated that the degree of succulence in *Salicornia herbacea* agg. was largely determined by the Cl^- ion though it was also increased by NO_3^- and Na^+, whereas the SO_4^- and Ca^{++} ions depressed it. Recent work in my own laboratory has shown that *Salicornia australis* will not tolerate an excess of sodium carbonate though it responds to excess NaCl and Na_2SO_4. STROGONOV (1962) has re-emphasized this differential effect of the Cl^- and SO_4^- ions, though he appears to be unaware of the earlier work by VAN EIJK. STROGONOV does, however, point out that, in the presence of excess Cl^-, plants develop halosucculent characteristics whereas in the presence of excess SO_4^- they develop haloxeric characters. The degree of halosucculence or haloxerism in plants should therefore depend upon the relative proportions of the different "formative" [1] ions in the cells. ARNOLD (1955) has taken this matter somewhat further and he believes it is the ratio of absorbed to free "formative" ions in the cells that is significant. Whilst this generalization may be true for many species it cannot be regarded as universal, because STROGONOV (1962) has shown that barley in the presence of excess Cl^- develops haloxeric features, so that there is species specificity as well as ion specificity.

STROGONOV (1962, 1964) has reported some very detailed studies on the effect of Cl^- and SO_4^- on the morphology of cotton, barley and *Salicornia*. With Cl^- ion alone involved, leaf thickness (succulence) reaches its maximum, whilst a mixture of SO_4^-/Cl^- has a greater effect than Cl^-/SO_4^- (the first ion is present in greater quantity than the second) (Table I).

TABLE I

Effect of Salinity upon Anatomy of Cotton leaves

Type of Salinity	Leaf Thickness	Thickness of Palisade Layer	Thickness of Mesophyll	% of control (total thickness)
Control	189.2 μ	83.6 μ	83.6 μ	100
Sulphate	250.8	118.8	110.0	132.5
Chloride	506.0	228.8	220.0	267.4
Control	270	111	107	100
Cl^+/SO_4^-	304	130	111	113
SO_4^-/Cl^-	346	148	143	129

The Cl^- ion upsets initiation and differentiation of leaf primordia in cotton so that their growth and elongation is inhibited. This depression of growth is a recognized phenomenon under saline conditions, and was noted many

1. Ions responsible for the development of halosucculence and haloxerism.

years ago by HARSHBERGER (1911) in respect of length and width of spikes of *Typha angustifolia* (Table II).

TABLE II

Salinity (sp. gravity)	Spike, mean length	Spike, mean width
1.0145	8.9 cm	1.9 cm
1.014	10.8	2.3
1.005	15.8	2.2

Until recently the amount of work involved in handling great masses of figures has curbed the collection of data and restricted their interpretation. The arrival of the Computor on the scene has materially altered the situation. WEIHE (1964), making use of an IBM machine has recently studied the relationship of environmental factors to certain features of a salt marsh community. Some 33 physiognomic characters were analysed in relation to some 22 abiotic habitat factors. He was able to show that there was a strong correlation between Hemicryptophytes, Nannophanerophytes and mean monthly precipitation. Shoot succulence was positively correlated with temperature, NaCl and Cl⁻, shoot woodiness (degree of lignification) in Nannophanerophytes was correlated with temperature and loss on ignition of the soil. It is evident that this kind of data treatment holds great promise for the future and researchers need not hesitate to collect as much data as they can. The present author (1966) has already pointed out that we need much more data collected in relation to species ontogeny and season.

VEGETATION AND SALT TOLERANCE

It has become increasingly clear (CHAPMAN, 1966) that the tolerance of vegetation to salinity is dependent very largely upon the kind of soil salinity (Cl^-, SO_4^-, etc.), upon the species variety or race being investigated and upon its state of development (whether juvenile, adult, flowering etc.). Appreciation of this conclusion enables one to understand the variations that are evident in published data on salt tolerance of species. The effect of kind of salinity was first noted by BATALIN in 1875, though he did not appreciate the reason for it. Other workers have also observed that crop plants grown in different geographical regions apparently changed their salt tolerance, whereas in fact it was the salinity that had changed.

The results of the extensive experiments carried out by STROGONOV (1962) have shown that plants grown under excess Cl^- have a greater tolerance towards salinity but are less tolerant to drought, heat and cold than plants grown under excess SO_4^-. In an attempt to explain Cl^- tolerance it has been suggested that it is associated with a binding of the Cl^- ions by albumens in the leaves of the plants, associated with an increased strength in the bond between the chlorophyll molecule and protein in the plastids, thereby protecting the biocolloids involved in metabolic processes against the actions of toxins. The binding of the water to the albumens is regarded as removing water out of the system and therefore prevents high concentrations of salts being obtained. This theory requires much more detailed study, because so long as any water remains as a solvent, and provided it is not saturated and energy is available to pump ions in against a concentration, there is no reason why high salt concentrations should not be reached. Control of the internal ion content is obviously important, as GREENAWAY (1962) has shown for various varieties of *Hordeum*, but this control is operated in more than one way, and binding of water to colloids is not necessarily the sole answer.

In the present state of our knowledge it would seem more important to study other methods of salt control, more particularly active methods of salt removal. This would include the mechanism of excess salt removal by means of glands, a process which appears to require energy (ARISZ et al. 1955). The present writer has pointed out (1960) that we need to know whether the mechanism actively affects only one ion, e.g. Na$^+$ or Cl$^-$, the other ion being associated in a purely passive manner. In view of the fact that salt crystals can be collected or dissolved off the leaf it would seem that exudation must take place largely in molecular proportions. SCHOLANDER et al. (1962) have made a preliminary study of this phenomenon in leaves of certain mangroves and they found that there is a diurnal cycle with *Aegialitis* and *Aegiceras*, minimal excretion taking place at night. No such diurnal cycle was observed in *Avicennia*. In all three genera about 90% of the soluble Cl$^-$ exuded was matched by molecular (NaCl) proportions of sodium and 40% was matched by K$^+$, giving ratios which are about the same as in seawater. One interesting feature that emerged was that the excreted material had a concentration 10–20 times that in the sap of the leaf cells, and this concentration was certainly not reached by transpiration evaporation because it continues while the leaf is surrounded by oil.

Tolerance towards salinity can also be achieved by the periodic removal of salt-saturated organs. This is probably necessary even in highly succulent plants such as *Salicornia*, *Allenrolfea* and *Halocnemum*, where sloughing off of some or all of the succulent tissue takes place at the end of the growing season. In *Salicornia australis*, at least, maximum osmotic pressures are reached in summer (40–50 atm), just before the succulent tissue starts to dry off. Other halophytes, such as *Juncus maritimus*, in which the salinity reaches maximum values (30–40 atm) at the end of summer (MASON, 1955), may also dispose of surplus salt by the loss of entire leaves. This is a feature of saline vegetation that so far has not received much study.

In the case of mangrove vegetation STROGONOV (1962) has suggested that vivipary, found in many mangrove genera, is important because it is argued that the gradual penetration of salt from mother plant to seedling brings about an increased salt tolerance in the latter. Whilst this argument may be valid for some species, e.g. *Rhizophora*, it is clearly not applicable to seedlings of *Avicennia nitida* (CHAPMAN, 1944) where the mean salinity of the seedling on the parent is between 1.5–2%, but almost immediately after being shed into the sea or mud it rises to above 6% and subsequently goes even higher. One can indeed argue that development of the seedling on the tree is only possible because salt is restrained from entering the seedling. PANNIER (1962) has confirmed the existence of a barrier tissue in *Rhizophora mangle* which lies between the cotyledonary body and the envolving tegument of the seedling. One may suspect that the micropylar haustorium in *Avicennia* seedlings may function similarly, though at present there is no experimental evidence to support the idea.

SALINITY AND VEGETATION GROWTH

The first impact of salinity upon growth of vegetation occurs at seed germination. All the work that has been carried out in this field shows that not only glycophytes but halophytes also germinate best under non-saline conditions, i.e. below 0.5% NaCl (CHAPMAN, 1960). In the case of halophytes some germination can be expected up to a concentration of 2% NaCl in the water. Further work, in the light of STROGONOV's (1962) contributions, would seem to be desirable in order to re-examine earlier conclusions (STOCKER, 1928; SCHRATZ, 1936) that salinity in parent habitat affects germination. Additional work is also required in order to separate out the extent of ion antagonism effect in relation to germination. As long ago as 1906 V. OSTERHOUT demonstrated that wheat germinated and grew better in a mixture of Na^+ and K^+ ions than in nutrient solutions containing either ion alone.

In some cases, especially that of halophytes such as *Suaeda, Salicornia, Aster tripolium, Aster subulatus*, it seems that inhibition of germination is primarily an osmotic effect, because when subsequently transferred to fresh water seeds germinate (SUSSEX in CHAPMAN & RONALDSON, 1958). This, however, is not always true. STROGONOV, for example, has shown that fully imbibed seeds of cotton do not germinate, and he has postulated a toxic effect of the Cl^- ion as being responsible. This ion is also reported as having a toxic effect on seeds of *Haloxylon*. Such toxicity does not occur in all plants because seeds of *Phaseolus aureus* are more affected by SO_4^- than by Cl^-. So far as plants are concerned the kind of salinity is therefore important as well as the actual concentration and one has in addition to consider each species separately.

Salinity, however, may not be the sole factor that affects germination, and there is clear evidence (CHAPMAN, 1960) that temperature and light (TSOPA, 1939) may also be concerned.

Once germination has occurred subsequent growth may be influenced by seasonal salinity changes or by occasional maximal values (REMANE & LASSIG, 1960). Growth rate of glycophytes falls off with increasing salinity, but in the case of obligate halophytes optimum growth conditions demand the presence of some salt, e.g. for *Aster tripolium* 0.12–0.25% NaCl, for *Salicornia stricta* (*herbacea* agg.) 2.5% (MONTFORT & BRANDRUP, 1927, 1928). In *Atriplex vesicaria* it has been shown that deficiency symptoms and death occur when there is less than 42 ppm of sodium chloride in the medium. In some species there is no doubt that the impact of salinity may vary with age of the seedlings, e.g. *Suaeda novae-zelandiae* (TURNER in CHAPMAN, 1960) and Cotton (STROGONOV, 1964), the seedling stage being more sensitive than the adult. Again, on naturally occurring saline soils growth and development changes with the kind of salinity (see section on morphology) (Table III).

TABLE III

Effect of Salinity on Growth

Salicornia herbacea	Height (cm)	Number nodes	Fresh weight	% of control
Control	14.5	18.7	1.33	100
Sulphate salinity	20.5	17.0	3.19	181.8
Chloride salinity	19.0	21.8	6.36	248.4
Suaeda glauca		No. branches		
Control	39.8	20.6	8.5	100
Sulphate	51.2	33.3	22.6	265.8
Chloride	30.6	15.6	10.7	125.8

In the case of glycophytes, such as cotton, tomato, barley, with increasing salinity the protoplasm retreats from the cell wall and the intercellular plasmodesmata are ruptured. This may be regarded as a protective mechanism which enables the plant to pass into a more or less dormant condition and so resist the unfavourable environment (STROGONOV & OKNINA, 1961; STROGONOV 1962, 1964). Since the process is reversible, permanent damage does not occur. It has also been observed that the phenomenon is more marked with chloride than with sulphate salinity.

The genus *Atriplex* would seem worthy of increasing study. BEADLE (1952) has studied germination in five species and BLACK (1961) has shown that *A. vesicaria* has an extremely high tolerance to salinity. In this species the main centres of accumulation are in the rapidly developing young leaves with low levels in the roots. Distribution of Na+ and Cl− within plants does vary, primarily in relation to the habitat, e.g. *Phragmites*, but ontogeny has also to be considered (CHAPMAN, 1942).

SALINITY AND WATER RELATIONS OF PLANTS

Although SCHIMPER put forward his theory of physiological drought in the last century there is still no really clear picture concerning the water relations of plants growing in saline soils. The effect of excess salt on plants can be explained in two possible ways:

(a) The increase in the osmotic pressure of the soil solution results in a reduced water uptake by the plants.

(b) There is a toxic effect exerted by the excess salts.

Over the years most of the emphasis has been directed on alternative (a) and it is only recently (STROGONOV 1962, 1964) that more attention has been directed to the second possibility. It is clear that individual ions do exert a specific effect upon the transpiration rate (v. EIJK, 1939) and it is equally clear that much more work is required on this aspect, particularly in relation to the proportions of free ions present in the cells. STROGONOV (1962, 1964), as a result of his experiments on cotton, has shown that for this genus at least plants in sulphate salinity have a lower water content and a higher transpiration rate than plants in equivalent chloride salinity. Other features in plants from sulphate soils, as compared with plants from equivalent chloride soils, are lower suction pressure, lower osmotic pressure, a lower protoplasmic viscosity and a lower degree of hydrophily of cell colloids, though the colloidal system itself is more stable. More of the energy absorbed by the leaves is expended on the increased transpiration rate and hence there is a greater demand on the soil water. With chloride salinization this is not the case and there is, indeed, little evidence for arguing that a condition of physiological drought exists where there is Cl-salinization. The development of haloxerism associated with SO_4^- salinization, however, suggests that SCHIMPER's theory may prove valid for this type of salinity.

The difference in plant response between chloride and sulphate salinization, taken in conjunction with all the varying effects of subsidiary ions and the antagonism of the two major ions, indicates why there is no regular pattern of behaviour discernible in the numerous data on the water relations of plants from saline habitats. Because of the multiplicity of factors involved it is in practice impossible to lay down hard and fast rules concerning the permissible salinity of any irrigation water (KELLEY, 1963).

There is one last comment I would like to make in connection with water relations of saline plants. It has been proposed by GENKEL (in STROGONOV, 1964) that the success of halophytes can be accounted for by the favourable conditions of the water-table, though this view must refer particularly to maritime halophytic vegetation. Whilst the presence of some species might be due to the existence of a high water-table, it is also equally evident that other species cannot tolerate waterlogging. This has been experimentally demonstrated for *Suaeda fruticosa* and *Halimione portulacoides* (CHAPMAN,

1960), whilst observations on a mangrove area at Auckland, where a drainage channel has been blocked by road-building operations, showed that *Avicennia marina* var. *resinifera* cannot tolerate continued water standing around the stems.

VEGETATION AND IONS

It has become increasingly evident that future work on saline vegetation must centre primarily around the impact of different ions upon the metabolism of the plants. There is abundant evidence that the inter-relations of sodium and potassium are particularly important. Both in halophytes and crop plants the presence of potassium greatly influences the quantity of sodium absorbed. A detailed study of the uptake of sodium and potassium by *Suaeda macrocarpa* has been reported in detail by BINET (1962), whilst BAUMEISTER & SCHMIDT (1962) have shown that growth in *Salicornia herbacea* and *Aster tripolium* is improved if a small quantity of potassium is added to the sodium chloride. In other plants, such as *Allenrolfea* and *Suaeda divaricata* the amount of potassium absorbed and accumulated can exceed the amount of sodium.

In the case of *Atriplex vesicaria* BLACK (1960) has suggested the existence of two different mechanisms for the absorption of alkali cations. The first is an Na^+ mechanism where potassium ions are able to compete when the sodium concentrations are low: this results in what BLACK has called "luxury" uptake levels of potassium. Secondly there is a potassium mechanism which is totally independent of sodium ion competition. The same species of *Atriplex* and allied species in Australia are capable of accumulating sodium and potassium even under conditions of low concentration. In this respect they behave like marine algae which also have the power to accumulate certain ions, such as iodine, from a dilute solution of the ion, in this case – sea-water. WALTER (1962) has proposed that plants capable of this phenomenon should be called xerohalophytes. The removal of sodium chloride from soils by plants has actually been suggested as one means of desalinising soils (BOYKO, 1964). CHANDRI et al. (1964) have calculated that a community of pure *Suaeda fruticosa* should remove at least 2,400 lb of sodium chloride per acre per year (= 2450 kg/ha). This would only be effective, of course, in conjunction with an irrigation programme.

Most of the work on ions up to the present has been directed at sodium, potassium and chlorine. The significance of the sulphate ion in relation to the chloride ion has been magnificently stressed by STROGONOV (1962, 1964), who has shown that excess sodium sulphate induces haloxerism and excess sodium chloride halosucculence. When the two ions are present together, as may often be the case, we are totally ignorant of the interaction of the two ions, that is whether there are separate pump mechanisms, and if so whether these pump mechanism are inter-related. Similarly we possess no information as to whether in the presence of excess sodium and magnesium sulphate there are separate pump mechanisms, and if so whether they are related to any special respiratory mechanism.

Apart from other aspects of differential ion activity already mentioned, there is the specific toxic effect of the chloride ion which has been postulated for some considerable time, though there has been no real understanding of how it could operate. Symptoms of salt poisoning in plants either take the form of chlorophyll bleaching or else of browning of parts of the leaves. STROGONOV (1964) maintains that the bleaching of the chlorophyll is associated with a decrease in the strength of the bond between the pigment and the plastid protein, but this decrease is reversible if the excess chloride disappears. In the necrotic areas, if there is no reversible change, it is believed that toxic organic substances accumulate. Such toxic substances, when transported through the plant, poison other normal cells and in this way a progressive necrosis develops. It should be emphasized that salt poisoning is most pronounced in true glycophytes and probably only occurs in halophytes with a low salinity tolerance. It must be absent in euhalophytes. Indeed, a study of the difference in cellular physiology between glycophytes and halophytes should throw considerable light on the mechanism of salt poisoning, and could well have a valuable impact upon the selection of suitable species for cultivation on reclaimed saline lands. It may also contribute to our understanding of salt adaptation.

There is some evidence that chloride toxicity is associated with disturbances of the normal nitrogen metabolism. In the case of maize it seems that there is an incomplete glycolytic breakdown of glucose so that there are sufficient acceptors available for an inorganic element like nitrogen. Both SO_4^- and Cl^- salinity bring about disturbances in the root nitrogen but in general those caused by chloride seem to be larger than those caused by sulphate. In barley, sunflower and *Vicia faba* increase in chloride salinity was followed by an increase of amino acids in the leaves (STROGONOV, 1964). Accumulation of the amino acids provides an additional substrate upon which enzymes can act and intermediate substances, not normally found in plants, can make their appearance. Two such substances are the aliphatic diamines, putrescine and cadaverine. In the presence of potassium deficiency putrescine has been shown to accumulate and to be responsible for the appearance of typical necrotic spots. STROGONOV and SHEVYAKOVA (1961) have shown that the application of putrescine to plants of *Vicia faba*, potatoes, barley, cotton and sunflower, results in the production of leaf necrotic spots similar to those associated with sodium chloride poisoning, and·subsequently the presence of putrescine was demonstrated in salt poisoned plants of *Vicia faba*. It could not, however, be found in salinized plants of sunflower so that there is a degree of specificity in the response. One cannot, in fact, argue from the above work of STROGONOV that the production of putrescine is the explanation of every case of chloride poisoning. It is very likely that other Cl⁻ containing organic compounds are also involved, though at present we know nothing of their nature. Work in the immediate future should concentrate upon the detection and isolation of further toxic compounds comparable to putrescine.

VEGETATION, PHOTOSYNTHESIS AND RESPIRATION

This is a field in which very little work has been carried out and it is one which should be thoroughly studied. Halophytes form parts of a wide variety of ecosystems and it is increasingly important that in some cases, e.g. mangrove forest and salt marsh, these systems be analyzed for their overall productivity and energy interchange relationships. This can only be achieved if we have some knowledge of the metabolic rates of the principal species. Work of this nature has been carried out by GOLLEY, ODUM & WILSON (1962) for a Puerto Rican mangrove forest. These investigators showed that wind velocity played a not inconsiderable part in the amount of carbon dioxide produced per minute, the maximum values being recorded at 15 l/min. and at a light saturation of 3000 foot candles. Work of this nature is very time consuming because the carbon assimilation of the foliage varies from apex to base of a tree and this has to be associated with the variation in available light at the different levels. In the case of the Puerto Rican forest it was shown that, allowing for the relatively crude methods of securing the data, the overall photosynthetic rate about balanced the respiration rate, and it was therefore concluded that the forest was not in a process of rapid dynamic change.

Investigations into the respiratory activity of aerating mangrove roots (pneumatophores) have been undertaken by TROLL & DRAGENDORFF (1931), SCHOLANDER et al. (1955) and the present writer (CHAPMAN, 1954), but although the results are of great interest they do not permit of any general application to an overall balance sheet.

It has also become evident that in the case of mangrove seedlings there is an anaerobic component to the respiration rate and the full implications of this in relation to habitat occupation have still to be worked out (CHAPMAN, 1962). In the meantime work is in progress to determine the effect of salinity upon such anaerobic respiration and also upon aerobic respiration. It will be seen that all the above work has centred around maritime halophytes and practically no information is available in respect of desert halophytes, nor is there any information available about the impact of different "kinds" of salinity (chloride or sulphate) upon photosynthesis and respiration.

CONCLUSION

The principal feature that emerges at the present time from a consideration of vegetation in relation to salinity is the inadequacy of our knowledge about the response of naturally occurring halophytes and of glycophytes to salinity induced by different ions. Data are required covering all aspects of morphology and physiology. It has become clear that plants respond differently to NaCl and Na_2SO_4 salinity and in respect of the former it would seem that Cl^- poisoning is a major feature. Here again, whilst there are clues to the kind of organic toxin that may be involved, until a wider range of plants has been investigated it would be premature to even postulate a mechanism. Finally it is also evident that intensive study of the effect of salinity upon photosynthesis and respiration is urgently required in order that some assessment can be made of actual and potential productivity of different saline areas.

SUMMARY

Classification of plants in relation to salt concentration can be on various bases, but in the past there has been neglect of the effect of the different ions and future progress must be directed towards this end. In the presence of excess Cl plants develop halosucculent characteristics whereas in the presence of excess Na_2SO_4 they develop haloxeric characters. The degree these two sets of characteristics are exhibited by any one species should therefore be related to the relative proportions of Cl and SO_4 ions available. Use of computer will, in future, enable workers to evaluate many more morphological features in relation to abiotic habitat factors.

It has become increasingly clear that the tolerance of vegetation towards salinity is dependent not only on the kind of saline ion (Cl, SO_4, CO_3) but also upon the species, variety or race under study and its state of development. Control of excess Cl^- can take place either biophysically within the cells or else by active removal of the excess through special glands or by periodic removal of the salt-saturated organs.

Seed germination of halophytes as well as of non-halophytes takes place most readily and effectively under non-saline conditions. Here again future work must be directed towards sorting out the extent and effect of ion antagonism and mechanism of inhibition, i.e. whether it is a purely osmotic effect or whether toxins are involved.

The effect of excess salt on plants can be due to either a toxic influence or to reduce water uptake by plants arising from the increased osmotic pressure of the soil solution. Insufficient attention has been paid to the first alternative. Whilst the success of halophytes has been ascribed to the favourable water table conditions, this is by no means always the case. Evidence is slowly accumulating for the existence of separate mechanisms for sodium and potassium absorption and this field needs much more study. No work appears so far to have been carried out in comparing absorption mechanisms for the Cl and SO_4 ions relative to each other. STROGONOV has demonstrated the existence of a toxic effect on some plants in the presence of excess chloride. It appears to be associated with disturbance of the normal nitrogen metabolism and the appearance of certain aliphatic diamines.

Until much more work has been carried out on the effect of salinity upon photosynthesis and respiration it will be impossible to prepare halophytic ecosystems in terms of their overall productivity and energy interchange relationships. Despite the intensive work of the past upon halophytic vegetation and the effect of salinity upon plants there are many major problems unsolved if we are to make effective use of saline lands.

RÉSUMÉ

La classification des plantes en relation à la concentration saline peut s'effectuer sur des bases différentes, mais dans le passé on n'a pas dédié beaucoup d'attention à l'effet des différents ions; c'est pourquoi à l'avenir le progrès devra poursuivre ce but. En présence d'un excès de Cl, les plantes présentent des caractéristiques halophiles, tandis qu'à la présence d'un excès de Na_2SO_4 elles montrent des caractéristiques halophobes. Le degré de ces deux caractéristiques, qu'on observe dans n'importe quelle espèce, doit donc être mis en relation avec les proportions relatives de ions de Cl et SO_4 disponibles. L'usage d'un calculateur permettra, à l'avenir, l'évaluation de caractéristiques morphologiques ultérieures, en relation avec les facteurs du milieu inorganique.

On constate toujours plus clairement que la tolérance de la végétation à la salinité dépend non seulement du genre du ion salin (Cl, SO_4, CO_3) mais aussi des espèces, variétés et races examinées et de l'état de développement. Le contrôle d'un excès de Cl peut être effectué soit biophysiquement dans les cellules, soit en enlevant l'excès par des tampons spéciaux, ou en enlevant périodiquement les organes saturés de sel.

La germination des semences des halophytes ainsi que des non halophytes se passe de la façon la plus rapide et efficace dans des conditions non salines. Par conséquent, à l'avenir il faudra s'orienter vers une classification en fonction de l'importance et de l'effet de l'antagonisme ionique et des mécanismes inhibitoires, c'est à dire, tendant à établir s'il s'agit simplement d'un effet d'osmose, ou si des toxines y sont impliquées.

Les conséquences de l'excès de sel sur les plantes peuvent remonter soit à une influence toxique, soit à une réduction de la capacité d'absorption de l'eau par les plantes, dûe à un accroissement de la pression osmotique de la solution du terrain. On n'a pas dédié assez d'attention à la première de ces possibilités. Par contre, le résultat satisfaisant des halophytes a été attribué aux conditions favorables de la nappe phréatique, ce qui toutefois ne correspond pas toujours à la réalité.

On accumule lentement les preuves de l'existence du mécanisme séparé de l'absorption et utilisation du sodium et du potassium; par conséquent ce domaine a besoin d'études ultérieures.

Aucune étude comparée n'a été effectuée jusqu'ici sur le mécanisme d'absorption des ions de Cl et de SO₄ dans leur rapport réciproque. Stro-gonov a démontré l'existence, dans certaines plantes, d'effets toxiques en présence d'un excès de chlore. Ce phénomène serait, à ce qu'il semble, associé à des troubles du métabolisme normal de l'azote et à la présence de certaines diamines aliphatiques.

Tant qu'on n'aura pas effectué des études sur les effets de la salinité sur la photosynthèse et sur la respiration, il est impossible de préparer des systè-mes écologiques halophytiques en fonction de leur productivité globale et des rapports énergétiques réciproques. Malgré les études faites dans le passé sur la végétation halophyle et les effets de la salinité sur les plantes, les problèmes les plus importants n'ont pas encore trouvé une solution, du moins en vue de l'utilisation pratique des terrains salins.

RIASSUNTO

La classificazione delle piante in relazione alla concentrazione salina può effettuarsi su diverse basi, ma nel passato é stato trascurato l'effetto dei differenti ioni, quindi nel futuro il progresso deve indirizzarsi a questo scopo. Alla presenza di un eccesso di Cl le piante presentano caratteristiche alofile mentre alla presenza di un eccesso di Na_2SO_4 esse presentano carat-teristiche alofughe. Il grado di queste due caratteristiche che si osservano in qualsiasi specie, deve quindi essere messo in relazione alle relative propor-zioni di ioni di Cl e di SO₄ disponibili. L'uso di un calcolatore permetterà ai ricercatori di valutare anche ulteriori caratteristiche morfologiche in rela-zione ai fattori dell'ambiente inorganico.

Appare sempre più chiaro che la tolleranza della vegetazione alla salinità dipende non solo dal genere dello ione salino (Cl, SO₄, CO₃) ma anche dalle specie, varietà e razze esaminate e dallo stato di sviluppo. Il controllo di un eccesso di Cl può essere eseguito sia biofisicamente nelle cellule sia rimuovendo l'eccesso con speciali tamponi o rimuovendo periodicamente gli organi saturi di sale.

La germinazione dei semi delle alofite come pure delle non alofite av-viene in maniera più veloce e più efficace in condizioni non saline. Ecco quindi che nel futuro ci si deve indirizzare verso una classificazione in funzione dell'entità ed effetto dell'antagonismo ionico e dei meccanismi inibitori, cioè, se è semplicemente un effetto di osmosi o se sono interessate anche le tossine. Le conseguenze del sale sulle piante possone derivare sia da un influenza tossica sia da un diminuito assorbimento di acqua da parte delle piante, dovuto ad una aumentata pressione osmotica della soluzione del terreno. Si é prestata poca attenzione alla prima possibilità; per contro

il risultato soddisfacente delle alofite é stato attribuito alle condizioni favorevoli della falda freatica, il che tuttavia non è sempre provato.

Si vanno lentamente raccogliendo prove per l'esistenza del meccanismo
separato dell'assorbimento e utilizzazione del sodio e del potassio, di conseguenza questo campo ha bisogno di ulteriori studi. Nessuno studio comparato é stato eseguito fin d'ora sul meccanismo di assorbimento degli ioni di Cl
e di SO₄, in rapporto fra loro. STROGONOV ha dimostrato l'esistenza in alcune piante di effetti tossici in presenza di eccesso di cloro. Questo sembra
essere associato ad uno scompenso del metabolismo normale dell'azoto e
alla presenza di certe diamine alifatiche.

Fino a che non si eseguiranno studi sugli effetti della salinità sulla foto-
sintesi e sulla respirazione, è impossibile preparare raggruppamenti ecologici alofitici in funzione della loro produttività complessiva e dei rapporti
energetici reciproci. Nonostante gli studi fatti nel passato sulla vegetazione
alofila e gli effetti della salinità sulle piante, i problemi più importanti non
sono ancora stati risolti, almeno in termini di utilizzazione pratica dei terreni salsi.

REFERENCES

ARISZ, W. H., I. J. CAMPHUIS, H. HEIKENS & A. J. VAN TOOREN. 1955. The secretion of the
salt glands of *Limonium latifolium* relze. *Acta bot. Neerl.* 4 (3): 322–338.

ARNOLD, A. 1955. Die Bedeutung der Chlorionen für die Pflanze. *Bot. Stud.* 2.

BAUMEISTER, W. T. & L. SCHMIDT. 1962. Über die Rolle des Natriums im pflanzlichen
Stoffwechsel. *Flora* 152: 24–56.

BEADLE, N. C. W. 1952. Studies in halophytes. I. The germination of the seed and establishment of the seedlings of five species of *Atriplex* in Australia. *Ecology* 33 (1): 49–62.

BINET, P. 1962. La répartition du sodium et du potassium chez *Suaeda macrocarpa* MOQ.
Physiol. Plant. 15 (3): 428–436.

BLACK, R. F. 1960. Effects of NaCl on the ion uptake and growth of *Atriplex vesicaria* HE-
WARD. *Austr. J. biol. Sci.* 3 (3): 249–266.

BOYKO, H. 1951. On regeneration problems of the vegetation in arid zones p. 62–80. *In* Les
bases écologiques de la régéneration de la végétation des zones arides. Union Int. of
Biol. Sciences Paris. Série B (colloques) No. 9.

BOYKO, H. 1964. Biological desalination. Report to Unesco. (mimeogr.).

BOYKO, H. 1966. Basic ecological principles of plant growing by irrigation with highly
saline or sea-water. p. 131–200. *In* H. BOYKO [ed.] Salinity and Aridity – New Approaches
to Old Problems. Dr. W. Junk, The Hague.

CHAPMAN, G. W. 1931. The cause of succulence in plants. *New Phyt.* 30: 119–127.

CHAPMAN, V. J. 1942. A new perspective in the halophytes. *Quart. Rev. Biol.* 17 (4): 291–311.

CHAPMAN, V. J. 1944. 1939 Cambridge University Expedition to Jamaica. *J. Linn. Soc. Bot.* 52:
407–533.

CHAPMAN, V. J. & J. W. RONALDSON. 1958. The mangrove and salt marsh flats of the Auckland Isthmus. *N.Z. D.S.I.R. Bull.* 125.

CHAPMAN, V. J. 1960. Salt Marshes and Salt Deserts of the World. L. Hill, London.

CHAPMAN, V. J. 1962. Respiration Studies of mangrove seedlings. I. *Bull. Mar. Sci. Gulf.
Carib.* 12 (1): 137–167.

CHAPMAN, V. J. 1962. Respiration Studies of mangrove seedlings. II. Ibid. 12 (2): 245–263.

CHAPMAN, V. J. 1966. Vegetation and Salinity p. 23–42. In H. BOYKO [ed.] Salinity and Aridity – New Approaches to Old Problems. Dr. W. Junk, The Hague.

CHANDRI, I. T., B. H. SHAH, N. NAGRI & I. A. MALLICK. 1964. Investigations on the role of Suaeda fruticosa FORSK in the reclamation of saline and alkaline soils in W. Pakistan plains. Plant & Soil 21 (1): 1–7.

EIJK, M. VAN. 1939. Analyse der Wirkung des NaCl auf die Entwicklung, Sukkulenz und Transpiration bei Salicornia herbacea, sowie Untersuchungen über den Einfluss der Salzaufnahme auf die Wurzelatmung bei Aster tripolium. Rec. Trav. Bot. Neerl. 36: 559–657.

GOLLEY, F., H. T. ODUM & R. F. WILSON. 1962. The structure and metabolism of a Puerto Rican red mangrove forest in May 1962. Ecol. 43 (1): 9–19.

GENKEL, P. A. 1954. Solenstoichivost᾿ rastenii i puti ee napravlennogo porysheniya (Salt tolerance of plants and its planned increase). Izdatel᾿stvo Akad. Nauk. SSSR. Timiry Chten. XII.

GREENAWAY, H. 1962. Plant Response to saline substrates I, II. Aust. J. biol. Sci. 15 (1): 16–57.

HARSHBERGER, J. W. 1911. An hydrometric investigation of the influence of sea water on the distribution of salt marsh and estuarine plants. Proc. Amer. phil. Soc. 50: 457–496.

IVERSEN, J. 1936. Biologische Pflanzentypen als Hilfsmittel in der Vegetationsforschung. Copenhagen.

KELLEY, W. P. 1963. Use of saline irrigation water. Soil Sci. 95 (6): 385–39.

MASON, G. W. 1955. Some aspects of the Halophyte Problem. M. Sc. Thesis Auck. Univ.

MONTFORT, G. & W. BRANDRUP. 1927. Physiologische und Pflanzengeographische Seesalzwirkungen II. Jb. wiss. Bot. 66 (5): 902–946.

MONTFORT, G. & W. BRANDRUP. 1928. Ibid. III. Ibid. 67 (1): 105.

OSTERHOUT, W. I. v. 1906. On the importance of physiologically balanced solutions for plants. 1. Marine plants. Bot. Gaz. 5: 42.

PANNIER, P. F. 1962. Estudio fisiológico sobre la viviparia de Rhizophora mangle L. Acta Cient. Venez. 13: 184–197.

POKROVSKAYA, E. I. 1954. Obmen veshchestr rastenii na zasolennykh pochrakh (Metabolism of Plants on Saline Soils). Avtoreferat Kandidatskoi dissartatsii. Moskva.

POKROVSKAYA, E. I. 1957. [Certain data on oxidation reduction process in halophytes]. Painiatn Akademika H. A. Maskowa. Shornik static, Moskva, 268–274 (in Russian).

REMANE, A. & G. LASSIG. 1960. Einwirkungen des Salzgehaltes auf Vermehrung und Wachstum bei Lemna minor. Kiel. Meeresfor. 16: 221–228.

SCHOLANDER, P. F., L. VAN DAM & S. SCHOLANDER. 1955. Gas exchange in the roots of mangroves. Amer. J. Bot. 42 (1): 92–98.

SCHOLANDER, P. F., H. T. HAMMEL, E. HEMMINGSEN & W. GAREY. 1962. Salt balance in Mangroves. Pl. Phys. 37 (6): 721–729.

SCHRATZ, E. 1936. Beiträge zur Biologie der Halophyten III. Prings. Jb. wiss. Bot. 83: 133–189.

STOCKER, O. 1928. Das Halophyten Problem. Ergebn. Biol. 3: 265–354.

STROGONOV, B. P. & E. Z. OKNINA. 1961. O Sostoyanii pokoya u rastenii v usloviyakh zasoleniya (The rest period of plants under saline conditions). Fisiol. rastenii 8 (1): 79–85.

STROGONOV, B. P. 1962. Fiziologicheskie osnovy solenstorchivosti Rastenii ± zdatel᾿stvo Akad. Nauk SSSR (1964, Physiological basis of salt tolerance of Plants, trsl. POLJAKOFF-MAYBER, A. & A. M. MAYER. Israel Prog. Sci. translat. Jerusalem).

TROLL, W. & O. DRAGENDORFF. 1931. Über die Luftwurzeln von Sonneratia L. und ihre biologische Bedeutung. Planta 13 (2): 311–473.

TSOPA, E. 1939. La végétation des Halophytes du Nord de la Roumanie en connexion avec celle du reste du pays. S.I.G.M.A. 70: 1–22.

WALTER, H. 1962. Die Vegetation der Erde. Vol. 1. Jena.

WEIHE, K. V. 1964. Beiträge zur Ökologie der Mittel- und westeuropäischen Salzwiesen-Vegetation (Gezeitenküsten). I. Methodik, Standorte und vergleichende morphologische Analyse. Beitr. Biol. Pflanz. 39 (2): 189–237.

Author's address: Professor Valentine Jackson Chapman, University of Auckland, P.O. Box 2175, Auckland, New Zealand.

Saline Soil Problems with Particular Reference to Irrigation with Saline Water in India

by

P. C. RAHEJA

Saline sodic soils exist in all parts of the world. Their distribution is, however, largely restricted to the extreme arid, arid and semi-arid zones where a high evaporation rate conduces to accumulation of salts in the surface layers. Lands of this type exist in South West Asia, comprising parts of Turkey, the Arabian Peninsula, Iran and Afghanistan; the Indo-Pakistan desert tract; Central Eurasia where it is cut off by the mountain ranges of the Pyrenees, the Alps, the Balkan, the Carpathians, the Black Forest Mountains, the Caucasian Mountains, the Ural, the Himalaya, the Tien-Shan Mountain, the Kunlin, the Stanovoi Peaks and the Yablonski Mountains, all acting as barriers against moisture-bearing winds from the oceans; North Africa comprising Morocco, Algeria, Tunisia, Libya and Egypt; the Sahara-Sahel region constituting the intermediate zone between the extreme arid desert and the Sudan; the Trans-American region extending from Oregon to Mexico and from the foothills of the Rocky Mountains to the Pacific coast in North America, and the coastal areas facing the Pacific ocean, Meso-America and the coastal region of Peru, of northern Ecuador, of South Central Chile and of North Western Argentine of South America; Southern and South West Africa; and finally Central Australia. In all these regions wherever agriculture is practised, saline lands have developed as the natural equilibrium between rainfall, vegetation and environment has been disturbed and high evapotranspiration established due to diversified cropping systems with or without irrigation. Wherever in arid regions irrigation has been provided, salinity has developed at a very fast rate.

In India the area under irrigation has steadily increased during the past seven decades and in the last fifteen years the potential of irrigation has developed with great speed. It is estimated that the maximum potential area in India, cultivated by irrigation, is 187 million acres. By 1966 about 45% of this potential is expected to be achieved. Currently over 80 million acres are irrigated from canals, from medium-sized storage reservoirs, and from dug out wells and drilled tube wells. These three types are termed large, medium and small sized irrigation works, respectively.

In recent years there has been a great deal of emphasis on the development of small sized irrigation works in arid areas and in consequence, the scope of exploitation of the groundwater resources has been very much enlarged. These groundwaters are, however, not all entirely suitable for irrigation. Quite a large proportion of them present more or less a hazard of salinity, and it is for this reason that the protective irrigation judiciously practised makes all the difference in the crop production in arid countries.

All along the western coast of India tidal lands are inundated during the monsoon season. In areas where seasonal streams discharge their waters into the sea, saline sodic flats, called *Ranns*, have developed. These seasonal streams deposit silt and clay on the beaches making them slushy during the rainy weather. During the summer season these deposits are baked by the sun and become hard. Where the stream flow does not occur the fine sands are saturated with sea brine and growing of crops is restricted there by the excess of salt in the soil solution when the tidal action has subsided in the post-monsoon period.

EXTENT OF SALINE SODIC SOILS

Irrigation has been in vogue in the Indo-Pakistan region since ancient times. Already during the Vedic period, that is from about 2000 to 600 B.C. the peasants dug wells and constructed inundation canals in order to irrigate their crops, and it may be of interest to note that until the end of this period there is no mention of a rise in salinity on the soil surface. Irrigation was gradually extended during the Hindu, the Afghan, the Mugal and the British periods. Simultaneously with the extension of irrigation in the Indus Valley and in the Jamna basin, which are semi-arid to arid areas, the salinity level of the soil rose conspicuously. By 1942–43, the saline lands in Western Pakistan already comprised about 700,000 acres. The explanation of this is that a sub-alluvial ridge is running from Delhi to Shahpur, and since the general flow of groundwater moves from north-east to south-west, this ridge impedes the free drainage of these areas and checks the flow of under-ground streams. In fact, by 1960, more than 3 million acres out of 12.5 million irrigated acres had already seriously been affected by salinization in the 16 canal districts of the erstwhile combined Punjab. Of these three million acres, 1.5 million have completely gone out of cultivation and additional salinization is taking place at a rate of 100,000 acres every year (Fig. 1).

A similar development can be observed in other parts of this subcontinent. At present about 20 million acres of saline lands already exist in the arid and semi-arid regions of North West India. The extent of these lands vary from state to state and some details may, therefore, be of interest.

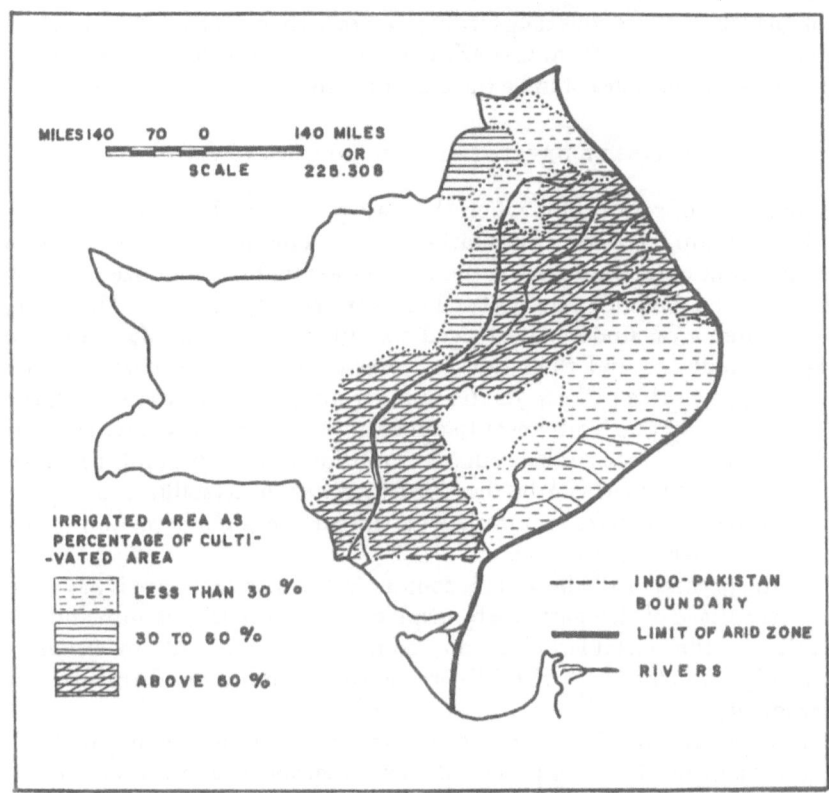

Fig. 1. Arid zone of India-Pakistan irrigation 1942–43 (after O. P. Bharadwaj).

About 1.5 million acres have been salinized to varying degrees in the Punjab. In Uttar Pradesh sodic saline, sodic and saline soils exist mainly in the western and central parts of the State and cover over 7 million acres in different stages of degradation. In the State of Maharashtra saline sodic soils have developed in the irrigated tracts. The salts present there are mainly sulphates and chlorides and most of the salts are concentrated in the B-horizon. Depending upon the water-table, the zone of salt accumulation occurs at 1 to 2 m depth. The surface soil develops a high degree of sodium saturation and such soils comprise 42% of the respective area. In the State of Gujarat tidal coastal saline lands occur in the Rann of Kutch and the Little Rann of Kutch. These sodic flats have a large extension and occupy an area of more than 8–10 million acres. Along the Western Coast of the State the so-called Khar lands are to be found on the beaches of the Arabian Sea, where they occupy an area of about 50,000 acres. The total soluble salts in these soils range from 3.04–3.60%, their pH from 8.60–8.76, and their $CaCO_3$ content from 3.2–6.6%. The exchangeable calcium is, however, low. In all other states of India the areas of saline soils are very limited.

SALINITY DEVELOPMENT UNDER IRRIGATION

Salinization of soils occurs when the Ca-clay or Mg-clay reacts with dissolved sodium salt in the soil solution. As the replacing power of Na^+ is 5.0 as against 1.6 of Ca^{++} and Mg^{++}, the Ca-clay is rapidly sodiumized and by dilution, the replacing power of sodium very rapidly increases. Therefore, in irrigated areas under waterlogged conditions, Ca^{++} and Mg^{++} ions are rapidly displaced and their salts are precipitated. Clay becomes highly sodiumized in this process. As the water from the surface evaporates, the salts are brought up from lower to upper horizon. When excessive concentration of a particular salt occurs beyond the saturation point, i.e. in an higher amount than can be dissolved in water, it crystallizes and these crystallized precipitates accumulate on the surface, a phenomenon which we know as salt efflorescence.

In the process of alkalinization, sodium in the base exchange is adsorbed on the surface of clay particles by displacing calcium, magnesium and potassium in the crystal lattice. In due course this excess sodium, in water (NaOH), combines with the CO_2 in the soil solution and forms sodium carbonate.

In some soils the "B" horizon is sodiumized due to the leaching down of the salts. In this leaching process the exchangeable sodium is replaced by divalent cations of Ca^{++} and Mg^{++} in the A horizon of the soil. The sub-soil then becomes highly salinized and, if there is enough clay present, this clay will be cemented and a sodiumized indurated layer will develop. Layers of this type are found in the Takyr soils of U.S.S.R., in the Thur soils of the Punjab and in the Usar soils of Uttar Pradesh.

A degradation of alkali soils takes place when H-ions replace the Na and Mg ions in them. Sodium carbonate is gradually leached out leaving the hydrogen clay in the soil. By this the pH value is lowered to 6.0. This reaction occurs in soils which have a low content of calcium carbonate. The whole process is known as solodization and such soils are called solods.

Sometime, however, after the degradation, a reaccumulation of salt or a regradation occurs as sodium ions flock in the soil (RAHEJA, 1965).

SALINITY IN IRRIGATION WATERS

In an arid environment the accumulation of salts in the soil can be observed at a much faster rate under irrigation than under non-irrigated condition. Therefore, by far the largest proportion of saline soils have been developed under the influence of irrigation. The two main sources of irrigation water are (a) surface-water and (b) ground-water. Flowing surface waters usually have a low content of salts as is evident from the data in Table I (LIVING-STONE, 1961).

TABLE I

Mean composition of river waters of the world in ppm

	HCO_3^-	SO_4^-	Cl^-	NO_3^-	Ca^{++}	Mg^{++}	Na^+	K^+	Sum
North America	68.0	20.0	8.0	1.0	21.0	5.0	9.0	1.4	142
South America	31.0	4.8	4.9	0.7	7.2	1.5	4.0	2.0	69
Europe	95.0	24.0	6.9	3.7	31.1	5.6	5.5	1.7	182
Asia	79.0	8.4	8.7	0.7	18.4	5.6	9.3		142
Africa	43.0	13.5	12.1	0.8	12.5	3.8	11.0	—	121
Australia	31.6	2.6	10.0	0.05	3.9	2.7	2.9	1.4	59
World	58.4	11.2	7.8	1.0	15.0	4.1	6.3	2.3	120

In most of the countries the sums of cations and anions in their rain-water are low and it is most unlikely that they may cause salinity when directly used for irrigation. About half of the value consists of HCO_3^-, possibly combined with calcium, which has an additional beneficial effect by its neutralizing sodium and magnesium salinity.

There-against groundwaters tend to have a high content of soluble salts and the comparison, presented in Table II for two specific areas, mirrors a general rule.

TABLE II

Average pH, total solids and E.C. of water of the Aligarh and Kanpur districts

Source of water	pH		Total Solids ppm		E.C. 10^6	
	Aligarh	Kanpur	Aligarh	Kanpur	Aligarh	Kanpur
Canal	8.0	8.5	165	325	236	466
Well (Ground)	8.1	8.6	501	943	715	1400
Drainage (Sub-soil)	8.8	8.7		1821	2602	2882

In this case the canal waters in the Gangetic plains have a lower total quantity of dissolved solids and, accordingly, a lower E.C. value than the groundwater. The E.C. values of the latter are three times the values of the former. The drainage waters are particularly highly charged with salts and the E.C. values exceed seven to eleven times those of the canal water (AGARWAL et al. 1957).

The mineral content of groundwater depends upon the kind of rock with which it comes in contact. Following the process of groundwater recharge, changes in the salt content of the rain-water or stream water occur by reduction processes, base exchange and concentration, and the cations in the water reach an equilibrium with the cations in the soil. "The reduction processes, mainly of biochemical nature, influence the concentration of SO_4 in the groundwater" (YARON & SHALHEVOT, 1965). The great variations in mineral content of various groundwaters, caused by the different lithological formations, with which they are in contact, are best shown by an example.

Water samples from surface wells located in various lithological formations were collected in India and analysed for total soluble salts, cations and anions. The results of 95 samples are reported in Table III.

TABLE III

Relationship between lithology of the aquifer and chemical composition of waters

Lithological formation	No. of samples	T.S.S. ppm			Dominant cations (25%)				Dominant anions (25%)			
		Min	Max	Mean	Ca	Mg	Na+K	Mixed	HCO$_3$+CO$_3$	SO$_4$	Cl	Mixed
Blown sand	16	212	5696	1918	1	1	12	2	4	2	6	4
Younger alluvium	12	513	9618	4293	–	1	10	1	–	1	10	1
Older alluvium	14	740	7974	3276	–	–	14	–	–	2	11	1
Malani volcanic	29	267	6300	2623	1	–	26	2	1	1	17	1
Jalore granite	12	305	5376	1524	–	–	9	3	1	–	5	6
Siwana granite	12	1595	7899	4747	–	–	2	–	–	–	–	–

T.S.S. = Total Soluble Salts.

The figures of Table III give a good picture of the large variations in the total amount of soluble salts in waters from various lithological formations. Jalore granite and blown sand formations have low mean values compared to those of the other formations. The highest mean value, based upon two samples, was found in Siwana granite and the next highest in an aquifer of Younger alluvium. In most of the water samples the predominant cation is sodium and the predominant anion is chloride. Out of sixteen samples of waters from the lithological formation of blown sand four were of the carbonate type and two were sulphate waters. In the other formations, the number of samples containing other anions than chloride is only small (CHATTERJI, 1963).

The pH values of irrigation waters of high E.C. and high S.A.R. (C) and those of waters of low E.C. and low S.A.R. (A) compare favorably and there is no significant difference. The sodium content and cation exchange capacity in the latter waters are both very low and, in consequence, their sodium adsorption ratio is also very low. These are sweet (fresh) waters and they are extensively used for irrigation on light soils without exhibiting any hazard of salinization of the soil.

On the other hand, waters of moderate E.C. and high S.A.R. values (B)

TABLE IV

Irrigation waters classified on the basis of E.C. and S.A.R. values

Sample No.	pH	E.C. in micromhos /cm at 25°C	Na meq/l	K meq/l	Ca meq/l	Mg meq/l	Total cations meq/l
A. Waters of low E.C. and low S.A.R.							
1.	8.4	650	3.04	0.07	1.41	2.22	6.74
2.	8.4	1166	5.86	1.64	0.86	2.06	10.42
3.	8.5	1368	9.56	0.05	1.62	2.11	13.34
4.	8.3	796	5.81	0.18	0.76	1.70	8.51
5.	7.5	546	3.33	0.24	1.45	2.69	7.71
B. Waters with moderate E.C. and high S.A.R.							
1.	8.8	3784	28.26	6.41	0.41	4.33	39.41
2.	8.9	3780	30.43	0.51	0.31	1.47	32.73
3.	8.9	2252	15.22	0.17	0.83	2.13	25.86
4.	8.7	3063	19.56	0.15	2.18	2.56	25.47
5.	8.8	3758	26.95	1.92	2.49	9.33	40.69
C. Waters of high E.C. and high S.A.R.							
1.	8.0	10771	69.56	0.21	11.19	18.10	99.60
2.	8.3	8751	60.86	1.84	8.14	25.13	95.97
3.	8.4	7208	63.04	0.85	5.10	10.73	79.72
4.	8.4	6307	43.48	7.92	4.27	11.17	66.84
5.	8.4	12117	65.21	8.97	5.64	32.74	112.56

Sample No.	Cl meq/l	SO_4 meq/l	CO_3 meq/l	HCO_3 meq/l	Total anions meq/l	SAR	Villages from which samples drawn
A. Waters of low E.C. and low S.A.R.							
1.	3.50	0.02	0.81	2.22	6.55	2.3	Bagra.
2.	3.45	0.62	2.02	4.47	10.56	4.8	Bhitola.
3.	6.00	trace	2.63	5.47	14.10	7.0	Dhaola.
4.	3.32	2.07	0.59	2.52	8.50	5.3	Tikhi.
5.	1.72	3.56	0.28	2.04	7.60	2.3	Gol.
B. Waters with moderate E.C. and high S.A.R.							
1.	20.00	13.43	1.73	5.94	42.10	18.4	Narnawa.
2.	18.00	4.48	2.53	10.25	35.26	32.4	Mandwala.
3.	13.00	2.87	1.23	5.25	22.35	12.5	Tilora.
4.	17.00	2.51	0.57	5.95	26.03	12.7	Sangana.
5.	22.50	5.28	3.84	6.88	38.50	11.33	Nagni.
C. Waters of high E.C. and high S.A.R.							
1.	94.50	trace	1.62	3.54	99.66	18.2	Debhawas.
2.	79.00	8.43	1.01	2.53	90.97	14.9	Muk.
3.	70.50	6.29	0.65	2.93	80.37	22.4	Kurakhera.
4.	50.00	14.38	0.57	2.46	70.21	15.7	Akoli.
5.	104.50	1.27	2.63	4.96	113.36	14.9	Num.

present already some hazards for irrigation. Sodium in the cation exchange complex is very high and the calcium content is low. Waters of this group are rich in HCO_3 anions and have, therefore a high alkalinity. These waters can only be used prudently for the irrigation of crops.

Although the pH values of the C category of waters are similar to those of Category A, the sodium and chloride contents are both very high. These waters present a very strong hazard in irrigation and they can be utilized only when salt resistant crops or grasses are grown.

Table V compares a large number of samples, altogether 236, from well waters and their analyses (MONDAL, 1964). The distribution of the groups of this category is based upon their E.C. and S.A.R. values.

TABLE V

Classification of different types of groundwaters [1]

Range of conductivity micromhos/cm at 25°C	Total No. of water samples	Per cent of waters	S.A.R. groups					
			0–10 (S_1)	10–18 (S_2)	18–26 (S_3)	26–34 (S_4)	34–42 (S_5)	Over 42 (S_6)
1	2	3	4	5	6	7	8	9
0–250 (C_1)	Nil	Nil	Nil	Nil	Nil	Nil	Nil	Nil
250–750 (C_2)	14	5.9	14 (100)	Nil	Nil	Nil	Nil	Nil
750–2250 (C_3)	67	28.4	54 (80.59)	7 (10.44)	4 (5.97)	2 (2.96)	Nil	Nil
2250–5000 (C_4)	70	29.6	33 (47.14)	20 (28.57)	12 (17.12)	3 (4.28)	1 (1.43)	1 (1.43)
5000–10,000 (C_5)	62	26.4	13 (20.96)	16 (25.80)	14 (22.58)	8 (12.90)	5 (8.06)	6 (9.69)
10,000–15,000 (C_6)	17	7.2	1 (5.88)	4 (23.52)	6 (35.29)	1 (5.88)	1 (5.88)	4 (23.52)
Above 15,000 (C_7)	6	2.5	Nil	Nil	2 (33.33)	2 (33.33)	1 (16.66)	1 (16.66)
	236							

1. Distribution of S.A.R. of the waters expressed as percentage of the total No. of water-samples under each conductivity class is given in parenthesis.

The majority of the waters of the groups C_1–C_4 and S_1–S_3 is suitable for irrigation. As the electrical conductivity (E.C.) increased, the proportion of irrigation waters in the S_3–S_6 groups increased simultaneously. Almost all waters, which had an E.C. higher than group C_5 were not suitable for irrigation. The waters of C_6 and C_7 classes together constituted 9.7% of the sampled wells. The percentage of waters which fall in C_3, C_4 and C_5 category form 84.4% of the total number of wells. Although these groups offer some hazard in raising crops in the summer season, as such crops have high water requirements, these waters are safely used for winter crops such as wheat, rape and barley.

The various classes of waters which can be used for irrigation are further classified as shown in Table VI, in which their distribution is based upon the amount of total soluble salts and the S.A.R. values (MONDAL, 1964).

TABLE VI

Distribution of groundwater according to soluble salts and S.A.R. values

S. No.	Type of groundwater	No. of samples	Percentage of total wells up to class IV (C_4) type	Area irrigated in sq. miles [1]	Percentage of the total area under these type of groundwaters
1	2	3	4	5	6
1.	C_2–S_1	2	2.13	42.48	19.21
2.	C_3–S_1	34	36.17	86.08	43.46
3.	C_3–S_2	8	8.51	15.36	6.95
4.	C_3–S_3	2	2.13	2.52	1.14
5.	C_4–S_1	24	25.53	44.12	19.96
6.	C_4–S_2	12	12.76	9.36	4.23
7.	C_4–S_3	8	8.51	4.08	1.85
8.	C_4–S_4	4	4.26	7.08	3.20
	Total 94		100.00	211.08	100.00

Within the E.C. classes C_2, C_3 and C_4 with combination of S.A.R. values of S_1, S_2, S_3 and S_4, the percentage of total wells and the percentage of the total area irrigated with these types of groundwaters have been worked out. The wells of which the conductivity range is 250–750 and their S.A.R. value ranges from 0 to 10 constituted only 2.13% of the total number of wells but supplied water for irrigation of 19.21% of the total irrigated area. C_3 type of waters, which have a very low S.A.R. value (S_1) constitute 36.17% of the wells and these wells irrigate about 44% of the total irrigated area. All these classes of waters together irrigate over 62% of the commanded area. Another class of waters which contribute considerably to the irrigation potential are C_4 (E.C. 2250–5000) and S_1 (SAR 0–10) types of waters. The area irrigated per well is less than that of the other two types of waters. The rest of the wells have only a low command as their waters offer too strong a salinity hazard for crop production.

GUPTA & ABICHANDANI (1965) studied the E.C. values of the well waters from six villages for changes in conductivity during the year 1964. The results are summarized as follows.

1. One square mile = 640 acres = 2.59 km².

TABLE VII

E.C. in micromhos/cm of irrigation water before and after the monsoon season

Name of Village	Season/Month of sampling				
	Pre-Monsoon		Monsoon	Post-Monsoon	
	April	June	August	Oct.	Nov.
Dantiwada	8,293	8,093	7,400	3,794	3,372
Caperda	11,349	10,959	10,175	10,848	9,885
Jelwa	6,111	5,901	—	4,972	5,319
Dhankri	7,857	7,587	—	7,232	7,559
Suraita	8,730	8,430	—	9,040	8,722
Shikarpura	17,460	16,017	12,750	4,520	2,326

Three of the villages were inaccessible during August and water samples could not be drawn from them for the determination of their electrical conductivity. The general trend of the results indicates that the level of salinity of the irrigation water is generally lower in the post-monsoon than in the pre-monsoon season since the recharge of the aquifer takes place in the post-monsoon season. The rainfall, in general, was deficient over the whole tract otherwise there may have been a substantial fall in the values in December. It may be added that the irrigation of wheat, lucerne and barley starts in November and continues till April. It is during this period that most of the irrigation waters have a lesser salt content than in the hot weather of the summer months.

GROWTH OF CROPS WITH SALINE WATER

Saline waters are frequently used for crop production in North Africa, in the Middle East and in the Indo-Pakistan desert region. Such irrigation waters are reported from Algeria to have 5000 mg/l total salt content, Tunisia 5600 mg/l, Israel 2300–6000 mg/l, and Western Rajasthan 3,800 to 11,200 mg/l (CHEVALIER, 1950; AUBERT, 1958; SHALHEVET & REININGER, 1964; BOYKO, 1959, 1966; MONDAL, 1964). Most of these waters are applied to sandy soils wherein the salts do not accumulate in the profile and are leached down with the rain-water in one to two years and this restores the salt balance in the soil.

The conditions are quite different in soils with a less sandy texture. Thus, for instance, the electrical conductivity of the saturation extract of the soil (1:2 ratio) was determined for soil samples drawn from various depths of a light and a medium textured soil with a lime kankar pan at 40–60 cm depth (GUPTA & ABICHANDANI, 1965). The comparative data of two sites for each of these textural classes are detailed in Table VIII.

TABLE VIII

E.C. in micromhos of 1:2 saturation extract of soil samples drawn in December and March from irrigated and non-irrigated fields

Texture of soil	Soil Depth cm	Period of sampling Irrigated 1963–64		Year of irrigating the crop	
				Fallow 1963–64	Irrigated 1964–65
		Dec. 1964	March 65	December 1964	March 65
Light soil	0– 20	614	663	1,005	1,224
	20– 40	670	714	960	1,428
	40– 60	647	918	1,005	2,040
	60–120	1,117	1,581	2,010	2,346
	120–200	1,563	1,377	2,122	2,040
Medium soil	0– 20	1,061	1,326	1,898	2,346
	20– 40	1,005	1,224	2,457	2,346
	40– 60	–	3,060	–	1,832

In the 1964–65 fallow plot, there occurred practically only a very small change in the salt balance of the soil in March compared with that of December. In the 1964–65 irrigated plot, however, the salinity level in the first metre was definitely increased by the irrigation with saline water. In the medium textured soil the salinity was appreciably higher than in the light soil.

In order to evaluate the effect of various levels of salinity on the yield of fodder maize BHUMBLA et al. in 1964 conducted a study in small field plots. The relevant data are summarized in Table IX.

TABLE IX

Effect of quality of irrigation water on the yield of maize green fodder

Treatments	1962–64 (Control = 2425 g/plot)				Mean	Remarks	
	C_1	C_2	C_3	C_4		Symbol	E.C. $\times 10^6$
S.A.R. 4	1889	1742	1406	936	1493	C_1	3,330
S.A.R. 8	1733	1648	1357	790	1382	C_2	5,600
S.A.R. 12	1740	1516	1342	989	1397	C_3	11,200
S.A.R. 16	1823	1310	971	905	1252	C_4	20,000
Mean	1795	1554	1269	905			
	S.E. for conductivity =		91.6 g/plot				
	L.S.D. @ 5%		= 179.6 g/plot				
	L.S.D. „		= 235.8 g/plot				

The control treatment gave a yield of 2425 g. As against this, yields of 1795, 1554, 1269 and 905 g were obtained with E.C. values of 3,330, 5,600, 11,200 and 20,000 respectively. With the rise in the E.C. values a direct fall in yield of maize green fodder could be observed. The influence of S.A.R. values was also very strong in reducing the yield, but the successive increase in sodium adsorption ratios had a less marked effect than the rise of the E.C. values.

Similar experiments have been conducted on wheat in order to study the effect of increasing values of electrical conductivity and of the sodium adsorption ratio of the soil (Table X).

TABLE X

Effect of salinity of irrigation water on the yield of wheat grain

Treatments	1961–62 (Control = 815 g)				Mean	1963–64 (Control = 644 g)				Mean
	C_1	C_2	C_3	C_4	in g	C_1	C_2	C_3	C_4	in g
S.A.R. 4	648	639	447	112	461	650	574	554	126	476
S.A.R. 8	750	764	537	154	551	589	600	639	114	485
S.A.R. 12	682	608	443	207	485	662	605	586	268	530
S.A.R. 16	756	657	354	151	479	669	548	523	317	514
Mean	704	667	445	156		642	582	575	206	
S.E.		27.6 g/plot					45.5 g/plot			
L.S.D. 5%		54.1 g/plot					89.3 g/plot			
L.S.D. 1%		74.5 g/plot					117.2 g/plot			

By the F-test the effect of conductivity was significant, while that of SAR was non-significant.

$C_1 = $ 3,330 E.C. $\times$ 10^6
$C_2 = $ 5,600 E.C. $\times$ 10^6
$C_3 = $ 11,000 E.C. $\times$ 10^6
$C_4 = $ 20,000 E.C. $\times$ 10^6

The differences due to the S.A.R. values are less marked within the range from 4 to 16 which means that the reduction in yield to almost half occurs when the S.A.R. values are less than 4. The rise in electrical conductivity had a very adverse effect. In both years the loss in yield was very small at an E.C. value of 3,330. The yield decreased, however, substantially with a further rise of electrical conductivity. In general, wheat could tolerate a salinity level of 5,600 E.C. value.

The growth behaviour of tobacco which is a non-salt tolerant plant, was studied by diluting sea-water and amending its composition by adding potassium and phosphorus salts to it (KURIAN et al., 1964). These plants were raised in earthen pots (25″ × 25″) containing a sandy soil of the following mechanical composition: Coarse sand – 68%; fine sand – 26%; silt and clay – 3%; carbonates – 3%. The sea-water was amended by addition of phosphorus and potassium salts. The results of this study are summarized in Table XI. The amendment of tap water showed, as expected, a highly significant increase in growth of the plants, but also the amendment of sea-water by addition of potassium and phosphorus, particularly at 10,000 ppm salt concentration dilution improved very much the growth of the plants and they compared very favourably with the series under tap water irrigation. It shows that the addition of potassium salts in proper ratios brings about a

TABLE XI

Growth behaviour of tobacco irrigated with sea-water dilutions

Treatments	Amendment Total K meq/l	Total H_2PO_4 meq/l	Height of plant cm	No. of leaves	Area of leaves cm^2
Control					
Tap water	—	—	14.62	9.40	38.21
Tap water amended	108.43	9.44	34.10	14.60	125.26
Sea-water dilutions					
10,000 ppm	2.10	—	10.90	6.60	28.78
10,000 ppm amended	88.88	7.50	16.88	9.60	54.08
15,000 ppm amended	3.15	—	11.10	5.40	24.66
15,000 ppm amended	133.28	11.25	12.10	6.60	29.14
L.S.D. 5%			6.13	1.85	19.82
L.S.D. 1%			9.36	2.53	27.03

balanced ionic condition which prevents the toxic effects of sea-water. Much better, but not yet published results have recently been achieved with wheat, irrigated with sea-water in dilutions up to 20,000 ppm and with other species even with pure sea-water from the Indian Ocean (36,000 ppm T.D.S.).

SUMMARY

Saline Soils constitute a major problem in many arid and semi-arid countries. In India alone saline soils cover about 20 million acres. Here, this salinity has largely developed
  (a) under irrigation from canals;
  (b) in inundated lands in depressions along the rivers and streams;
  (c) by tidal encroachment in coastal areas; and
  (d) by irrigation with highly saline waters on clay containing soils.
    In the canal-irrigated areas of the arid and semi-arid regions, the salinity problem has been created by weathering of sodium bearing minerals in the soil profile and by a rise of the watertable within the capillary fringe. In the inundated lands, unless subjected to flushing, the sodium salts have accumulated by being washed into them from the surrounding areas. These spotty solonetz soils are the consequence of impeded drainage in depressions, due to the formation of a cemented stratified layer in the sub-soil. Tidal saline lands, on the west coast of India, develop salt encrustation after the monsoon season. The content of sodium and magnesium salts differs greatly between the various localities and varies from 3–15%.
    In the arid zone, over 30% of the groundwaters show high S.A.R. values, a high electrical conductivity, a low calcium content and high chloride

values. Sodium accumulates in the top soil horizon unless it is leached down the profile by precipitation. The waters of low electrical conductivity (500–1000) and low S.A.R. values (2.3–7.0) are mostly sweet. There is a slight hazard involved in irrigating medium-heavy soils with irrigation waters of moderate E.C. values (2250–3780) and high S.A.R. (10–32) values. Rotations, with 2–3 years fallows, have to be adopted to allow a leaching of the sodium salts into the sub-soil by monsoon rains. Most of the waters in the groups of C_1–C_4 and S_1–S_3 types are suitable for irrigation on light soils.

There is a significant decrease in the yield of maize and wheat with an increase in the E.C. values of the soils, the greater salt sensitivity of these two important crops being with maize. A successive increase in sodium absorption ratios of the soil had less effect on crop yields than the increase in the E.C. values.

Sea-water diluted to a concentration of 10,000 ppm and amended by the addition of phosphorus and potassium salts at 7.50 and 88.88 meq/l levels, substantially increased the yield of tobacco leaves in pot culture experiments.

RÉSUMÉ

Les terrains salins occupent aux Indes une superficie de près de 20,000,000 acres. La salinité y est très répandue à cause de l'irrigation au moyen de canaux, dans les zones inondées et les dépressions le long des fleuves et des cours d'eau, à cause de la marée montante le long des côtes et de l'irrigation avec de l'eau fortement salée.

Dans les régions arides ou semi-arides irriguées, le problème de la salinité est dû aux altérations atmosphériques du minéral de sodium dans les couches du terrain et à la montée du niveau de la nappe phréatique au sein de la couche capillaire.

Dans les zones inondées, moins sujettées aux grandes poussées de l'eau, les sels de sodium se sont accumulés à cause du délavage des terrain aux alentours. Ces "solonetz maculés" sont déterminés par la difficulté des drainages dans les dépressions dues aux stratifications imperméables dans le sous-sol.

Les terrains salins de marée, le long de la côte occidentale de l'Inde donnent lieu à des incrustations salines fort sensibles après la saison des moussons. Le contenu en sels de sodium et de magnesium peut subir des variations allant de 3 à 15%. Dans les terrains arides plus de 60% de l'eau phréatique est caractérisé par des valeurs très hautes de S.A.R., par une forte conductance électrique et par une faible teneur en calcium et une haute teneur en chlorure.

L'accumulation du sodium a lieu à la superficie du terrain, à moins qu'il n'y ait lieu un délavement vers les terrains profonds causé par la tombée des pluies. Les operations de drainage qui s'enscrivent à la culture des plantes qui demandent de l'eau un niveau élevé, tels que le riz, le *Trifolium alexandrinum* et la canne à sucre, entraînent une amélioration de la salinité dans les terrains irrigués.

L'enfouissement des engrais verts a donné d'excellents résultats avec la *Sesbania aculeata* pour la bonification de ces terres. Les amendements avec le gypse sont nécessaires dans les terrains sodiques-salins ou sodiques avec une forte teneur en argile.

La bonification des "solonetz maculés" est très difficile à moins que l'on n'emploie en même temps le drainage et l'amendement du sol. Les terrains salins de marée ne peuvent pas être facilement bonifiés, à moins que la composition du sol ne présente de 10 à 15% de carbonate de calcium. Des techniques spéciales de reboisement ont été mises à point pour l'aménagement de cultures industrielles de *Prosopis juliflora* et de *Casuarina equisetifolia* et pour créer des paturages de plantes chénopodes alophytes et de diverses espèces erbacés.

L'irrigation avec l'eau saline est possible lorsq'elle est réalisée rationellement et de façon à assurer un bon équilibre des sels dans le terrain où poussent les racines, tenant compte des propriétés physiques et chimiques du terrain, de l'assolement et des facteurs de l'aridité. On a choisi pour ces terres les variétés les plus résistantes de millet, sorgho et blé.

I terreni salini occupano un'area di circa 20,000,000 di acri, in India. La salinità si è diffusa largamente a causa dell'irrigazione per mezzo di canali, nelle zone inondate nelle depressioni lungo i fiumi ed i corsi d'acqua, a causa del risalire della marea nelle zone costiere e dell'irrigazione con acqua fortemente salata.

Nelle zone appartenenti alle regioni aride o semiaride irrigate, il problema della salinità è stato determinato dall'alterazione atmosferica dei minerali di sodio negli strati del terreno e dall'alzarsi del livello della falda freatica nell'ambito dello strato capillare.

Nelle zone inondate, meno soggette a grandi flussi d'acqua, i sali di sodio si sono accumulati a causa del dilavamento delle zone circostanti. Questi solonetz maculati sono la conseguenza della difficoltà di drenaggio nelle depressioni dovute alla formazione di stratificazioni impermeabili nel sottosuolo.

I terreni salini di marea, nella costa occidentale dell'India, producono notevoli incrostazioni saline dopo la stagione dei monsoni. Il contenuto di sali di sodio e di magnesio varia dal 3 al 15%. Nella zona arida più del 60% dell'acqua freatica presenta alti valori di S.A.R., una forte conducibilità elettrica, un basso contenuto di calcio ed un alto contenuto di cloruro.

Il sodio si accumula nell'orizzonte superficiale del terreno, a meno che questo non sia dilavato dalle precipitazioni verso i profili profondi. Le opere di drenaggio che seguono la coltura di piante che necessitano acqua alta, come il riso, il *Trifolium alexandrinum* e la canna da zucchero, migliorano la salinità nei territori irrigati.

Il sovescio con la *Sesbania aculeata* ha dato ottimi risultati nella bonifica di tali terreni. Gli ammendamenti col gesso sono necessari in terreni sodici salini o sodici con forte contenuto di argilla.

E' difficile bonificare i solonetz maculati, a meno che non si usi contemporaneamente un sistema di drenaggio e di ammendamento del suolo. I terreni salini di marea non possono essere bonificati con facilità, a meno che essi non presentino dal 10 al 15% di carbonato di calcio nel profilo del suolo.

Sono state sviluppate tecniche speciali di rimboschimento per l'impianto di colture industriali di *Prosopis juliflora* e di *Casuarina equisetifolia*, e per favorire pascoli di chenopodiacee alofitiche e di varie specie erbose.

L'irrigazione con acqua salina è possibile dove essa è applicata in maniera razionale in modo tale da assicurare un ottimo equilibrio di sali nella zona radicale, tenuto conto delle proprietà fisico-chimiche del terreno, delle rotazioni colturali e dei fattori di aridità. Per dette zone sono state scelte delle varietà resistenti di miglio, sorgo e grano.

REFERENCES

AGARWAL, R. R., C. L. MEHROTRA & C. P. GUPTA. 1957. Spread and intensity of soil alkalinity with canal irrigation in Gangetic alluvium of Uttar Pradesh. *Indian J. agric. Sci.* 27 (4): 363–373.

AUBERT, G. 1958. Discussion on paper of Dr. H. HEIMANN The irrigation with saline water and the toxic environment. Potassium Symp., 223.

BHUMBLA, D. R., J. S. KANWAR, K. K. MAHAJAN & BHAJAN SINGH. 1964. Effect of irrigation with different sodium and salinity hazards on the growth of the crops and the properties of the soil. *Proc. Symp. Problems Indian Arid Zone, Jodhpur.*

BOYKO, H. 1966. Salinity and Aridity – New Approaches to Old Problems Mon. Biol. XVI, p. 1–22, Dr. W. Junk N.V., Den Haag.

BOYKO, H. 1967. Salt-Water Agriculture, *Scient. Amer.*, March 1967, p. 89–95.

BOYKO, H. & E. BOYKO. 1959. Seawater Irrigation – a new line of research on a bioclimatic Plant-Soil-Complex. *Int. J. Bioclimat. Biometeorol.* III, part II., Sec. B. 1, 1–24.

BOYKO, H. & E. BOYKO. 1966. Experiments of Plant Growing under Irrigation with Saline Waters from 2000 mg/litre T.D.S. (Total Diluted Solids) up to Sea-water of Oceanic Concentration, without Desalination. Mon. Biol. XVI, p. 214–282, Dr. W. Junk N.V., The Hague.

CHATTERJI, P. C. 1967. Some chemical types of ground waters from Siwana Development Block region (Barmer district) of Western Rajasthan. *Quart. J. Geol. Min. & Metallurgy Soc., India* (in press).

CHEVALIER, G. 1950. *Revue franç. de l'oranger* 213: 115–124.

GUPTA, I. C. & C. T. ABICHANDANI. 1965. Unpublished data.

KURIAN, T., E. R. R. IYENGAR, M. R. NARAYANA & D. S. DATAR. 1964. Effect of sea water dilutions and its amendments on tobacco. *Proc. Symp. Problems Indian Arid Zone*, Jodhpur.

LIVINGSTONE. 1961. Chemical composition of Rivers and Lakes. *U.S.D.A. Geol. Survey Paper* No. 440, pp. 64.

MONDAL, R. C. 1964. Quality of ground waters in Siwana, Jalore and Saila Development Blocks. *Proc. Symp. Problems Indian Arid Zone*, Jodhpur.

MONDAL, R. C. 1965. Unpublished data.
RAHEJA, P. C. 1965. Aridity and Salinity, a survey of soils and land use. p. 43–127. *In* H.
    BOYKO [Ed.] Salinity and Aridity. Mon. Biol. XVI, Dr. W. Junk, N.V. The Hague.
SHALHEVET, J. & P. REININGER. 1964. *Israel J. agric. Res.* 14 (4): 179–187.
YARON, B., & J. SHALHEVOT. 1965. Quality of irrigation water, Chapter VII *In* International
    Sourcebook on Irrigation and Drainage in Relation to Salinity and Alkalinity, F.A.O./
    UNESCO. Pub.

Author's address: Professor Dr. P. C. RAHEJA, Project Manager, Lake Nasser Water
Resources Research and Development Project, Near East Regional Offrce, FAO, Cairo,
UAR.

Behaviour of Salty Soils under Irrigation in Puglia

by

NICOLA FICCO

Reference is made to some experiments carried out in the Plateau of Apulia on soils of a silty character (silt content: 80.95%).

The waters for the irrigation came from an artesian layer utilized by means of wells drilled in battery. Their total salinity was 1.5‰.

Before the experiments were started, salinity was found to vary in the first 50 cm of the soil in ascending sense beginning from the ground level.

The experimental program was focussed towards the variables: doses of irrigation and fertilization. Accordingly, the following schemes were mainly applied to each crop:

I_1 = normal irrigational treatment;
I_0 = normal irrigational treatment plus two waterings to permit the leaching of the salts;
C_a = PKN fertilization in the following proportion: 4:1:2;
C_b = PKN fertilization in a proportion of 6:1.5:3.

Each scheme was repeated three times with a "randomized" system.

The dimensions of the plots which had an area of 365 m² (8.5 × 43) each, were established in such a way as to allow the use of machines in operations of cultivation.

CROPS CULTIVATED AND IRRIGATION TREATMENT

The experiments were started in 1962 and were completed in 1964. The crops cultivated were as follows:

  – in 1962: tomatoes – fodder maize, sorghum for fodder, alfalfa;
  – in 1963: sunflower, fodder sorghum, grain maize, tomatoes, alfalfa, and fodder maize;
  – in 1964: fodder sorghum, grain maize, and fodder maize.

Since the soil was of such a heavy character, particular care was devoted to the nursing of the cultures especially in order to prevent any commencement whatsoever towards the formation of cracks which might have affected considerably the specific water volumes of the irrigations. The anti-parasitic treatments were carried out primarily against pyralidid of maize and against the red spider of tomatoes.

After the first irrigation, all crops, showing stagnant water on the plots, caused by inadequate levelling of the land, had to receive timely care and nursing. As regards the cultivation of alfalfa, several zones where the germination of the seed did not appear uniform, had to be sown again.

In fixing the rotations of irrigation, we had to take care of the modalities of delivery by the collective irrigation plant which uses those artesian waters in whose perimeter falls the zone where the tests are carried out. Said zone constitutes part of a district which was recently transformed into an irrigated farm district in pursuance of the Land Reform Law.

The salty land belonging to the district is part of an area which for the Plateau of Apulia, according to an estimate by BOTTINI, amounts to approximately 10,000 ha. At the time of the conversion of this land, steps were taken to provide them with hydraulic benefits by means of slightly curved ridges with small lateral ditches. The latter failed to offer sufficient guarantees for satisfactory yields because of their inadequate physico-mechanical make-up and due to the presence of a phreatic layer at a depth of about 3 m with maximum deviations up to 1.93 m from the surface. The special conditions of the soil and of the brackish underground layer of the land subjected to irrigational treatment were taken into account in selecting the crops. The fertilization applied was of the normal type (Scheme C_a: mineral perphosphate 25 kg/ha – potassium sulphate 100 kg/ha – calcium nitrate 50 kg/ha) and of the higher type (Scheme C_b: the afore-mentioned quantities, indicated under C_a, were augmented by 50%).

The methods of irrigation applied were as follows: lateral infiltration from furrows for tomatoes, sunflower and grain maize, lateral infiltration from canals for fodder maize and fodder sorghum, and submersion for alfalfa.

The variation in the irrigation treatment consisted in the number of waterings. On some plots where we wanted to observe the irrigation effect upon the leaching of the soil, we applied a larger number of irrigations for the same crop. We preferred making a few more waterings with a view to facilitating the leaching of the salts from the arable condition rather than increasing the single doses of irrigation since the heavy character of the soil would undoubtedly have exercised an unfavorable influence on the crops.

The following table shows the most significant data concerning variations in the saline content of the soils, concerning the exchangeable sodium and concerning the pH reported at the beginning of the experiments and after their conclusion. As regards the exchangeable sodium, the absolute figures are indicated in milli-equivalents whereas the calculation of the Ca and Mg ions for the establishment of the relationship of sodium absorption (SAR $= Na^+ : (Ca^{++} + Mg^{++}) / 2$) is still pending.

Wishing to illustrate graphically the variations in the saline content of the soil in function of the number of irrigations (Schemes I_0 and I_1) and of the fertilization (Schemes C_a and C_b), the following curves would be obtained on the basis of diagrams the abscissas of which are the takings of the samples whereas their ordinates are salinity expressed in grams per thousand. The values of the salinity, reported in the ordinate, represent the average of the partial values shown in the table and referred to the plots subjected to the same irrigational treatment and fertilization, as specified hereunder.

TABLE I

| Dates of samples | Salinity | |
	Avg values reported scheme I_0 & I_1	Avg values reported scheme C_a & C_b
9.XI.1962	7.120–7.920	7.918–7.741
24.IV.1963	3.496–2.930	3.553–3.152
19. V.1964	2.690–2.164	2.507–2.650
21. X.1964	2.720–2.736	2.849–2.902

TABLE II

| Plot No. | Experimental scheme | Soil Salinity | | | | Exchangeable Sodium | | pH | |
		9.XI.62	24.IV.63	19.V.64	21.X.64	1962	1964	1962	1964
1. 2. 3	I_0A	3.65	3.75	2.44	2.87	6.99	7.90	8.70	8.79
5. 6. 7	I_0B	8.35	4.73	1.63	1.97	10.60	6.87	8.43	8.90
9.10.11	I_1A	5.56	4.06	2.90	1.83	11.30	5.79	8.36	8.32
13.14.15	I_1B	5.56	3.36	3.93	4.64	9.60	9.47	8.78	8.30
17.18.19	I_0A	7.82	6.24	4.86	3.95	5.30	3.75	8.35	8.17
21.22.23	I_0B	2.41	3.37	0.99	3.05	4.94	8.29	8.15	8.25
25.26.27	I_1A	8.33	5.17	2.00	4.22	10.10	7.04	8.30	8.23
29.30.31	I_1B	8.26	2.93	4.83	1.53	5.40	6.18	8.28	8.29
33.34.35	I_0A	11.81	2.13	1.31	1.97	9.77	4.92	8.06	8.18
37.38.39	I_1B	13.65	1.26	0.69	0.96	18.30	3.24	7.80	8.05
49.50.51	I_0A	10.14	1.58	1.02	2.03	9.55	5.95	7.97	8.08
53.54.55	I_0B	4.86	0.81	0.72	1.10	5.21	4.14	8.07	8.10
57.58.59	I_1A	9.36	4.85	3.83	4.29	6.17	5.32	8.35	8.10
61.62.63	I_1B	6.90	3.43	2.75	3.21	4.34	6.90	7.70	8.20
65.66.67	I_0A	8.55	2.72	2.59	1.50	5.56	6.31	7.83	8.20
69.70.71	I_0B	8.60	3.29	2.95	4.72	3.17	6.60	7.84	8.08
73.74.75	I_1A	7.61	2.48	2.16	3.85	4.35	5.49	8.03	8.03
77.78.79	I_1B	6.86	2.53	2.83	2.15	4.78	4.63	8.13	8.20
85.86.87	I_0B	6.77	2.84	2.32	3.02	5.91	6.53	8.36	8.55
89.90.91	I_1A	6.35	2.55	1.96	1.98	3.43	5.32	8.01	8.41
93.94.95	I_1B	5.19	2.61	2.86	2.67	5.01	5.78	7.90	8.41
Total		156.59	67.05	51.57	57.51	149.78	126.42	171.30	173.84
Avg		7.45	3.19	2.45	2.73	7.13	6.02	8.15	8.27

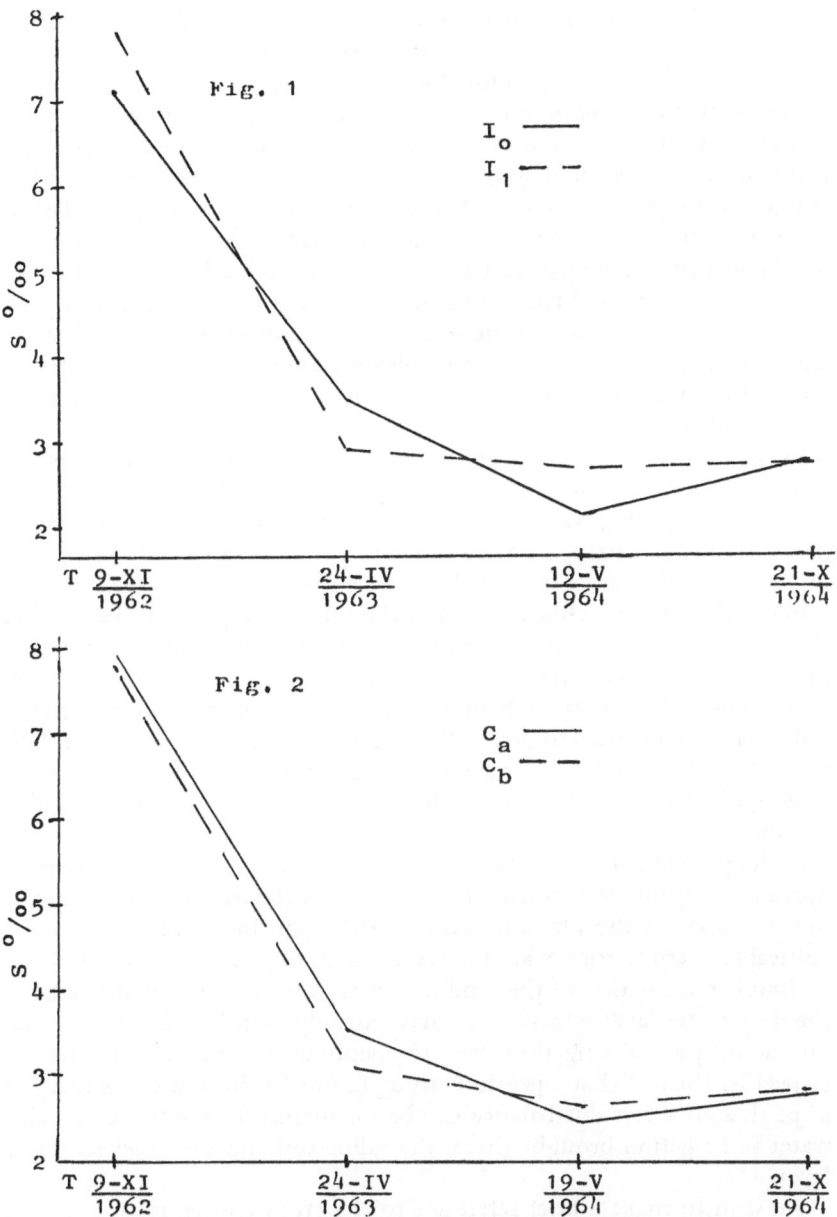

Fig. 1. Variation of soil salinity in function of the number of waterings.
Fig. 2. Variation of soil salinity in function of fertilization.

It seems obvious, however, that in order to ascertain the effect of irrigation on the soil, we have to consider the drop in salinity rates between the first and the second sampling and for this we have to take into account, in addition to the washing action of the irrigation waters, the influence of winter rains, the amount of which was recorded at about 200 mm for the period under review. A plausible reason for the noteworthy decline in the salinity of the soil between the first and the second sampling can be explained by the fact that the soils under observation, which were irrigated for the first time, appeared to be more sensitive to leaching than after repeating the irrigational treatments with a certain regularity. At any rate, the percentage of salinity in the soil remained almost stationary after the second irrigational treatment as regards the plots treated under Scheme I_1, whereas the rate of salinity underwent certain fluctuations for the plots treated under Scheme I_0.

As regards variation of salinity in function of the fertilization schemes, this does not appear to exercise a decisive influence on salinity itself if we are to judge by the proceeding of the relative curves of the values extrapolated from the analyses carried out. The pH data show a slight displacement of the soil reaction in the field of alkalinity.

In conclusion, the saline content of the soil was found to be reduced at the end of the experiments to one third of that found at the beginning of the experiments. However, it is a question whether it would not be necessary to check if the balance between irrigation water (total salinity 1.5‰), soil (total salinity down to 40 cm 3‰) and phreatic layer (the verified salinity of which is 3.28 g/l), will remain the same in the future or whether it will undergo variations and rise to such an extent as to injure normal cultivations.

Under present conditions it is not deemed necessary to carry out drainage operations capable of removing the products of the leached salts from the soil and prevent the phreatic layer to rise again by capillarity into the cultivable stratum, for we are interested to attempt the realization of the hydraulic reclamation of the land in this region through the drainage of the deep water layer which, as we have already mentioned, is being used for the purpose of irrigation. We are encouraged in our attitude in this respect by the fact that a previous study of our Institution offers justified hope that an acceptable balance can be maintained between the brackish water of irrigation brought there, the saline soil and the brackish water layer.

We wish to make a brief reference to the crops chosen from amongst those which had already shown themselves resistant to salinity and which have given satisfactory yields mainly in plots treated under Schemes I_0, C_b, with the exception of tomatoes which suffered severe damage from an attack of the red spider (see page 161 ff).

SUMMARY

The author refers to experiments made during three years on typical heavy salty soils of the Tavoliere di Puglia with the purpose of studying the influence of irrigation and mineral fertilization on salty soils.

The results obtained with the following different treatments have been compared:
- normal irrigation;
- normal irrigation with two more waterings for leaching. In both cases using on some plots mineral fertilization in the rapport PKN = 4:1:2, and on others PKN = 6:1.5:3.

Summing up, from the observations made, it appears that the irrigation has a positive influence for leaching the soil at the first irrigation treatment, and afterwards the salinity in the soil is stabilized on values that permit the crops to give good yields.

About the fertilization, it was found that it did not have influence on the salinity of the soil.

RÉSUMÉ

L'auteur se rapporte aux épreuves faites pendant trois années sur les terrains salins du Tavoliere di Puglia dans le but d'étudier l'influence de l'irrigation et de l'engrais sur les terrains salins.

On a établi une comparaison des résultats suivants:
irrigation normale
irrigation normale avec deux arrosages supplémentaires pour consentir de lessiver les sels; on a employé dans les deux cas des parcelles avec engrais mineral dans le rapport PKN 4:1:2 et d'autres dans un rapport PKN 6:1.5:3.

D'après ces observations, il résulte que l'irrigation exerce un effect positif sur le lessivage du terrain au moment de la première irrigation et après que la salinité du terrain s'est fixée sur des valeurs consentants d'obtenir une bonne récolte.

On a remarqué que l'engrais n'avait aucune influence sur la salinité du terrain.

RIASSUNTO

L'autore si riferisce alle prove condotte durante tre anni sui terreni salsi del Tavoliere di Puglia, con lo scopo di studiare l'influenza dell'irrigazione e concimazione sui terreni salsi.

Sono stati messi a confronto i risultati ottenuti con le seguenti tesi:
Trattamento irriguo normale;
Trattamento irriguo normale più due adacquamenti per consentire il lisciviamento dei sali; ed in tutti e due i casi usando in alcune parcelle

concimi minerali nel rapporto PKN 4:1:2 e in altre un rapporto PKN
6:1.5:3.

Dalle osservazioni fatte risulta che l'irrigazione ha una influenza positiva
per il lisciviamento del terreno al primo trattamento irriguo e dopo che la
salinità del terreno si sia stabilizzata su valori che permettono di ottenere
buoni raccolti.

E' stato osservato che la concimazione non aveva alcuna influenza sulla
salinità del terreno.

Author's address: Dr. Nicola Ficco, Ente per lo Sviluppo dell'Irrigazione e Trasformazione Fondiaria in Puglia e Lucania, Bari, Italy.

Experiments in the Improvement of Saline and Alkaline Soils of the Zebra Plain (Morocco)

by

ALAIN RUELLAN

Situated in eastern Morocco, in contact with the Rif and the Middle Atlas range on the left bank of the Moulouya, 35 km from the Mediterranean, the Oued Zebra plain forms part of the great watershed of the Lower Moulouya which will embody nearly 80,000 ha of lands under irrigation; the Zebra plain itself will be irrigated over an area of some 10,000 ha.

The Zebra plain is a vast synclinal which throughout the quaternary was filled in by alluvions and colluvions, usually sandy-argillaceous and nearly always strongly calcareous. Protected against marine influences by the small Riffian chain of the Kebdana, its climate is of the Mediterranean sub-arid type; the average winter rainfall is 300 mm and the mean annual temperature 19°C, but winter frosts are no exception there and in summer the temperature frequently exceeds 40°C; thus the climate, despite the proximity of the sea, is markedly continental. Finally, the natural vegetation which at present covers this plain is a sort of steppe consisting essentially of wormwood (*Artemisia herba alba*), jujube (*Ziziphus lotus*) and plants belonging to the Salsolaceae family.

THE SOILS OF THE ZEBRA PLAIN

The Zebra Plain has been subjected to a detailed pedological survey, having been mapped to a scale of 1:20,000 with the co-operation of J. L. GEOFFROY and CH. MASSONI.

Within the framework of the French classification (AUBERT, 1963) the vast majority of soils found in this plain are brown isohumic sub-tropical soils, more or less ancient as determined usually by the age of the quaternary deposits on which they developed.

The principal physical and chemical characteristics of these soils are as follows:

Chalk content: nearly all the soils are strongly calcareous from the surface downwards; the content may be 10–25% in Horizon A. There is moreover nearly always a chalk accumulation horizon which usually begins at a depth of 40–50 cm. But this accumulation horizon varies enormously from one type of soil to the other; its thickness ranges from 30 to over 150 cm; its chalk content from 30 to 90%; the forms assumed by the chalk in this accumulation horizon are very diverse: spots, granules, nodules, incrustations, crusts, slabs.

Texture: sandy-argillaceous on the surface. But beginning from a depth of 30–40 cm and down to 100–120 cm these soils nearly always show a horizon that is definitely more argillaceous and rubefied: in this horizon the clay content may vary between 35 and 55%. The varieties found are generally illite, chlorite, a little kaolinite and here and there a little Montmorillonite.

Organic matter: these soils have a fairly low content in organic matter, but it is deeply distributed (isohumic distribution): there is 2–2.5% on the surface and 0.5–0.7% at a depth of 50 cm. This organic matter is in an advanced stage of evolution with a carbon/nitrogen ratio equal to or below 10.

Structure: polyhedric to nut-shaped on the surface, more or less fine, sometimes very fine polyhedric in depth, in the clay and chalk accumulation horizon, where the structure of these soils is very well developed. Their structural stability, however, is always very weak, especially in the B horizons; according to HÉNIN's (1960) method the instability indexes are 2–10 on the surface and 10–20 in depth.

Salinity and alkalinity: the Zebra Plain soils are to a very great extent salinated and alkaline. This salinity and alkalinity, however, is hardly ever apparent on the surface, but only from a depth of 30–50 cm and lower.

To take salinity first, the conductivities of horizons B and C oscillate between 8 and 25 mmhos; this corresponds to 3–10‰ total salts, which are essentially sodium chloride. Surface conductivity never exceeds 2–3 mmhos.

Now as regards alkalinity: one first finds that from the surface the soils show high pH figures: pH water 8.4–8.7 and pH KCl 7.4–7.6. Nevertheless despite these pH figures the surface horizons never contain more than 5% sodium on the absorbent complex. By contrast there is often much magnesium: 30% and more.

In the depth, starting from 30–40 cm, pH water is 8.5–8.6 when the soils are salty, but if salinity is low it oscillates between 9.0 and 9.5. Figures for pH KCl range between 7.7 and 8.2. On the adsorbent complex of these B and C horizons the sodium content frequently exceeds 10%, but this is not generally the case and it is not rare to find that for pH figures higher than 9 the adsorbent complex has a low sodium content. By contrast there is always much magnesium: 30–60%.

It should finally be stressed that there is at present no phreatic layer in the Zebra plain; thus the salinity and alkalinity of the soils originated either from ancient pedogenesis, or from alluvial and colluvial mother rocks that could have been deposited in a saline and alkaline condition.

THE PROBLEMS POSED IN RECLAIMING THE SALINE AND ALKALINE SOILS
OF THE ZEBRA PLAIN

The soils of the Zebra Plain, strongly saline from a depth of 40–50 cm, are thus also strongly alkaline; it would appear, however, that this alkalinity, proved beyond doubt by the high pH figures, is not due only to the presence of sodium on the adsorbent complex.

In order to understand the causes of this alkalinity, we undertook detailed laboratory studies of the adsorbent complex, of the soil solution and of the chalk in these soils. Available space does not permit inclusion of details of the tests carried out and results obtained. Some of these results have already been published (RUELLAN 1963, 1964). We shall content ourselves with summarizing the essential conclusions as hereunder:

(1) The alkalinity of the soils of the Zebra plain may be due:

To the presence of small quantities of magnesium carbonate not exceeding 2–3%; we were able to verify several times that this factor alone could easily account for the existence of pH figures between 9.0 and 9.5;

To the presence of 10–15% of sodium on the adsorbent complex.

To the presence of both factors.

(2) It would seem on the other hand that the presence of large quantities of magnesium on the adsorbent complex in no way accounts for these high pH indexes.

Thus reclamation of the soils of the Zebra plain through irrigation poses a certain number of problems on the salinity and alkalinity plane.

First problem: that of washing away the salts, which must be eliminated to a depth of at least 1 m. Can this desalination be done easily and quickly? What method should be used?

Second problem: if this desalination is done too quickly, faster than dealkalization, will it not permit the dispersion of the alkaline clay, the destruction of this very unstable structure, the creation of an impermeable sub-superficial horizon? It should further be pointed out that alkalization is certainly not the factor responsible for structural instability; texture and organic matter also play an important role.

Third problem: is de-alkalization itself possible, since it requires diminution of the sodium percentage on the adsorbent complex simultaneously with diminution of magnesium carbonate percentage?

It is moreover indispensable to verify whether it is really necessary to eliminate this magnesium carbonate in order to secure worthwhile (vegetal) crops, i.e. is it really necessary, particularly with a view to improving the assimilation of mineral elements, to lower the pH of the soils to a figure near to 8.0?

Finally, a fourth series of problems: what irrigation methods, irrigation doses, fertilizers, cultivation techniques, vegetal yields should be applied on these soils – first to improve them and next to secure the best productivity from them?

In order to study and attempt to resolve these problems as a whole, it has naturally been necessary to set up an experimental station, which has now been working for more than five years. Many results have already been obtained: in the following pages, after rapidly describing the water used for irrigation and the principal tests carried out, we shall sum up these results in the field of evolution of soil salinity and alkalinity.

The irrigation water

By means of two dams that great Moroccan river, the Moulouya, is used for irrigation of the Lower Moulouya perimetre.

The principal chemical features of this water are as follows:

Its salinity is not negligible, oscillating between 700 micromhos, a minimum which it attains at the end of the winter and 2,000 micromhos, a maximum reached at the end of the summer. This salinity is therefore fairly strong and in zones already under irrigation we often found considerable salt accumulations – up to 12‰ – at the top of the furrows. After a few years' irrigation, if no precautions are taken, soil salinity may attain 2 to 4‰ in the first 20–30 cm.

As regards the composition of this salinity, it is very favourable for the improvement of alkaline soils. Actually the principal characteristic of the Moulouya water is its high content in earthy-alkaline cations and sulphate anions. In milli-equivalents, the calcium percentage in relation to the cation total is most frequently around 40%; the figure for magnesium is 30% and the sulphate percentage in relation to total anions usually ranges from 40 to 50%. Chloride and sodium content is therefore always rather low, with the sodium adsorption ratio oscillating between 1.0 and 2.5. We may therefore describe the Moulouya water as "de-alkalizing." But for soil improvement in the Zebra plain it has one drawback: it is over-rich in magnesium and we were able to verify through laboratory and terrain tests that its use caused an increase of magnesium on the adsorbent complex of the soils.

Tests carried out

The following principal tests are already carried out or still under way at the Zebra Plain experimental station:

Soil desalination by the following two methods:

"Crash" desalination through irrigation in a very large single dose, and

Progressive desalination by washing at each irrigation.

Soil de-alkalization by the following methods:

Utilization of the Moulouya's waters only;

Surface application of solid gypsum;

Addition of gypsum in irrigation water;

Application of organic matter in the form of manure and green fertilizers;

Application of greatly diluted sulphuric acid in irrigation water.

Comparison between various irrigation methods and particularly between irrigation by sprinkling and gravitational methods of irrigation, as
regards their action on the evolution of the physical and chemical properties of the soils.

Research in cultivation.

Finally, a variety of cultures are tested in order to study simultaneously
their behaviour and their influences on the evolution of the soils. So far the
following cultures have been effected: cotton, sugar beet, sweet pepper
(*niora*), alfalfa, Alexandrine clover, turnip, sugar cane, potato, sunflower.

As a whole these tests are carried out on two of the plain's principal soil
types: a deep-lying sub-tropical brown isohumic soil with very marked
clay accumulation and in which chalk accumulates in the form of spots
and granules, dosing only 25 to 40%, and a shallow sub-tropical brown isohumic soil in which the chalk accumulation, which limits depth starting
at 30–50 cm, is a chalk shell 30 to 60 cm thick with a chalk dosage of 55–90%.

Finally, all these tests have always been preceded and accompanied by
many laboratory tests.

Results obtained in soil desalination and de-alkalization.

(1) Desalination. This may on the whole be carried out easily and quickly.

On deep soils, we obtained the following results: By increasing irrigation
doses 20–30% on the water requirements of the cultures, soil desalination to
a depth of more than 1 m is effected in two years; salinity is stabilized at
between 1 and 1.5‰, i.e. a conductivity of 1–2 mmhos; that is certainly the
minimum conductivity which it is possible to obtain, in view of the salt
content of the irrigation water.

This result may be secured more quickly by means of a 50% increase in
the irrigation doses.

It is obtained immediately through application of a single irrigation dose
of 5000 m³/ha.

On shallow soils, despite previous breakage of the chalk shell, salt removal was shown to be definitely slower, particularly for encrusted horizons and for those situated under the shell. There is nothing surprising in
this in view of the fact that in these soils the general permeability is fairly
low and that water cannot circulate quickly except by way of the sub-
horizontal cracks of the chalk crust and the vertical cracks caused by
breakage.

(2) De-alkalization. De-alkalization, that is lessening of pH figures to a level

around 8.0, naturally proved – in view of the causes of alkalinity – a much trickier job.

It is first of all necessary to stress a very important point: if the maximum desalination which could be obtained for a deep soil is definitely adequate to permit cultivation of most plants, even those which are very sensitive to salts, this desalination is nevertheless not total; and it is thanks to the preservation of a slight salinity, which prevented clay dispersion and subsoil impermeabilization, that sufficient drainage was maintained – a drainage that permitted the beginning of de-alkalization after most of the salts had gone.

As regards the deep-lying soils, the results which we have so far obtained are – briefly summarized – as follows:

(a) To eliminate the sodium from the adsorbent complex is an easy task. By simultaneous study of the evolution of the adsorbent complex and particularly of the evolution of the composition of the saturation extract, we were able to follow sodic de-alkalization very closely; without any soil improvement, one obtains in 18 months, through a 40% increase in irrigation doses, an almost total sodium elimination to a depth of over 1 m. This de-alkalization is accompanied by a slight improvement of the structural stability of the deep argillaceous horizons. One nevertheless finds that, despite sodium elimination, on the one hand, pH figures for horizons B and C remain high: 8.7 to 9.0; on the other hand, the sodium has been partially replaced on the adsorbent complex by magnesium; finally, the structural stability remains very weak.

(b) Surface applications of solid gypsum are almost entirely without effect; they dissolve much too slowly.

(c) Laboratory attempts to percolate the soils with gypsum-enriched Moulouya water showed that sodic de-alkalization is speeded up: there is nothing surprising in this.

There is a considerable desorption of magnesium: the desorbed quantities are noticeably greater than those which can originate from the adsorbent complex only, the gypsum therefore certainly causes a very slow dissolving of the magnesium carbonate; at the end of the experiment, the pH of the earth sample is 8.6–8.7, therefore inferior to that obtained after sodium elimination.

(d) On the terrain, adding gypsum to irrigation water has so far produced noticeably less significant results.

We successively added to the irrigation water 0.3 g/l gypsum, then 0.6 g/l, then 1 g/l. Yet despite precautions of all kinds the gypsum dissolves very badly and actually only very little, and especially reaches the experimental lots very irregularly. As a result, after four years of tests, the results are highly contradictory:

On some lots, pH figures actually fell to 8.6–8.7 according to the laboratory tests and cultivation yields improved fairly markedly;

On the other hand, on other lots no difference was apparent between the control lots and those irrigated with gypsum-treated water.

Anyway this method does not seem to provide the solution: it is expensive and complicated, as against increases in yield that remain mediocre. We are nevertheless continuing the tests.

(e) In the laboratory we have also tried much diluted sulphuric acid. The results are definite: percolation of a sample of earth alkalized with Moulouya water to which is added 0.5 or 0.2 g/l of sulphuric acid causes rapid dissolution of the magnesium carbonate and one pretty quickly gets pH figures of 8.2–8.4 which are more or less normal for calcareous soils.

(f) On the terrain, irrigation tests with water containing 0.2 g/l of sulphuric acid have only just started. We therefore have no results to give. As it seems to us, however, the results of this experiment may be expected to condemn this method. In fact the use of sulphuric acid, which presents many dangers, is very expensive, and it seems unlikely that the increase in the cultivation yields would be such as to cover the costs. It should in fact be stressed that the yields which we obtain at present are far from negligible: 1.5–2 t/ha for Egyptian cotton, 80–100 t/ha (in the green) for alfalfa; yet other cultures (particularly sugar beet) produce mediocre results and especially all cultures still produce very irregular results.

(g) Applications of manure on a considerable scale have so far produced no results either in the pH field or in yields. But it is certainly still too early to draw conclusions: the action of organic matter, if any, can only appear after a very long interval.

Thus if the desalination of the soils of the Zebra plain poses no problems and if on the other hand after five years' irrigation structural instability has not caused the impermeabilization of the subsoil, it appears by contrast that full de-alkalization is very difficult to obtain. But the problem must now be approached from a different angle: is such de-alkalization really indispensable? Would not other cultivation practices such as the application of organic matter, of mineral fertilizers in abundance, of oligo-elements, and choice of judicious crop rotation serve to counteract the harmful effects of these high pH figures? It is in this direction that we now propose to orientate our researches.

SUMMARY

The soils of the Zebra Plain are saline and alkaline from a depth of 30–50 cm. The soils under irrigation can easily be desalinated. Alkalinity, by contrast, which is due simultaneously to the presence of sodium of the adsorbent complex and to the presence of small quantities of magnesium carbonate, is difficult to eliminate. Thanks to the favourable composition of the irrigation water, which is rich in calcium sulphate, sodium elimination is rapid. By contrast, the dissolution of the magnesium carbonate, which is slightly accelerated by adding solid gypsum to the irrigation water, is very slow, and one fails to get pH figures lower than 8.6. Only very strongly diluted sulphuric acid makes possible the attainment of a pH figure of 8.2; but this method seems too expensive. Anyway it is not certain that the elimination of magnesium carbonate is indispensable.

RÉSUMÉ

Les sols de la plaine du Zebra sont salés et alcalisés à partir de 30–50 cm de profondeur. Le dessalage des sols, sous irrigation, est facilement réalisable. Par contre, l'alcalisation, qui est due à la fois à la présence de sodium sur le complexe adsorbant et à la présence de faibles quantités de carbonate de magnésium, est difficile à éliminer. Grâce à la composition favorable de l'eau d'irrigation, riche en sulfate de calcium, le départ du sodium est rapide. Par contre, la dissolution du carbonate de magnésium, qui est légèrement accélérée par l'apport de plâtre dans l'eau d'irrigation, est très lente, et on n'arrive guère à obtenir des pH inférieurs à 8.6. Seul l'acide sulfurique très dilué permet d'atteindre un pH de 8.2; mais c'est une méthode qui semble trop coûteuse. De toutes façons, il n'est pas certain que l'élimination de ce carbonate de magnésium soit indispensable.

REFERENCES

AUBERT, G. 1963. La classification des sols. La classification Pédologique Française. *Cahier de Pédologie O.R.S.T.O.M.* 3: 1–7.
HENIN, S. 1960. Le profil Cultural. 320 p.
RUELLAN, A. 1960. Les sols salés et alcalisés de la plaine du Zébra. *Soc. Sci. Nat. Phys. du Maroc*, Trav. Sect. de Pédol., 13–14: 157–164.
RUELLAN, A. 1963. Etude Pédologique de la plaine du Zébra. O.N.I. Maroc 320 p. (Ronéo).
RUELLAN, A. 1964. Les sols salés et alcalisés en profondeur de la plaine du Zébra; premiers résultats d'une expérimentation destinée à étudier leur amélioration et leur évolution sous irrigation. Commun. au 8ème Congrès-International de la Science du Sol. Bucarest.

Author's adress: Alain Ruellan, Office de la Recherche Scientifique et Technique Outre Mer (ORSTOM), 14 rue de Douaumont, Rabat, Morocco

Freshwater Organisms in Biotopes Influenced by Sea-water [1]

by

MEERTINUS P. D. MEIJERING [2]

The island of Spiekeroog, where the investigations were made on which I wish to report here, is situated within a relatively uniform chain of alluvial sea sand islands off the Dutch and German North Sea coast. Spiekeroog island is about 5 km distant from the East Frisian mainland. The geological profile of the East Frisian coastal region shows a thick ribbon of deep sea sand deposited within the tidal strip on marine clay layers. From these sea-sand layers sand was driven up into a dune-belt forming the northern part of the islands, while the southern part contains relatively flat sand-marshes. The east-west extension of the thus structured kernel of Spiekeroog covers 4 km, the north-south extension 2 km. The sandmarsh lies within the reach of storm-floods which occur mainly during the winter months. On Spiekeroog only 1.7 km² can remain free of salt water in the event of extremely high floods.

Generally the waters of the dune-belt can be reckoned to the limnetic type of water which we can learn from REDEKE's map of Dutch waters (see: REMANE & SCHLIEPER, (1958), p. 9). These are bubbles of fresh water containing extraordinarily few salts and which are able to maintain themselves under the islands amidst pure sea-water. The freshwater bubble under Spiekeroog's waterworks, lying 300 m south of the open North Sea, reaches down to a depth of 63 m. The formation of these freshwater bubbles has not been fully clarified even today, but recently further investigations have been started. The phenomenon is all the more astonishing, because the islands are very unstable in their form and extension. To demonstrate this I should like to show here the overlap of two maps of Spiekeroog constructed by TECHOW. The west contour shows Spiekeroog in 1738, the east in 1951.

1. Dedicated in gratitude to Professor Dr. W. E. ANKEL, Giessen, West Germany, on his 70th birthday, 7.8.1967.
2. I am indebted to Dr. H. BOYKO for his kind invitation to offer the following lecture to this symposium, and to UNESCO and the World Academy of Art and Science for financial support.

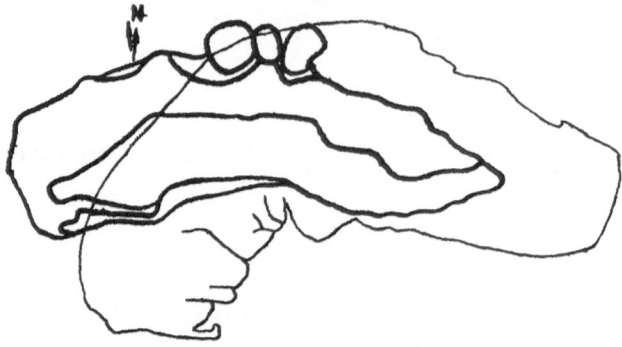

Fig. 1. Overlap of two maps of Spiekeroog.
Thick lines: 1738   Thin lines: 1951

On a fifty-year-old eastern part of the island the depth of the freshwater
bubble was measured and a value of 45 m was found. These amounts may
demonstrate how fast freshwater reservoirs can be formed under the
islands. The rapid formation of these freshwater accumulations can be
explained partly in the fact that sea-sand under the islands contains no
small particles of organic or inorganic origin which would be able to
accumulate marine salt ions. This phenomenon has been frequently
stressed by BOYKO & BOYKO (1959). After the disastrous flood of February 16th,
1962, we saw that the masses of sea-water which broke into the dune-belt
sank down rapidly into deep underground layers. About one year after the
flood we could no longer prove abnormally increased salt contents within
reach of our Hermann Lietz-School drinking-water pumps, which take
water from a depth of 8 m. Thus the freshwater bubbles can regenerate
themselves very quickly when they suffered from encroachments of sea-
water.

Let us now restrict our attention to the uppermost layer of the freshwater bubble. We can see that the principles mentioned above hold for this
biological important layer, and this even to a greater extent. Usually we
find small periodical waters on the surface of Spiekeroog, most of them in
the transitional zone between the dunes and the sandmarsh. Since the
groundwater-table is frequently very low in this sandy type of soil, the
possibility arises that the dry bottoms of ponds can be washed out by rainfall in addition to the general tendency of salt water to sink away. The
regeneration of fresh water after the flooding of the ponds with sea-water
can often be accellerated in this way. Thus we have a very unstable type of
water in respect of salt contents, which I have been investigating in collaboration with J. B. REDFERN since 1961.

Most of the waters of Spiekeroog are of artificial origin. They are or were used by the islanders as drinking-water ponds for people as well as cattle or as water reservoirs for the fire brigade. They have all been dug in recent historical time, and so we are often very well informed about their age. Only a few of them are of natural origin, and are shallow depressions washed out by floods in the sandmarsh. I do not think that any one of them is older than 200 years, and only very few may be older than 100 years. We are thus able to introduce a historical factor into our zoogeographical reflections, which is useful for ecological interpretations.

In our investigations on the waters of Spiekeroog we have great help from the excellent map material available on this island in particular. Our basis is a vegetation map published by WIEMANN & DOMKE (1959), which enables quick topographical inquiries and in addition to this gives us a very good impression of the vegetation surrounding our ponds; from this an impression of life conditions can be derived. The text to this map is still in preparation by Dr. WIEMANN in Hamburg.

The main object of our investigations on Spiekeroog are Cladocera. 25 ponds of the island have a relatively stable population of these animals. During a four year period of investigation only 4 of these ponds were never reached by sea-water. They are situated in the valleys of the inner dunes. Thirteen more ponds lie in the dune valleys; they were, however, reached once on February 16th, 1962, the highest flood in the history of the island which occurred just within the time of our investigations. Sea-water masses broke some dune chains and filled up the valleys behind them. The remaining 8 ponds lie in the sandmarsh, more or less exposed to storm-floods. They are partly in a moor, partly in the higher sandmarsh of the *Sagino maritimae – Cochlearietum danicae*, or in the zone of the *Junco – Caricetum extensae*, where floodings occur more frequently. The salt regions of the lowest part of the sandmarsh have many ponds, but these are only inhabited by Cladocerans as an exception.

From November 1962 to July 1963, BRANDES worked on the salt contents of all these ponds. His findings gave us an impression not only of the rapid regeneration of fresh water in the ponds which I have already mentioned but also of the extremely unstable salt concentrations in the ponds of the sandmarsh. The following table contains the data of some typical ponds which may demonstrate these facts.

On looking at this table we must keep in mind that on February 16th, 1962, sea-water with a salt concentration of 32‰ swept over all these ponds with enormous vehemence. The height of the sea-water-table may have been up to 3 m! Only pond d was not reached by the flood. We can use this pond, then, as a standard for the following regeneration of fresh water in other ponds. During the first 9 months after the flood the salt content in all the ponds diminished to at least β-oligohaline values. Since Cladocera appeared in s in October 1962, this must have also been the case in this pond. The

TABLE I

Pond	Saltcontent in ‰			
	25.11.62	24.3.63	12.5.63	17.6.63
d	0.08	0.07	0.08	0.09
o	0.26	0.06	dry	dry
a	0.35	0.13	0.10	dry
w	1.43	0.08	0.12	0.21
v	2.39	0.29	0.30	0.51
u	0.51	0.71	0.79	1.57
q	1.21	0.04	0.07	dry
s	17.04	0.53	0.69	dry

high value here had its origin in a subsequent inundation which occurred during the last week of October 1962. The best regeneration of fresh water happened in the shallow ponds o and a in the dune-belt. Here BRANDES found limnetic values as soon as November. In general the lowest values were measured in March 1963, 13 months after the flood, when most of the ponds were limnetic and the rest almost limnetic. Even pond s could be reckoned to this type of water, being almost limnetic after a period of only 4 months of regeneration. The ponds in the dune region and pond q which is quite near to the dune type of ponds even showed normal values in March 1963; this is similar to the non-flooded pond d. The small rise in the salt contents towards the following summer was caused by evaporation. Some of the shallow ponds became dry at that time. Only pond a is shaded by bushes. Here the process of regeneration went on until the pond was dry.

TABLE II

Rainfall 1962/63 on Spiekeroog in mm		Mean rainfall on Norderney in mm
1. 3.62–1.12.62	470.5	526
1.12.62–1. 4.63	115.1	178
1. 4.63–1. 7.63	158.9	134

To render the whole complete I can give a table of rainfall during the periods of time between BRANDES' measurements. For comparison the mean values of the East Frisian island of Norderney are listed beside the actual values of Spiekeroog. During the first 13 months after the flood rainfall was below normal, especially during the winter months of 1963. This fact could have helped to restore fresh water so quickly in the shallower ponds such as the newly flooded pond s, since it became almost dry in February 1963; accordingly the bottom of the pond could be washed out before the pond was filled with new water from rainfall. Deeper ponds, however, may profit more from heavy rainfall. Finally it may be clear that all these processes leading to the regeneration of fresh water are only understandable from the high permeability of the subsoil. Otherwise salt ions would be accumulated and dissolved once more in the water by rainfall.

After these remarks on the biotope let us now turn to the Cladocera, which form the leading living-element in the ponds described. Following KAESTNER (1959) we look upon Cladocera as a suborder of Onychura which form the subclass of the Phyllopoda together with the order of Notostraca. Finally the Phyllopoda belong to the class of Crustacea. Probably the best known species of Cladocera is *Daphnia pulex*, one of the six species inhabiting Spiekeroog. The structure of this 3 mm long animal may be perceptible from the picture in KAESTNER (1959, p. 751). For our purposes the reproductive organs are most important. Two long ovaries run parallel to the middle part of the intestine. The caudal vaginal outlet leads to a fold between the abdomen and the shell, the so-called brood pouch. Usually in this brood pouch a greater number of more or less developed young can be found which go back to parthenogenetic subitan eggs. But under the influence of external factors it is possible that resting-eggs are produced which need fertilization. These resting-eggs, one or two in number, are often deposited into ephippia, as in *Simocephalus vetulus* (see BROOKS (1959), p. 616), another species inhabiting Spiekeroog island. These resting-eggs are able to withstand winter cold or dryness. When new suitable life conditions occur, females develop out of these resting-eggs, so starting a new period of parthenogenetic reproduction. According to this scheme population cycles occur on which research has been done for a hundred years.

For a long time the theory was defended that the beginning of gamogenetic reproduction, i.e. the production of resting-eggs, was obligatory, and so fixed to some inherent scheme of time. This view was always contradicted, but it finally exerted itself. We know today that suitable external conditions such as optimal temperature, sufficient food-supply, sufficient life space per animal, and optimal salt contents of the medium can maintain parthenogenetic reproduction for an unlimited time. Only a marked decrease in one of these factors below the optimum causes bisexuality and the production of resting-eggs, bringing to an end what we call a generation cycle. In most cases it is a decrease in the food-supply in connection with diminishing space per animal caused by the rapid multiplication in the population which induces gamogenesis. With MORTIMER (1936) we call this a crowding-effect.

Ecologists usually divide cladocerans into mono-, di- and polycyclic species. I am not against these expressions as long as nothing more is meant than the actually observed issue of a cycle in accordance with the rhythm of seasonal variations in the environment. But we cannot link an imagination of inherent rhythms with these expressions as are found very often in organisms, especially plants. It is of great importance to confirm this if we are to understand the appearance of limnetic cladocerans in marine border regions. Sea-water floodings occur irregularly, and so they are able to alter the scheme of seasonal variations in the environment of a pond thoroughly. Organisms with an inherent scheme of time, with seasonally fixed cycles of activity, would, now and then, come into collision with intolerable salt contents; and this would frustrate their development at an indispensable time. This would mean that such a biotope could not be settled constantly.

The evaluation of our observations after several floodings on Spiekeroog, especially after the tremendous flood of February 1962, made it clear that the resting-eggs of all the cladoceran species on the island can resist sea-water up to North Sea concentrations. This holds even for *Simocephalus vetulus* and *Chydorus sphaericus* which are restricted to the dune-belt. According to this all the species reappeared after the flood. But it was typical that the reappearance of the different species occurred in the various ponds at very different times, this being related to the rapidity of freshwater regeneration in the ponds. In pond a where, as we saw, fresh water was restored very

TABLE III

Species	Pond	Date of reappearance
Daphnia pulex	a	28. 5.1962
,,	w	4. 5.1963
Simocephalus vetulus	w	4. 5.1963
Chydorus sphaericus	a	28.10.1962
,,	w	4. 5.1963
Moina rectirostris	s	19. 9.1962
,,	u	29. 8.1963
Daphnia magna	s	24.10.1962
,,	u	2. 6.1963
Macrothrix hirsuticornis	u	24.10.1962
,,	s	31. 3.1963

quickly, *Daphnia pulex* reappeared as the first species already on May 28th, 1962, only 3 months after the flood. In pond w, however, we did not observe the first specimens of *Daphnia pulex* sooner than May 4th, 1963, since there the regeneration of fresh water took more time. The first specimens of *Simocephalus vetulus* reappeared in w on May 4th, 1963, 15 months after the flood, when we already believed that the species had been exterminated on the island. *Chydorus sphaericus* was found in pond a on October 28th, 1962, but in pond w not earlier than on May 4th, 1963. We found *Moina rectirostris* before the second inundation in pond s on September 19th, 1962, the same species in pond u, however, not before August 29th, 1963. A population of *Daphnia magna* was active in s on October 24th, 1962, but it was refound in u for the first time on June 2nd, 1963. Finally the arctic form *Macrothrix hirsuticornis* reappeared in u earlier than in s; this was on October 24th, 1962, and March 31st, 1963, respectively, because a new flood reached s at the very beginning of the winter when *Macrothrix* is active. These few examples suffice to show that resting-eggs can wait for much more than a year until suitable salt conditions are present in connection with optimal temperatures.

In these two physiological abilities: the resistance of the resting-eggs in water of a high salt content and the ability to link the beginning of a generation cycle with the beginning of suitable conditions in the environment apart from certain seasonal dates, in other words, the ability of cladocerans to insert their own time scheme actively in the irregular rhythm of the environment, go a long way in explaining their capacity to conquer this extreme biotope. An inherent seasonally fixed rhythm would doubtlessly exclude them from our island.

Apart from the resistance of resting-eggs we have to consider the resistance of active animals. The six species inhabiting Spiekeroog are not the same in this respect. Provisionally I should like to use the values of the highest salt contents in which new Cladocera were found after the flood as a measure of the animals' resistance. These amounts mark a physiological limit of great ecological importance. This does not, however, take into account what the uppermost limit of tolerable salt contents is for active animals in slowly evaporating water.

TABLE IV

Daphnia magna	2.53 ‰
Macrothrix hirsuticornis	2.53 ‰
Daphnia pulex	1.71 ‰
Moina rectirostris	0.60 ‰
Chydorus sphaericus	0.35 ‰
Simocephalus vetulus	0.12 ‰

From table IV we can see that four of six species are able to begin their cycles in brackish water of the ß-oligohaline type. Only two species have to wait for limnetic conditions. Accordingly we found the different species spread over different zones on the island. On the sandmarsh, *Daphnia magna* and *Macrothrix hirsuticornis* can spread as far as ponds of the *Puccinellietum distantis*, while only exceptionally *Daphnia pulex* and *Moina rectirostris* can follow as far as the *Junco – Caricetum extensae*. *Chydorus sphaericus* is present in all the ponds of the dune-belt and as an exception in the shallow ponds of the highest parts of the sandmarsh in the vicinity of the dunes. *Simocephalus vetulus* is, however, restricted absolutely to the dune-ponds. These ponds may be reached by sea-water two or three times in a century, but this need not be the end of the species if resting-eggs are present.

It is quite obvious that the flooding of a pond containing an active population of Cladocera will cause the immediate death of all the animals. This holds for all the species on the island. But a flood only becomes dangerous, when it happens at the beginning of a cycle, i.e. when all the animals are already out of the resting-eggs and are busy producing subitan eggs. Then no resting-eggs are present at all. The likelihood that this will happen is different from pond to pond, and it depends on their exposition to floods. For the survival of a species on the island it is of great importance how many ponds this species was able to settle in. We have seen that the rhythm of activity varies from pond to pond, and so a higher flood finds the species in different stages of the cycle. These differences are the results of the different rapidity of freshwater regeneration after a previous flood. Thus it may happen that a population in one pond is exterminated, while another pond containing resting-eggs saves the species. From this latter pond the species can reconquer the first. After a following flood just the opposite may be the case. So the species may spread again after each restriction from centres of resting-eggs in various localities.

Daphnia magna and *Macrothrix hirsuticornis* settle in the same type of pond. Since *Macrothrix* prefers low temperatures for parthenogenetic reproduction, we find young populations mainly during the winter months, when the chance of a storm-flood is much greater than during the summer. *Daphnia magna*, however, prefers higher temperatures. Accordingly newly active populations are to be found in the summer months, when the danger of floods is very small. From these differences in temperature optima the fact results that we could find *Daphnia magna* regularly or as an exception in 28 places, *Macrothrix hirsuticornis* in 16. So *Daphnia magna* is a more stable element on the island than *Macrothrix*. *Simocephalus vetulus* on the other hand will doubtlessly profit from greater safety in the dune-belt. But the restriction to this territory is caused by the fact that the salt resistance of this species is very small. In addition to this the populations act nearly synchronously in the five ponds where we have found the species. Thus they might be exterminated by an extremely high flood, if it were to occur at the beginning of a cycle. But this chance is small, since *Simocephalus vetulus* prefers the summer months for its activity.

To get an impression of the general salt resistance of a species on the island we have to consider several matters: the resistance of active animals in brackish water, the resistance of resting-eggs in sea-water, the number and diversity of the actually inhabited ponds, and the optima of temperature in connection with the likelihood of floods. Chemical research alone cannot draw the whole picture of the settlement of cladocerans in the semi-marine region of the island; we have also to consider the complicated relations in space and time between the organisms and their environment. I have been able to demonstrate only a small section of these relations here. A more complete description of our results is, however, in preparation.

A Frisian island offers very good possibilities for the kind of research, which I have mentioned here. Because of the limited size of the islands it is possible to investigate the fauna and flora in relation to life conditions as one complex. Since the islands are geologically young and still changing in form and extension, we have the opportunity of learning when the minima of life conditions enabling a species to settle on the island are fulfilled. In particular in the case of Cladocera we must keep in mind that migratory birds bring resting-eggs of various species to the island. Until now only six species have successfully invaded Spiekeroog island. In two cases we had the opportunity of proving that the invasion of a species failed. We have the impression of a great experiment of nature. A careful evaluation promises more knowledge of the physiological capacities which help organisms to penetrate these extreme biotopes.

SUMMARY

Freshwater ponds on the Frisian islands are frequently subject to seawater-flooding. Since the underground of the islands contains almost pure sea-sand, the heavy water of high salt-contents usually sinks into deeper underground layers instantly, and fresh water can be restored by rainfall. The result is a region of brackish waters which change their salt-contents permanently, following an irregular scheme of time.

These ponds are inhabited mainly by Cladocera, a group of small Crustacea. Population cycles occur in Cladocera, insofar as parthenogenetic reproduction alternates with the production of gamogenetic resting-eggs. The salt-resistance of active animals suffices for them to bear salt contents up to ß-oligohaline values or less, but the resistance of resting-eggs helps the species to survive in pure sea-water.

The beginning of a population cycle is marked by the development of females from resting-eggs. This happens when suitable life conditions, including suitable salt-contents, are present. Since there is no inherent scheme of time in Cladocera, population cycles can follow the irregular scheme of desalinations. So Cladocera can actively coordinate their periods of parthenogenetic reproduction with periods of low salt contents in different ponds.

Resting-eggs can be transported by birds and cattle. A possible extermination of a species in one pond or another caused by the seawater-flooding of an active population need not be the end of the species on the island, since a resettlement from neighbouring ponds may occur which were not reached by the flood or which contained resting-eggs at the time of inundation.

The consideration of geographical connections, the chemistry, temperature and other conditions of the environment, together with the changings of resistance in different physiological stages of the organisms give us some impression of the complex of methods, which help freshwater-organisms to settle in the extreme biotopes of brackish regions.

RÉSUMÉ

Les étangs d'eau douce des îles Frises font souvent l'objet d'inondations d'eau de mer. Etant donné que le sous-sol des îles contient du sable marin presque pur, l'eau qui est lourde à cause de la haute teneur en sel, d'habitude pénètre immédiatement dans les couches les plus profondes du soussol, tandis que l'eau douce peut être apportée par la pluie. Il en résulte une région d'eaux saumâtres qui changent continuellement leur teneur en sel à des temps très irréguliers.

Les étangs sont habités principalement par des Cladocères, une famille de petits crustacés. Les Cladocères sont sujets à des cycles de reproduction, car la reproduction parthenogénétique est alternée avec la production d'oeufs gamogénétiques qui sont pondus. La résistance aux sels des animaux vivants leur permet de tolérer les teneurs en sel jusqu'aux valeurs ß-oligoalines ou moins, mais la résistance des oeufs pondus permet à l'espèce de survivre dans l'eau de mer à l'état pur.

Le début d'un cycle de reproduction est marqué par la naissance des femelles des oeufs de déposition. Cela arrive quand se produisent des conditions de vie convenables, y compris des teneurs en sel adaptées. Les Cladocères n'ont pas un cycle biologique caractéristique, par conséquent les cycles de reproduction peuvent suivre les schémas irréguliers du dessalement. C'est ainsi que les Cladocères réussissent à coordonner activement leur périodes de reproduction parthenogénétique avec les périodes à plus faible teneur en sel dans plusieurs étangs.

Les oeufs pondus peuvent être transportés par les oiseaux ou par le bétail. Si une espèce est exterminée dans un étang déterminé à la suite d'un débordement d'eau marine qui amène à l'élimination de toute une population active, cela ne signifie pas que nécessairement toute l'espèce a disparu de l'île, car il se peut que dans les étangs avoisinants, qui avaient été frappés par l'inondation ou qui contenaient des oeufs de déposition dans cette periode, une nouvelle population se développe.

Si l'on passe à considérer les liens existant entre la géographie, la chimie, la température et les différents états physiologiques des organismes, nous avons un tableau de tout le processus favorisant l'établissement des organismes d'eau douce dans les biotopes extrêmes des régions saumâtres.

RIASSUNTO

Gli stagni di acqua dolce delle isole Frisie sono spesso soggetti a inondazione di acqua marina. Dato che il sottosuolo delle isole contiene sabbia marina quasi pura, l'acqua pesante per l'alto contenuto salino di solito penetra immediatamente negli strati più profondi del sottosuolo, mentre l'acqua dolce può essere rifornita dalla pioggia. Come risultato abbiamo una regione di acque salmastre che mutano in continuazione il loro contenuto di sali, e ciò avviene senza regolarità di tempo.

Gli stagni sono abitati principalmente da Cladoceri, una famiglia di piccoli crostacei. Nei Cladoceri si verificano dei cicli di riproduzione in quanto la riproduzione parthenogenetica si alterna con la produzione di uova gamogenetiche che vengono deposte. La resistenza ai sali degli animali viventi è sufficiente a far sì che essi possano tollerare i contenuti salini fino ai valori ß-oligoalini o meno, ma la resistenza delle uova deposte permette alla specie di sopravvivere nell'acqua di mare pura.

L'inizio di un ciclo di riproduzione è contraddistinto dalla nascita di femmine dalle uova di deposizione. Ciò avviene quando si verificano convenienti condizioni di vita, ivi compresi gli adatti contenuti salini. I Cladoceri non hanno un ciclo biologico caratteristico, perciò i cicli di riproduzione possono seguire gli schemi irregolari della desalinazione. In tal modo i Cladoceri possono coordinare attivamente i loro periodo di riproduzione parthenogenetica con i periodi di bassi contenuti salini in vari stagni.

Le uova deposte possono essere trasportate dagli uccelli o dal bestiame. Se una specie viene sterminata in un determinato stagno a causa di uno straripamento di acqua marina che ha spazzato via una popolazione attiva, ciò non significa necessariamente la fine della specie nell'isola, dato che può darsi che negli stagni vicini, che erano stati colpiti dall'inondazione o che contenevano uova di deposizione in quel periodo, ricominci una nuova popolazione.

Se consideriamo lo stretto rapporto che unisce la geografia, la chimica, la temperatura e le altre condizioni ambientali con i mutamenti di resistenza nei vari stadi fisiologici degli organismi, abbiamo un quadro dell'insieme di processi che favoriscono lo stabilirsi degli organismi di acqua dolce nei biotopi estremi delle regioni salmastre.

REFERENCES

BOYKO, H. & E. BOYKO. 1959. Seawater irrigation – a new line of research on a bioclimatological plant-soil complex. *Int. J. Biometeor.* 2: 1–24. (Sect. B1).

BRANDES, J. W. (unpublished). Über die Salzgehalte der Tümpelwässer auf Spiekeroog. Hermann Lietz-Schule, Spiekeroog.

BROOKS, J. L. 1959. Cladocera. *In* EDMONDSON, W. T. Freshwater Biology. 2nd Edit. John Wiley & Sons Ltd., New York.

KAESTNER, A. 1959. Lehrbuch der Speziellen Zoologie I, 4. Lieferung. Gustav Fischer Verlag, Stuttgart.

MORTIMER, CL. H. 1936. Experimentelle und cytologische Untersuchungen über den Generationswechsel bei Cladoceren. *Zool. Jb.* Abt. allg. Zool. Physiol. 56: 323–388.

REDEKE, H. C. (cit. after REMANE & SCHLIEPER).

REMANE, A. & C. SCHLIEPER. 1958. Die Biologie des Brackwassers. Schweizerbart'sche Verlagsbuchhandlung, Stuttgart.

TECHOW, E. (unpublished). Versuch, den Grad der Genauigkeit der Horst'schen Karte von 1738 durch Blattüberdeckung mit modernen Karten zu bestimmen. Hermann Lietz-Schule, Spiekeroog.

WIEMANN, P. & W. DOMKE. 1959. Vegetationsübersicht der ostfriesischen Insel Spiekeroog (1:5000). Staatsinst. allg. Bot., Hamburg.

Author's address: Dr. Meertinus P. D. Meijering, Limnologische Flußstation des Max-Planck-Instituts für Limnologie, 6407 Schlitz, Hessen, West-Germany.

PART IV: DESALINATION

Editorial Remarks

This part of the International UNESCO-WAAS-Italy Symposium on saline irrigation with and without desalination was dedicated to the problems of desalination and, in particular, to the extensive work carried out in this field in Italy.

This session was under the chairmanship of Professor Carlo SALVETTI, former President of the Board of Governors of the International Atomic Energy Agency (I.A.E.A.) and Vice-President of the National Committee for Nuclear Energy (C.N.E.N.).

Introductory Address

by

CARLO SALVETTI

Ladies and Gentlemen,

First and foremost I wish to thank the Organizing Committee of this Symposium and in particular the National Academy of Agriculture, its President and Vice-President for the honour which has been bestowed upon me, asking me to preside at to-day's meeting of which I am very proud.

Being a physicist, I am a layman in the field of growing plants under saline irrigation, but the problems to be discussed at today's session are in direct connection with the nuclear energy applications and this justifies my presence here.

As it has often and correctly been observed, immense stretches of land of our globe are acutely feeling the lack of sweet water, in domestic use, in industry and agriculture.

As you very well know, the absence or presence of such water is an important factor and in most cases, has a direct bearing on the progress of many undeveloped areas which are doomed or would be condemned because of their aridity, if the deficiency of this precious liquid is not remedied.

The only solution to this problem – as agreed in many cases – is the supply of drinking water by desalting brackish water: many desalting units of moderate output, i.e. a few thousand cubic meters per day, are already in use in many parts of the world.

The problem we are facing to-day is to increase the output of such units and at a lower cost; to improve them in order to build units capable of producing thousands of cubic meters of drinking water at a minimal cost; the general opinion is that in this field too, nuclear energy can play an important and decisive role.

Large efforts in this direction are made in many countries and many international agreements such as that between USA and Russia recently signed, are a proof of the worldwide interest in this activity and in this method of generating fresh water.

The problems of water desalting have also been dealt with in Italy for some time: it is with real pleasure that I wish to point out the work and efforts of Senator MEDICI, especially while he presided at the National Committee for Nuclear Energy. I wish to stress his efforts in Italy to encourage and boost research for water desalting through the use of nuclear energy.

As you know, many areas in our country will have to employ such new processes, in order to ensure – in the near future – the drinking water needed to maintain an adequate standard of well-being. The National Council for Research and many national industries are tackling the problem raised by desalting techniques.

The National Nuclear Committee, in collaboration with University Institutes, and a few industries, have been tackling the problem of application of Nuclear Energy for this new peaceful purpose. In this connection I should like to mention particularly the University of Pisa and the industries of Bombrini-Delfino-Parodi, the Montecatini Group and the Società Sori.

It is also worth mentioning that plans for a steam producing nuclear reactor using an organic liquid are about to be completed. This reactor has proved convenient and thus could be used at a competitive cost, in smaller dimensions than those now under study.

The specific purpose of this new study is to see whether such a particular reactor – using organic fuel – can be used to set up desalting units of small dimensions, compared to the giant types with which it is now proposed to exploit nuclear energy as a source of heat.

I wish to point out that for such particularly large installations which are now under study in the USA, huge financial investments are required. They can therefore be set up in areas only where special conditions prevail.

According to our opinion there is a definite interest to look into the matter of building desalting units of medium size, based on nuclear energy, on nuclear reactors, for the production of drinking water for medium urban centers, or for very particular uses even in the agricultural sector.

The two main lectures to be held prior to the excursion to the nuclear reactor will deal with different techniques and different sources of energy.

Engineer Dr. DI MENZA, sponsor and Director of our project, which uses nuclear energy in desalting sea-water, will, I believe, submit to you valid arguments to uphold the causes for which this particular type of reactor has been chosen and will indicate the prospects opened to us.

In the other lecture, Professor Giorgio NEBBIA of the University of Bari will speak about the conversion and transformation of brackish waters on which he has conducted experiments since 1953. Assisted by the National Council for Research, he has been studying ways of using sun rays for distillation of brackish waters in order to supply drinking water to small communities.

Of late, he has been looking into the economic problems of water and water desalting; I wish to add here that the Technology Institute of the University of Bari devoted much effort in experimenting, in particular, on membranes and inverted osmosis.

I am convinced that this morning's meeting will give us an opportunity to listen to interesting arguments on the latest techniques in water desalting and discuss their relative problems, thereby enabling us to reach, if possible, the most appropriate solutions.

It is in this hope that I have the pleasure to open the proceedings of this session.

Economics of the Conversion of Saline Waters to Fresh Water for Irrigation

by

Giorgio Nebbia

Humanity is facing an always increasing scarcity of water of satisfactory quality (what we call *fresh water*), not only for domestic and industrial needs, but also for agriculture.[13]

In various areas the possibility of considering saline underground waters and sea-water as a source of fresh water for domestic and industrial uses through desalination is today already a reality and is everywhere considered with increasing attention.[23]

In the present paper, after a short reference to the theoretical minimum energy for obtaining fresh water from sea-water, an analysis is presented of the cost of desalinated water and of the factors from which we may expect some reduction of these costs in the future.

The problem if and where desalinated water may be used for irrigation is also discussed.

THE MINIMUM ENERGY REQUIREMENT FOR DESALINATION

In order to have a basis for the evaluation of the cost of fresh water and of the various components of such cost it is opportune to consider the minimum energy requirements for obtaining fresh water from saline waters; this value depends on the thermodynamic properties of the water to be desalinated.

The minimum energy requirement for obtaining a weight unit of pure water from an infinitely large amount of sea-water has been analyzed in various papers[8, 12, 21] and has been reported as 0.64–0.68 kcal/kg, figures frequently indicated as 2.8 kWh/1000 gallons[21] and 2.98 kWh/1000 gallons,[8] respectively.

In this Laboratory a critical analysis has been carried out on the available data on the thermodynamic properties of average sea-water[14] and our figure for the minimum energy requirement for the above process is somewhat lower, i.e. 0.61 kcal/kg. (2.7 kWh/1000 gallons) at 25 °C.

If some concentration of the sea-water occurs during the fresh water production the minimum energy requirement increases. If saline waters, having a total dissolved salts content less than that of sea-water, are desalinated, the minimum energy requirement is less than the above quoted figure.

Anyway the real existing desalination plants consume an amount of energy which is at least 70 times greater than the theoretical minimum, and this energy consumption is one of the causes for the high costs of desalinated water.

THE COST OF DESALINATED WATER

The production of fresh water from saline or sea-water is subjected to the general laws of any industrial production of commodities.

The desalination is essentially an energy-consuming process and therefore the cost of fresh water obtained by desalination depends on two main parameters: (a) the plant cost, and (b) the energy consumption and cost.

The analysis of the cost for the various desalination processes has been treated in various papers [9, 17-22] that have proposed optimization procedures for identifying the minimum possible cost of fresh water.

In a general form the water cost equation may be represented as:

$$C = \frac{C_p\, a + C_M}{P} + C_E\, e \qquad [1]$$

where:

C: water cost, Lire/m³ *
C_p: plant cost, Lire
a: capital interest and amortization rate, per year
C_E: energy cost, Lire/kcal
e: energy consumption, kcal/m³
C_M: operation, maintenance, labor and other costs, Lire/year
P: water production, m³/year

C_p and e are related in that the lower the energy consumption, the higher the exchange surface and the higher the plant cost. In a first approximation it is possible to express C_p in the simple form:

$$C_p = A + \frac{B}{e}$$

---

* In the present paper Lire indicates Italian Lire. The conversion to the figures in U.S. dollars, which are perhaps more familiar to some of the readers, may be made on the basis of an official exchange rate of 625 Lire/dollar, although this conversion leads to only approximate results. Thus 100 Lire/m³ corresponds to 0.61 U.S. dollars/1000 U.S. gallons and to about 200 dollars/acre-foot; 1 Lire/kcal corresponds to 0.40 dollars/1000 Btu.

where A is expressed in Lire and B is a parameter representing the part of the plant cost which is related to the energy consumption and B is therefore proportional to the exchange surface and is expressed in Lire. $\dfrac{m^3}{kcal}$.

Equation [1] then becomes:

$$C = \frac{\left[A + \dfrac{B}{e} \right] a + C_M}{P} + C_E \, e \qquad [2]$$

In order to obtain the minimum unit water cost it is necessary to differentiate Equation [2]; one obtains:

$$\frac{dC}{de} = -\frac{aB}{P\,e^2} + C_E = 0 \qquad \text{if } e = \sqrt{\frac{aB}{P\,C_E}}$$

The minimum water cost, then, will have the form:

$$C_{min} = \frac{a\,A + \dfrac{a\,B}{\sqrt{\dfrac{a\,B}{P\,C_E}}} + C_M}{P} + C_E \sqrt{\frac{a\,B}{P\,C_E}} =$$

$$= \frac{a\,A + \sqrt{a\,B\,C_E\,P} + C_M}{P} + \sqrt{\frac{a\,B\,C_E}{P}} \qquad [3]$$

This result indicates that any decrease in the energy cost, C_E, and in that part, B, of the plant cost which depends on the specific energy consumption, e, (surface of the heat exchangers), causes a decrease of the water cost which is not linear but depends on the square root. Thus a reduction of 50% of any of these parameters causes a reduction of less than 30% in the water cost.

A sensible reduction of the water cost may be obtained through a reduction of that part, A, of the plant cost which does not depend on the specific energy consumption, for instance through a reduction of the cost of construction materials.

The equations for the water costs are furthermore complicated because the plant cost depends also on the water production and on other variables (e.g. it increases as energy cost increases), but the present approximation can be used for a semiquantitative analysis.

The above analysis is valid both for the distillation and the electrodialysis process (in this case the main energy consumption is due to electric energy instead of heat) and has a more general validity in the field of desalination processes since an increase of efficiency and a decrease of specific energy consumption require in general more complicated and expensive plants.

In order to obtain the order of magnitude of the water cost, C, calculated according to Equation [2], the results for two *multiflash* distillation plants are presented, on the basis of the cost data available from the literature [13].*

Case 1

$P = 5,000 \text{ m}^3/\text{day} = 1,500,000 \text{ m}^3/\text{year}$

$$C_p = A + \frac{B}{e} = 400,000,000 + \frac{30 \cdot 10^{12}}{e} \text{ Lire}$$

$a = 0.10/\text{year}$

$C_M = 24,000,000 \text{ Lire/year} = 16 \text{ Lire/m}^3$

e, kcal/m³	50,000	75,000	100,000	
C_p, Lire	1 billion	0.8 billions	0.7 billions	
$C_E, \dfrac{\text{Lire}}{\text{kcal}} \cdot 10^3$	C, Lire/m³			
0.5	108	107	113	*
1.0	133	145	163	
1.5	158	182	213	

By applying Equation [3] in the conditions of the line marked by an asterisk, one may verify that the optimum energy consumption, e, is 63,000 kcal/m³ and the minimum water cost is 106 Lire/m³.

Case 2

$P = 200,000 \text{ m}^3/\text{day} = 60,000,000 \text{ m}^3/\text{year}$

$$C_p = 8,000,000,000 + \frac{500 \cdot 10^{12}}{e} \text{ Lire}$$

$a = 0.10/\text{year}$

$C_M = 600,000,000 \text{ Lire/year} = 10 \text{ Lire/m}^3$

e, kcal/m³	40,000	50,000	60,000
C_p, Lire	20.5 billions	18 billions	16.4 billions
$C_E, \dfrac{\text{Lire}}{\text{kcal}} \cdot 10^3$	C, Lire/m³		
0.4	60	60	61
0.6	68	70	73
0.8	76	80	85

* One may compare the real energy consumption of the existing distillation plants (40,000–100,000 kcal/m³) with the minimum energy requirement (610 kcal/m³).

The plant of Case 1 has a production in the range of the already available units; the plant of Case 2 has dimensions which will be achieved in the next five years and for which projects are already being prepared.

The low figures of energy cost may be obtained in dual-purpose plants, i.e. plants in which the energy to the desalination plant is supplied as exhaust steam from a thermoelectric power plant, operated by fossil or nuclear fuels.

The sale of the electric energy may pay a part of the operation cost of the desalination plant and the thermal energy for such a plant may be accounted for the rather low costs, in the order of 600 Lire/1000 kcal or less.

The importance of dual purpose plants has been treated in various papers [3, 13, 17, 25] and the perspectives for the use of nuclear energy for such plants have also been the subject of various economic analyses.[2, 4, 15]

At least two projects for large dual purpose desalination and power plants are under development for Israel [16] and Southern California [6]; the production will be of some hundred thousands m³ per day of fresh water from the sea.

The results of such analyses show that in the near future a cost of 50 Lire/m³ (0.30 dollars/1000 gallons) for fresh water obtained by desalination can be reached with the developments already known.

In addition new processes may be developed that will cause some further small decrease of the water cost.

WATER DESALINATION AND REGIONAL DEVELOPMENT

The conversion of saline water to fresh water is an engineering achievement essentially intended for the help to arid countries.

The first desalination plants have been built to give water to communities and industries in desert areas where no other source of water was available.

Presently the great desalination plants in project can be considered as means for integrating the existing, but scarce, resources of natural waters as a part of large development programs for regions populated by millions of people. Such plants will produce power together with water and the amount of power and water to be produced must be carefully planned considering the already existing power and water facilities.

The impact expected from desalination in regional development of particular areas has been treated in various works.[13, 16, 23]

The desalinated water must be distributed in large aqueducts like natural waters today; it can be expected that the main users of desalinated water will be industries and communities that need water with particular qualities of salinity and purity and that are prepared to pay relatively high costs, being already accustomed to pay for water between 10 and 30 Lire/m³ (0.06 to 0.18 dollars/1000 gallons), and in some cases even higher costs.[1, 10, 13, 24, 26]

The main market for the desalinated water lies in those more and more numerous areas in which an insufficient amount of water is a limiting factor for development.[23]

PERSPECTIVES FOR THE USE OF DESALINATED WATER IN AGRICULTURE

The market for desalinated water in agriculture is by far less bright than the one outlined above.

In the first place in general irrigation does not need high quality water, the main limiting factor being the salinity which must not exceed limits that, however, are two to five times those tolerated for industrial and domestic water, with the exception of few salinity sensitive crops, such as citrus.

Secondly the cost of water for irrigation is in general very low and less than 10 Lire/m³ (0.06 dollars/1000 gallons, or 20 dollars/acre-foot); often such cost is less than 2 Lire/m³ and farmers cannot accept costs much higher than these.

In such conditions the gap between 50 Lire/m³, an optimistic figure for desalinated water, and 10 Lire/m³, a pessimistic figure for irrigation water, seems too wide for hoping a reasonable use of desalinated water for irrigation on a large scale in the next decade or two.

However there are particular cases in which the desalinated water and its cost may be accepted in agriculture.

One of such cases is that of Guernsey,[23] one of the Channel Islands, where economy is based on horticulture, with almost 500 ha of glass-houses in which primarily tomatoes and flowers are grown. In 1960 a distillation plant, with a production of 2000 m³/day, has been constructed and the desalinated water is partly used for the irrigation of the high-value crops, also at the present high cost of more than 200 Lire/m³. The cost charged by the Water Board to the consumers is, however, maintained within reasonable economical limits.

It is possible that other more or less numerous cases exist in which desalinated water may economically be used for irrigation at least of high-value crops.

It may be noted that where saline groundwaters exist, the water obtained by distillation, having no salt content, may be added to such saline waters to reduce their salinity to acceptable limits, thus increasing the total amount of available low salinity water.

When saline waters are present in the ground, however, the electrodialysis desalination process presents some advantages. In this process the energy consumption depends on the amount of salts removed from the saline water, differing in this respect from the distillation process in which the energy consumption is independent from the salinity of the treated water. When electrodialysis is applied just for reducing a little the salinity of natural waters the energy consumption is low and the water cost is relatively low, too.

Israel offers a very interesting example of a country in which better citrus crops may be obtained through irrigation with water obtained by partial desalination of the existing natural waters in the ground. An analysis of the economics of the agricultural applications of desalination in Israel has been recently published.[11]

This analysis shows that sensible savings are realized if the citrus crops are irrigated with desalinated water and desalination units are tested in a few kibbutz-settlements.

Desalinated water may also give an indirect contribution to agriculture offering better living conditions to workers in the field; this is done in Israel [11] and the Soviet Union [5] with electrodialysis units, fixed or mounted on trucks.

CONCLUSION

Although the desalination processes cannot presently, and in a near future, give a contribution to agriculture offering low cost water for irrigation, their importance in the conservation and development of the living and industrial level in a great part of the world is expected to increase year by year.

This is due to the fact that the cost of natural waters for domestic and industrial use, and also for irrigation is expected to increase with the increase of the demand and the depletion of the more accessible and cheap sources presently used, and that the cost of desalinated water tends to decrease following technological improvements and new inventions.

The population statistics show that the whole world must eventually depend on desalination for a substantial part of its water requirements.[7]

It is to be hoped that this fact will be realized by engineers, manufacturers, public administrators and the public itself because most of our future availability of water will depend on the technological and economical developments of the art of desalination and on its acceptance as one of the supply systems for water which is a commodity, an industrial product and, at the same time, a most important public service.

REFERENCES

1. E. ACKERMAN & G. O. G. LÖF. 1959. Technology in American Water Development. The Johns Hopkins Press, Baltimore.
2. S. BARON. 1964. Economics of Reactors for Power and Desalination. *Nucleonics*, 22, (4): 67–71.
3. BECHTEL CORPORATION. 1964. Cost Studies of Large Multistage Flash Saline Water Conversion Plants. Office of Saline Water Research and Development Progress Report No. 116.

4. CATALYTIC CONSTRUCTION COMPANY. 1964. A Study of Desalting Plants (15 to 150 mgd) and Nuclear Power Plants (200 to 1500 MWt) for Combined Water and Power Production. Office of Saline Water Research and Development Progress Report No. 124.
5. Chemical Engineering. 1965. p. 25 (Jan. 2).
6. Chemical Engineering. 1965. 72 (16): 46–48 (August 2).
7. R. COLAS. 1962. De l'éventualité et de la necessité de fabriquer de l'eau douce à partir de l'eau de mer. Dechema Monographien 47: 5–15. Verlag Chemie, Weinheim.
8. B. F. DODGE & A. M. E. ESHAYA. 1960. Thermodynamics of Some Desalting Processes. Advances in Chemistry, Series 27: 7–20.
9. R. B. EVANS & M. TRIBUS. 1965. Thermoeconomics of Saline Water Conversion. I&EC Process Design and Development 4: 195–206.
10. J. HIRSCHLEIFER, J. C. DE HAVEN & J. W. MILLIMAN. 1963. Water Supply Economics, Technology and Policy. The University of Chicago Press, Chicago.
11. R. MATZ. 1965. Desalination of Sea and Brackish Water. The Present Status of the Art in Israel. p. 87–108. In Acqua dolce dal mare. Tamburini Editore, Milano.
12. G. W. MURPHY. 1956. The Minimum Energy Requirements for Sea Water Conversion Processes. Office of Saline Water Research and Development Progress Report No. 9.
13. G. NEBBIA. 1965. Il problema dell'acqua e la trasformazione delle acque salmastre in acqua dolce. Cacucci Editore, Bari.
14. G. NEBBIA, E. M. PIZZOLI & O. DE MARCO. Alcune proprietà fisiche dell'acqua del mare tale e quale e concentrata. (in press).
15. OFFICE OF SCIENCE AND TECHNOLOGY. 1964. An Assessment of Large Nuclear Powered Sea Water Distillation Plants. Washington.
16. Report of the United States-Israel Desalting and Power Team. 1964. Washington.
17. R. S. SILVER. 1962. A Review of Distillation Processes for Fresh Water Production from the Sea. Dechema Monographien 47: 19–42. Verlag Chemie, Weinheim.
18. R. S. SILVER & W. K. J. WEIR. 1961. Fuel, Heat and Energy Aspects of Sea-Water Distillation. J. Inst. Fuel. p. 514–521.
19. H. C. SIMPSON & R. S. SILVER. 1963. Technology of Sea Water Desalination. National Academy of Sciences. National Research Council p. 387–413. (Publ. No. 942).
20. M. TRIBUS & R. EVANS. 1962. Thermo-economic Considerations in the Preparation of Fresh Water from Sea Water. Dechema Monographien 47: 43–71. Verlag Chemie, Weinheim.
21. M. TRIBUS & R. EVANS. 1963. Thermoeconomics of Sea Water Conversion. Department of Engineering, University of California, Los Angeles, (Report No. 62–53).
22. M. TRIBUS, R. ASIMOW, N. RICHARDSON, C. GASTALDO, K. E. ELLIOTT, J. CHAMBERS & R. EVANS. 1960. Thermodynamic and Economic Considerations in the Preparation of Fresh Water from the Sea. Department of Engineering, University of California. Los Angeles, (Report No. 59–34).
23. UNITED NATIONS. 1964. Water Desalination in Developing Countries. New York.
24. U.S. DEPARTMENT OF AGRICULTURE. 1955. Water. The Yearbook of Agriculture 1955. Washington.
25. U.S. DEPARTMENT OF THE INTERIOR. 1963. Memorandum Report on Combination Electric Power and Sea Water Desalting Plants in Arid Low Cost Fuel Areas. Washington.
26. Water Resources, Use and Management. 1964. Melbourne University Press, Melbourne.

Author's address: Professor Dr. Giorgio Nebbia, Laboratorio per lo Studio delle Fonti di Energia, Istituto di Merceologia, Università degli Studi, 70122 Bari, Italy.

The Application of Nuclear Energy to Sea-water Desalination

by

RAFFAELE DI MENZA

For some time now, one of the most discussed problems, especially in technical circles, has been the problem of supplying sufficient quantities of fresh water to the arid zones of our planet.

Until some months ago, lack of fresh water was mostly associated with underdeveloped areas where additional quantities of water seemed to be required, in the more or less distant future, according to development plans now under way.

However, something which occurred this last summer brought to light the problem of the ever diminishing availability of fresh water in a most dramatic way. It happened, significantly, in a city which epitomizes, without doubt, the most complete symbol of modern society's development and progress: New York. In this city, as indeed in the State which includes it, and in New Jersey (the latter renowned until recently for the good quality of its fresh water) there has taken place, during the last few months, a continuous and alarming decrease of fresh water availability for all purposes (domestic, industrial and agricultural); so much so that the competent authorities have been compelled to introduce drastic regulations in order to prevent any wastage of this ever more precious liquid. In the State of New York a first step has been taken towards partly alleviating this situation by initiating the establishment of a nuclear power desalination plant – which will be briefly described further on.

An equally alarming situation prevails also in several other zones of the United States. It should be noted, however, that even if the quantity of fresh water normally available for each inhabitant there tends to decrease, thus arousing concern as mentioned before, the availability of this liquid nevertheless remains several times – in some cases more than ten times – larger than that which has prevailed for years in other parts of our planet.

From studies and research carried out up to now it has emerged that, in order to satisfy the ever-growing hydric needs of many regions, it will be necessary to resort to sea-water desalination – and some progress has already been made in this direction.

Naturally, the problem which has to be solved today is that of refining the desalting techniques to a point at which the product can be obtained at the least possible cost and anyhow at a lower cost than that prevailing today. On the basis of current knowledge this would seem to be feasible in the very near future, and, as we have seen, remarkable efforts are being made towards the achievement of this goal.

I should point out that whenever fresh water is referred to in this paper it is generally taken to mean water for domestic, industrial and agricultural purposes. The quality of the water produced depends entirely on the desalination process adopted; we shall deal mainly with the source of steam necessary for the operation of the plant.

The most advanced modern desalination processes require either thermal or electric power; naturally, therefore, the use of atomic power has also been investigated for this particular application. As a matter of fact, for some time now, nuclear power has been gaining ground as an ever more cheap and promising heat source for the production of electric power.

In many areas today, nuclear plants are in a position to compete economically with conventional ones. The consistent and rapid advances made until now in the technology and running-in of the reactors, as well as envisaged future developments, permit us to hope that very soon these machines will not only be able to compete with but will become cheaper than the conventional sources of high temperature heat for production of electric power.

Whilst, however, in order to attain reasonable efficiency for this particular application it is necessary to produce steam at the highest possible temperatures – 250°C and higher – for the desalination of sea-water by distillation methods (which are today the most convenient), steam is required at about 150°C only. Because of this difference in temperature, the nuclear reactors may be employed either for the combined production of both electric power and fresh water, or for the production of fresh water only.

In the former, currently designated as double purpose plants, the steam produced is first utilized at high temperature and pressure in the turbine and, subsequently, at low temperature and pressure in the desalting plant. These multiple purpose systems, besides making for better utilization of the steam, have the advantage over those producing only electric power in that they can afford to eliminate some components such as, for instance, the turbine condenser and the related cooling water circulation system. Moreover, they make possible the creation of larger plants which, as will be shown later, require a lower capital investment.

On the other hand, they may present some disadvantages too – mostly operational ones – because the demands of fresh water and of electric power are frequently opposed; and the maintenance requirements of the desalination plant as well as other parts of the installation are not always in accord, thus leading to a waste of energy.

An example of a multiple-purpose nuclear reactor is that commissioned, some weeks ago, by the State of New York from American Machine and Foundry. This nuclear reactor will supply power for the production of about 4,000 m³ fresh water per day, 2,500 kw electric power and radioactive isotopes for industrial purposes. The cost of the plant, according to first available estimates, should be about 2,500 million Lire [1] and that of the water 60 Lire/m³. The latter, as can be seen, is comparable to that prevailing in some zones of Southern Italy.

The other method of utilizing nuclear energy for desalination is to employ a reactor for producing steam only – possibly at a low temperature – to be used directly in the distillation plants. Plants of this kind, called single-purpose plants, could be applicable in those areas where there is no additional request for electric energy and where fresh water has to be supplied in such large quantities as to justify, economically, a nuclear plant (e.g. as seems to be the case in Sicily today).

Understandably, in order for this particular application of the reactors to be economically viable, special machines would be required for the production of low temperature steam only, to be utilized directly in the distillation plants. These machines, of course, because of the lesser performance required of them, should be simpler and hence less costly than those now required for the production of electric power.

However, a reduction of operational temperature and pressure in the majority of the reactors on the market today does not necessarily imply a substantial reduction in the capital cost of the plant; thus it is more advisable to have double-purpose machines in sizes preferably beyond 1,000 thermal kw.

The aforementioned propositions are not altogether true for reactors cooled and moderated by organic liquid, which have not been widely developed up to date. Indeed, for this type of reactor, because of the characteristics or properties of the organic liquids used, the low temperature operation allows substantial modifications of the installation, especially where the choice of materials is concerned, thus reducing the capital cost and that of the thermal energy produced.

The Italian National Nuclear Energy Committee, making use of the experience acquired through the Organic Reactor Programme, has under study, in

1. In the present paper Lire indicates Italian Lire. The conversion to the figures in US dollars may be made on the basis of an official exchange rate of 625 Lire/dollar, although this conversion leads to only approximate results. Thus 1 million Lire corresponds to $ 1600.–, 100 Lire/m³ to $ 0.61 /1000 US gallons and 1 Lire/kcal to $ 0.40/ 1000 Btu.

collaboration with the University of Pisa and some national industries, a reactor of the type described above – for desalination only. At present we have not quite completed its economic evaluation in order to compare it with the double purpose and conventional plants, especially in relation to areas particularly suited to installation of single purpose plants. As we have said before, the energy required for converting salt water into fresh water can be supplied by conventional plants as well as by nuclear reactors.

As a rule, for relatively small projects, steam generators burning conventional fuel are more economical than nuclear reactors (as, by the way, in the production of electric energy). With increased dimensions, however, the situation is reversed. An example of a conventional installation is the desalination plant recently commissioned at Taranto, by the ITALSIDER Works, whose output is about 4,000 m³ fresh water per day, activated by waste steam derived from conventional installations.

It should be noted that, even in the case of conventional burners being used, if it is desired to produce only fresh water, the cost of steam is rather high because use has to be made of components which have been constructed for operating at high temperatures and pressures and therefore for the production of more valuable steam. It therefore follows that, in this case too, it is more suitable to make use of multiple purpose plants.

The sizes up to which conventional generators are more convenient than nuclear ones depend chiefly on the spot cost of the conventional fuel, on the capital outlay, on the entire investment and on other considerations relating to the site chosen.

From recently concluded studies in Israel and in the United States, it would appear that, in the zones under examination, for combined production of electric power and fresh water, nuclear plants become competitive for installations ranging from 200–400 thermal kw. By the way, it seems that the required additional fresh water foreseen in Puglia and other Italian regions is such as to call for larger sized plants than the abovementioned.

On the basis of experience acquired thus far in the field of utilization of nuclear energy for heat production, the larger the reactors the lower is their capital cost and correspondingly the cost of the energy produced.

In fig. 1 there are indicated some data of capital costs of nuclear and conventional power stations for production of electric power. These data are compiled from power stations actually built or from estimates and offers on an international scale.

Of course, what was previously said is valid also when the reactors are used in combination with desalination plants.

In the United States, Israel, England and other countries remarkable efforts are being devoted to the development of a technology of desalination and to the adaptation of nuclear reactors to this new peaceful application. In Italy too, the CNR, the CNEN and some important national

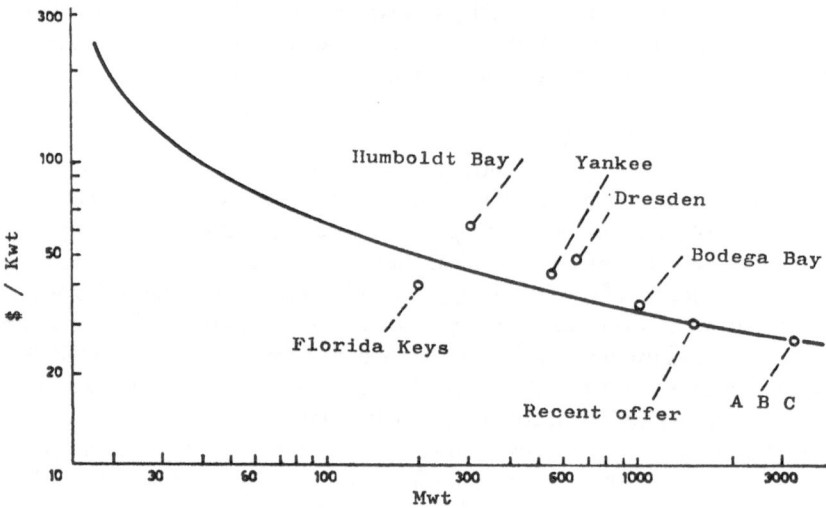

Fig. I. Capital costs of nuclear power stations in US $/thermal kw in relation to thermal potency in Mw.

industries are giving consideration to and confronting the problem of desalination.

In the United States a comparative study has recently been carried out, for the Government and the Atomic Energy Commission, of economic and technical evaluations in respect of desalination plants of various sizes. These are combined with steam generators of 200, 600, 1000 and 1500 thermal kw, powered with conventional or nuclear fuel and used for the combined production of electric energy and fresh water or fresh water only. We deem it useful to present some of the most significant results obtained because, in our opinion, they are sufficiently realistic to indicate what can be achieved in the next four or five years with this new and important application of nuclear energy.

The study is based on the assumption that the plant is to be situated on a given site in South California where, as it is known, there is an ever growing lack of fresh water.

Moreover, it has been assumed – bearing in mind those developments in nuclear technology and in desalination techniques which can be fairly confidently foreseen as attainable in the immediate future – that the plant will start operating in 1970, that it will be of 30 years' duration and will reach an efficiency rate of 80%. The different plants considered were in each case optimized as far as steam utilization is concerned, thus establishing the most economically viable ratios between quantities produced of electric energy and of water.

Amongst today's most developed reactors there were considered the Boiling Water type, the Pressurized Water, the Spectral Shift Control and the Heavy Water type.

The costs were estimated on the basis of information obtained from American constructors and from several design organizations. Wherever possible, costs were related to those of reactors already sold and which will start operation in the period 1968-70.

Conventional plants burn, as fuel, natural gas, the approximate cost of which is 860 lire per million calories. (This is lower than the cost obtainable in Italy.)

Both the above-mentioned nuclear and conventional generators produce steam ranging from 260–345°C (500–653°F).

The chosen water producing plant is of the multistage flash distillation type which can produce desalted fresh water, with salt content up to 1 ppm. The maximum admissible temperature of the brine has been taken as 125°C; the formation of deposits on the thermal exchange surface is controlled by the addition of sea-water with 120 ppm sulphuric acid, followed by vacuum de-aeration.

The cost of equipment construction has been calculated on the basis of current estimate procedure and on the engineering analysis of results hitherto obtained in the plants of San Diego, California.

TABLE I

Dual purpose plants for producing electric power and fresh water.
Fixed annual charges 7%, electric power sold at Lire 2.2/kWh.

Thermal Mw	200		600		1000		1500	
	Conv.	Nucl.	Conv.	Nucl.	Conv.	Nucl.	Conv.	Nucl.
Net electric power Mwe	25.40	21.58	83.11	74.45	144.07	135.24	217.98	214.98
Fresh water produced m³/day	135000	112000	345000	329000	525000	435000	785000	560000
Cost of water in Lire/m³	90	108	72	64	70	56.5	66.5	48

TABLE II

Dual purpose plants for producing electric power and fresh water.
Fixed annual charges 7%, electric power sold at Lire 3.7/kWh

Thermal Mw	200		600		1000		1500	
	Conv.	Nucl.	Conv.	Nucl.	Conv.	Nucl.	Conv.	Nucl.
Net electric power Mwe	30.36	23.34	97.16	85.89	164.63	146.20	246.32	224.61
Fresh water produced m³/day	101000	110500	245000	231000	381000	332000	587000	440000
Cost of water in Lire/m³	84	102	59.5	53	58	42	54	36

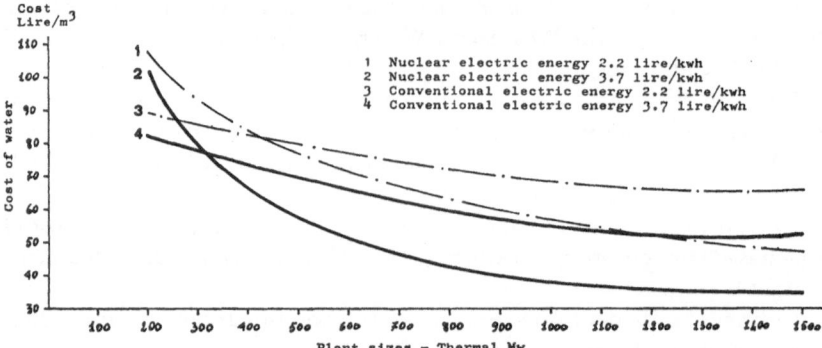

Fig. 2. Cost of water in Lire/m³ in relation to the plant size in thermal Mw for dual purpose plants, fixed annual charges 7%.

As can be seen from Tables I and II and fig. 2, if the electric energy produced is sold at 2.2 lire per kwh (a very low price which must be considered as a goal for the 'seventies) the cost of water, at the plant, would range from 90 lire/m³ for the 200 thermal kw installation with conventional fuel to 48 lire/m³ for that of 1500 thermal kw with nuclear fuel.

The production of water is, respectively, 135,000 m³ per day or 1.5 m³/sec for the 200 kw installation and 560,000 m³ per day or 6.5 m³/sec for the larger plant.

By selling electric energy at 3.7 lire per kwh, a more realistic price, the costs of water are of course reduced and they range from 84–36 lire/m³. As can be seen, even when adding the distribution costs, they are not at all prohibitive and are, I would say, even suitable for some zones.

The assumed yearly fixed charges of 7% include the amortization of the whole plant in thirty years with an interest rate of about 5.75% (or amortization in twenty years with interest rate of 3.5%).

The curves also show that nuclear plants become competitive with conventional ones at sizes ranging from 300–400 thermal Mw. For larger installations, the nuclear plants produce thermal energy and, hence, fresh water at a lower cost. This is mainly because, whilst for small installations the capital investment of conventional steam generators is so much lower than for nuclear ones as to cancel out the advantage of the lower annual cost of nuclear fuel, with the gradual increase in size the differences between capital outlays decrease progressively; at the same time the difference in yearly fuel costs – to the nuclear plant's advantage – is such as to reverse the whole situation.

In fact, for the installations considered in Table II with a credit for electric energy of 3.7 lire per kwh, at 200 thermal Mw, the conventional steam generator has a capital outlay of 860 million lire with an annual fuel consumption of 1,220 million, whilst the nuclear reactor has a capital outlay of 9,700 million lire (more than ten times the former) and consumes, annually, fuel for 580 million. On the other hand, at 1500 thermal kw the capital investment is 4.43 billion lire for the conventional steam generator as against 19 billion lire (less than five times the previous one) for the reactor, with yearly fuel costs of 9.15 billion for the conventional and 3 billion lire for the nuclear.

From fig. 2 it can also be noted how the cost of water progressively decreases with the expansion of installations up to 600–700 thermal Mw plants, gradually levelling out afterwards. This is explained by the fact that, below the said sizes, the increments of capital cost required for every single unit of production are very high particularly for nuclear reactors, while they diminish rapidly over 600 thermal Mw.

In the study under discussion, plants producing water only were also examined. The nuclear reactors (or the conventional steam generators) are not modified for operation at lower temperatures, but they produce steam at temperatures from 260°–345°C (500–653°F) which must then be reduced to about 150°C before it can be used in the distillation plant. Obviously, such a procedure does not ensure the most efficient and economical exploitation of the expensive machines used.

Investigation shows that under these conditions, for plants below 600 thermal Mw, when credit for electric energy is low (under 4 lire/kwh) and capital cost is high (yearly fixed charges of over 8%), the water produced by these single-purpose installations costs less than if derived from double-purpose plants.

In our opinion, and according to studies carried out in Israel, it is even more favourable for plants producing water only, if the reactor could be modified and simplified in order to produce low temperature steam which can then be used directly, without any waste in the distillation plant. A preliminary answer to this problem should be possible from a study of the organic reactor (ROVI) now under way at CNEN, which we mentioned previously. At any rate, this is a highly interesting matter which we ourselves have been studying in detail for some time and it is now being investigated in other countries such as the United States and Israel.

It should be noted, in passing, that if we were to take into consideration nuclear reactors of the most advanced type – which, however, have not yet been proved on a commercial basis – the cost of electric energy and of water would be lower still. But this is a topic which it will only be possible to discuss with any degree of soundness in the more or less near future.

Before concluding, it would be worthwhile examining, in more detail, the two plants whose capacity is most nearly what would be required to satisfy the hydric needs of some Italian regions which are affected by the decreasing supply of fresh water. I refer to the 200 and 600 thermal Mw plants, for which the data are taken from the catalytic report.

TABLE III

Dual purpose plant of 200 thermal Mw, fixed annual charges 4%

Electric power sold at Lire/kwh	2.2		3.72		4.95		6.20	
	Conv.	Nucl.	Conv.	Nucl.	Conv.	Nucl.	Conv.	Nucl.
Total cost of plant in billion Lire	22	36	18.5	33.5				
Cost of destillation plant in billion Lire	17	17.5	13.5	15				
Annual fuel cost in billion Lire	1.22	0.57	1.22	0.57				
Annual cost of steam and electricity generators in billion Lire	0.31	0.55	0.31	0.55				
Annual cost of water destillation plant in billion Lire	0.5	0.51	0.47	0.49				
Net electric power produced, Mw	25.54	19.87	29.77	23.61				
kg fresh water produced per kg steam consumed	20.39	20.41	17.45	18.82				
Fresh water output in 1000 m³/day	122	126	110	117				
Fresh water cost in Lire/m³	73	77	65	71	58	65	50	59

TABLE IV

Dual purpose plant of 200 thermal Mw, fixed annual charges 7%

Electric power sold at Lire/kwh	2.2		3.72		4.95		6.20	
	Conv.	Nucl.	Conv.	Nucl.	Conv.	Nucl.	Conv.	Nucl.
Total cost of plant in billion Lire	23	33.5	17	32.8				
Cost of destillation plant in billion Lire	18	15	11.8	14.5				
Annual fuel cost in billion Lire	1.22	0.58	1.22	0.58				
Annual cost of steam and electricity generators in billion Lire	0.31	0.55	0.31	0.55				
Annual cost of water destillation plant in billion Lire	0.53	0.48	0.47	0.48				
Net electric power produced, Mw	25.4	21.6	30.4	23.3				
kg fresh water produced per kg steam consumed	22.42	17.82	16.59	17.67				
Fresh water output in 1000 m³/day	135	112	101	110.5				
Fresh water cost in Lire/m³	90	108	84	102	76	96.5	71	90

In tables III and IV we find the most important data relating to a 200 thermal Mw plant, with conventional burner and with nuclear reactor.

The total cost of the conventional plant is in the region of 20 billion lire, of which about 15 billion lire, i.e. 75%, is for the distillation plant. Production of fresh water ranges from 100–135 thousand m³/day at a cost of from 65–90 lire per m³, ex plant.

The plant with the nuclear reactor has a much higher total cost – about 35 billion lire – of which about 15 billion, i.e. approx. 50%, is for the distillation plant. Production of water is more or less equal to the other but costs more: from 71–108 lire per m³ ex plant.

TABLE V

Dual purpose plant of 600 thermal Mw, fixed annual charges 4%

Electric power sold at Lire/kwh	2.2		3.72		4.95		6.20	
	Conv.	Nucl.	Conv.	Nucl.	Conv.	Nucl.	Conv.	Nucl.
Total cost of plant in billion Lire	51	57	36.5	43.5				
Cost of destillation plant in billion Lire	42.5	31	28	17				
Annual fuel cost in billion Lire	3.65	1.27	3.65	1.27				
Annual cost of steam and electricity generators in billion Lire	0.36	0.72	0.36	0.72				
Annual cost of water destillation plant in billion Lire	1.0	0.87	0.82	0.68				
Net electric power produced, Mw	82.2	79.9	95.6	89.1				
kg fresh water produced per kg steam consumed	19.13	15.49	14.68	10.89				
Fresh water output in 1000 m³/day	347	293	270	209				
Fresh water cost in Lire/m³	57	43.5	46	34	36			

Tables V and VI present the same data for a 600 thermal Mw plant.

The investment required for the conventional power station is still lower than that for the nuclear reactor but there is not such a great difference as in the previous case.

The cost of the distillation plant represents a percentage of the total investment slightly lower than that relating to the 200 Mw installation. The quantity of water produced ranges from 200–350 thousand m³ per day at costs which reach a minimum of 35 lire per m³ and which are, in any event, lower in the nuclear plant.

TABLE VI

Dual purpose plant of 600 thermal Mw, fixed annual charges 7%

Electric power sold at Lire/kwh	2.2		3.72		4.95		6.20	
	Conv.	Nucl.	Conv.	Nucl.	Conv.	Nucl.	Conv.	Nucl.
Total cost of plant in billion Lire	48.5	61.5	31.5	46.5				
Cost of destillation plant billion Lire	40.0	35.5	23	20				
Annual fuel cost in billion Lire	3.6	1.25	3.6	1.25				
Annual cost of steam and electricity generators in billion Lire	0.36	0.72	0.36	0.72				
Annual cost of water destillation plant in billion Lire	1.0	0.95	0.76	0.73				
Net electric power produced, Mw	83.0	74.5	97.2	86.0				
kg fresh water produced per kg steam consumed	19.02	17.28	13.31	12.10				
Fresh water output in 1000 m³/day	343	326	244	232				
Fresh water cost in Lire/m³	72	64	59.5	53	47	41		

Whereas the conversion ratios adopted – i.e. the ratios of quantity of water produced to quantity of steam – are very high considering the present state of technology in the smaller plant,[1] for the 600 Mw plant sufficiently realistic data were assumed which would be easily attainable in 1970.

From fig. 3 and fig. 4 we can further observe how the costs of water are affected by variations in the credits for electric power produced. In Italy we are undoubtedly within the high cost bracket of the kilowatt-hour and, consequently, within the low bracket cost of water.

It should be further noted how the yearly fixed charges and, hence, for given amortization terms, the interest rates on capital influence the cost of the product. For example, in the 600 thermal Mw plant if we pass from 7% fixed charges (30 years amortization, 5.75% interest rate on capital) to 4% (30 years amortization, about 2.5% interest rate on capital), for a kilowatt-hour credit of 4 lire, the cost of water passes from 50–32 lire per m³.

The interest rate of 2.5%, from a strictly commercial point of view, is low, but it could be obtained through the Government's intervention for an enterprise of such vital importance which would bring to many of the

1. This explains the somewhat inflated quantities of water produced in the different cases.

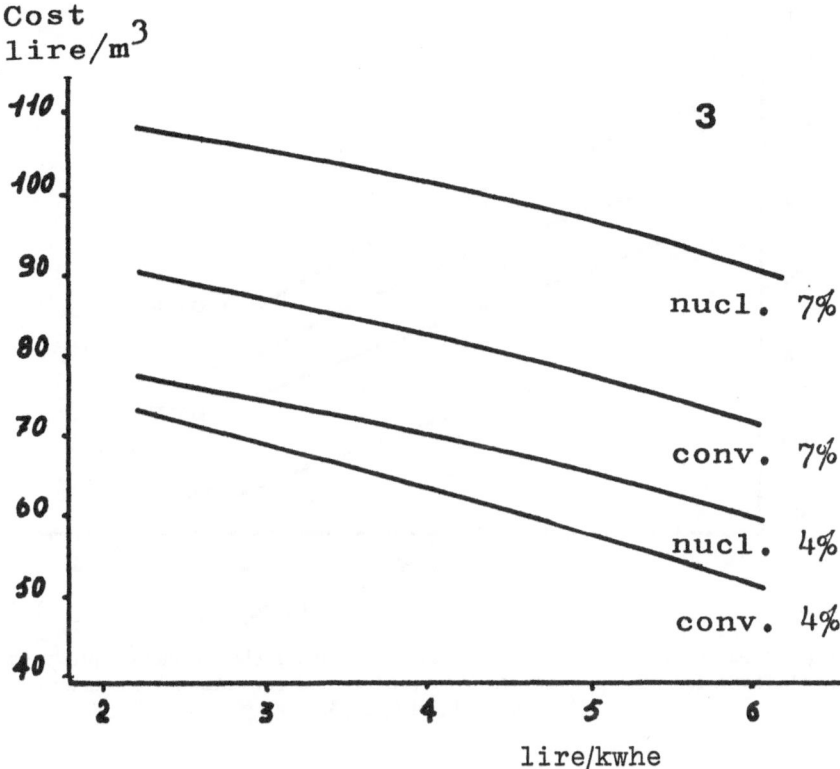

Fig. 3. Cost of water in Lire/m³ in relation to credits for electric energy produced in Lire/kwhe for dual purpose 200 Mwt nuclear and conventional plant, annual fixed charges 4 or 7%.

depressed areas of our country the water necessary to economic development and social welfare.

The study just examined will certainly not be applicable in its entirety to the situation prevailing in our arid zones, but it provides data which, in a comparative way, are highly indicative (technically and economically) of the direction to be followed in the solution of desalination problems.

We are not in a position to say much about the prevailing situation in the arid zones under development in Italy as regards the existing natural hydric resources, the feasibility of their exploitation and the relative timetables and costs. It is probable, however, that in some zones desalination is imperative and in this connection two facts can be pointed out at once.

We have seen from the above-mentioned study that nuclear power stations start to become competitive with the conventional plants, for double-purpose production, for installed power of the order of 300–400 thermal Mw. The assumed conventional fuel cost was 860 lire per million

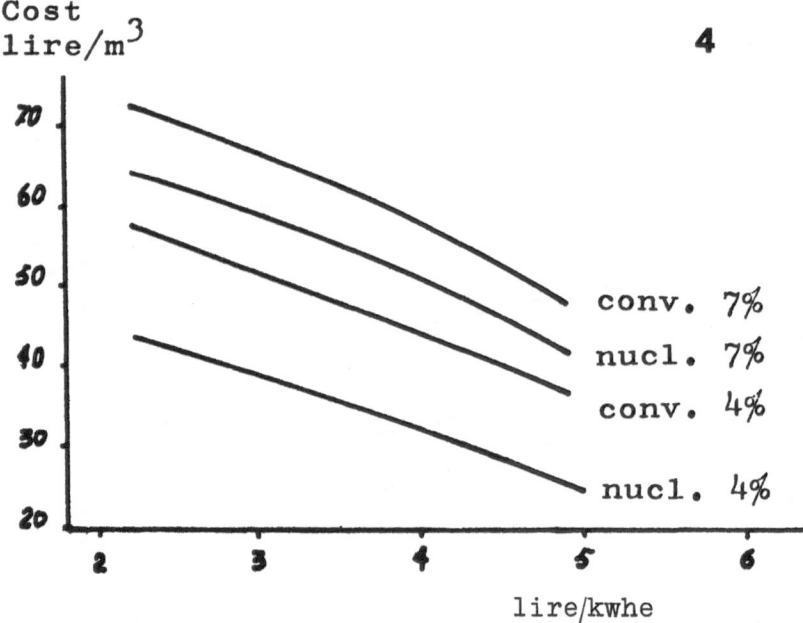

Fig. 4. Cost of water in Lire/m³ in relation to credits for electric energy produced in Lire/kwhe for dual purpose 600 Mwt nuclear and conventional plant, annual fixed charges 4 or 7%.

calories. In Italy the conventional fuel is generally much more expensive (about 1.5 times). This means that the competitive limit of the nuclear reactors becomes even lower thus making it probable that, for Italy, double purpose nuclear power stations or even single purpose plants, in some zones, for production of fresh water only may conveniently be adopted.

The other fact to be pointed out is that, in choosing the site and the size of the plants to be installed, the utmost attention will have to be paid to the problem and cost of the distribution of the water produced. In fact, because of the particular situation of our aqueducts it might be more suitable, all things considered, to install in some regions not a giant plant but several small plants which will produce water in the areas of consumption.

Most certainly, the recently initiated studies – especially those of the National Research Council – and the results of Conferences and highly qualified symposia such as this, will shortly put at our disposal all the data needed for adequately attacking, also in Italy, the problems relating to the supply of sufficient quantities of fresh water for domestic, industrial and agricultural purposes.

SUMMARY

After some general considerations on the ever increasing shortage of fresh water felt in some of the most progressed areas as well as in the under-developed countries, the possibility of using the nuclear instead of the conventional energy for the distillation of sea-water, is being examined.

Information is briefly given on a study, at present under way at CNEN, regarding the evaluation of a reactor moderated and cooled by organic liquids, to be used to produce low temperature steam for only desalination purposes.

Finally the results of a study recently concluded in the United States are illustrated and discussed. From them an idea can be obtained of the desalination plants, nuclear or conventional, that could be built in the near future. Particular care is paid to those plants that could be of interest for application in Italy.

RIASSUNTO

Dopo alcune considerazioni di carattere generale sulla sempre più crescente e sentita carenza della disponibilità di acqua dolce in alcune zone delle più progredite come in quelle sottosviluppate, si esamina la possibilità di sfruttare per la conversione delle acque marine l'energia nucleare in concorrenza con quella convenzionale.

Un breve accenno è fatto ad uno studio attualmente in corso al CNEN su un reattore a liquido organico da usare per sola produzione di vapore a bassa temperatura atto alla desalinazione dell'acqua del mare.

Vengono infine presentati e discussi i risultati di uno studio comparativo tecnico ed economico recentemente concluso in USA, dal quale si può ricavare un'idea di quelli che dovrebbero essere i complessi di desalinazione con sorgenti di calore convenzionale o nucleare che potranno essere realizzati nel prossimo futuro. In particolare vengono considerati impianti di dimensioni tali da poter sopperire ai fabbisogni addizionali di acqua dolce di alcune regioni italiane.

REFERENCES

Water Desalination in Developing Countries. 1964. United Nations, New York.

Di Menza, R. 1965. Progetto CNEN di un Reattore Nucleare Produttore di Vapore per la Desalinazione delle Acque Marine. Relazione presentata alla Tavola Rotonda sulla Desalinazione delle Acque Marine e Salmastre. Palermo.

Civilian Nuclear Power. 1962. Report to the President by U.S. Atomic Energy Commission.

Program for Advancing Desalting Technology 1964. Report to the President of USA from the Department of Interior and U.S. Atom Energy Commission.

Aschner, F. S., S. Yiftah et al. 1965. Feasibility of Nuclear Reactors for Sea Water Distillation.

DI MENZA, R., S. PITTORI, L. PIERAZZI & P. VALANT. 1964. Alcune Alternative per l'Utilizza-
    zione dei Reattori a Liquido Organico per Desalinazione di Acque Saline. CNEN
    Giornate dell'Energia Nucleare. Milano.
Large Nuclear Powered Sea Water Distillation Plants. 1964. Office of Science and Tech-
    nology.
RAMEY, J. T., J. K. CARR & R. W. RITZMANN. 1964. Nuclear Reactors Applied to Water
    Desalting. Third Geneva International Conference on the Peaceful Uses of Atomic
    Energy.
A Study of Desalting Plants (15 to 150 MGD) and Nuclear Power Plants (200 to 1500 MWt)
    for Combined Water and Power Production. 1964. Catalytic Construction Co. and
    Nuclear Utility Services, Inc.

Author's address: Eng. Raffaele Di Menza, CNEN, Via Prisciano 28, Roma, Italy.

Remarks on Some Agrarian Economic Questions Connected with Desalination

by

CARLO BARBERIS

According to my opinion the estimates made in connection with the cost of a cubic meter of desalted water – as estimated today by Prof. NEBBIA and Prof. DI MENZA – should lead us to an entirely optimistic state of mind in connection with the ever-increasing use of desalted water for irrigation and agriculture in general.

I am of the opinion that even with the present costs we find ourselves in a quite advantageous position with regard to the use of desalted water; but, naturally, this depends on many factors that I wish rapidly to review before this Assembly.

I have here a table indicating, for a certain number of crops, the highly interesting answer: What is the maximum cost per cubic meter of water, that each single culture can bear, in order to be irrigated?

These figures have been gathered by Dr. DEGAN of the International Center of Irrigation of Pioggia (Verona); they have an indicative value, but taking into consideration their source, they are absolutely valid as a basic orientation, and the standards set by Dr. DEGAN are of high interest to us. The figures are valid for the spray system of irrigation (for normal flood irrigation systems these figures are 30 to 40% lower, proportionally to the quantity of wasted water).

Taking into consideration the water reserve, irrigated cultures can pay for water consumption up to a maximum of about 10% of the product value for water used, i.e. up to a maximum value of 10% of the gross value of irrigation water produced. It is well known – and this belongs to the economic aspects of fertilizers – that water used for irrigation, determines a certain increase in production which has its proper market value; consequently this increase in output – referred to the cubic meter of water – constitutes the cost of transforming irrigation water.

Therefore, this 10% represent precisely the maximum limit of water transformation cost that the irrigated areas can, economically speaking,

TABLE I

Average cost of water production per m³ for irrigation [1] *(in Italy) and "limit" values for maximum tolerable cost of water*

Cultures	Production increase per ha in million Lire	Seasonal water consumption; m³/ha	Water production gross cost per m³ in Lire	Cost limit per m³ of water in Lire
(1) Flowers & hot-houses	10–20 [2]	5,000–8,000	2,000–2,500	200–500
(2) Intensive vegetable garden & citrus	2–4 [2]	4,000–6,000	500–660	50–66
(3) Tobacco	0.5–1.5	1,000–3,000	250–500	25–50
(4) Apple-trees, pear-trees	0.4–1.2	1,000–4,000	400–300	20–40
(5) Peach trees	0.3–1.2	1,000–3,000	300–150	30–40
(6) Vineyards, Olive-groves	0.3–0.6	1,000–2,000	300	30
(7) Artichokes (Autumn-Winter)	0.2–1.0	1,000–4,000	200–250	20–25
(8) Tomatoes, peppers, melon, cotton	0.5–8.0	2,500–4,000	200	20
(9) Vegetables, cabbage	0.2–0.3	1,000–2,000	150–200	15–20
(10) Spring & summer pot-herbs	0.2–0.25	2,000–3,000	80–100	8–10
(11) Sugar beet, corn & potatoes	0.06–0.25	600–5,000	50–100	5–10
(12) Medicinal plants	0.05–0.2	1,000–8,000	50–25	5–2.5
(13) Permanent meadows	0.1–0.25	4,000–10,000	25	2.5

pay, and here I wish to repeat, on condition that a spray system of irrigation is used, or other systems of similar efficiency.

The table puts at our disposal data of particular interest: for example, flowers can pay for irrigation water up to Lire [3] 200–250/m³; intensively cultivated vegetables and citrus fruit up to Lire 60 and more; apple trees and pear trees can bear cost of water up to Lire 50; tobacco can support up to Lire 50, peaches up to Lire 30/40; artichokes up to Lire 25; tomatoes, peppers and melons and so on, not more than Lire 20/m³ of water; other low income cultures like meadows with practically no commercial value can only pay a few liras for each cubic meter of irrigation

1. Annexed Note by Dr. DEGAN
The average production cost of a cubic meter of irrigation water for a definite crop per ha – on a standard farm – varies in different zones of Italy, but normally within certain limits.

With identical cultures and soils, this cost decreases from North to South due to the fact that in the South larger quantities of seasonal watering is required for obtaining the same increase in production.

2. The reference here is to the entire saleable crop, since these cultures are practically never produced without irrigation.

3. 100 Lire = US $ 0.16.

water. These data have a practical value for the grand average of unit productions, under an average farm operation pattern.

For the above mentioned reasons, it seems to us evident that where crops have to be concentrated on a certain stretch of land or a region, it is convenient to use desalted water for irrigation, if no other sources are available.

Taking into consideration the progress of water desalting techniques, it is worthwhile mentioning that in the near future the use of the mentioned data will be of a great help, whenever there will be a choice between using desalted water or using so-called "traditional" sources, mainly rain-waters gathered in ponds or reservoirs.

As a principle – following calculations I have been able to make during the recess of the Session – and using figures deriving from the reports of Messrs. NEBBIA and DI MENZA, in particular from the latter, it seems to me that the ratio is as follows: the cost of desalting water today is thrice that of water obtained by traditional means, especially, water supplied from artificial reservoirs.

But again it is worth mentioning that water desalting is also to be considered from an other interesting angle: we are given to understand that the construction cost of a reservoir will increasingly weigh on the price of a cubic meter of water, because most of the suitable sites have already been used, and the possibilities of constructing reservoirs are limited.

If – on the other hand – we consider the use of water desalting we notice that due to the progressive use of nuclear energy, there is a definite tendency to lower the production costs and besides that, water desalting possibilities are unlimited.

In my opinion such considerations should give this Convention food for thought: let us do some constructive thinking about the opportunity of a comprehensive plan for future investment: in the erection of water desalting units and towards a financial long-term policy concerning irrigation plans.

Author's address: Prof. Carlo Barberis c/o IFAGRARIA, Università di Roma, Via Dora 2, Roma, Italy.

APPENDIX

The UNESCO-WAAS-Italy Symposium

on Saline and Sea-water Irrigation with and without desalination
(Rome 5–9 September, 1965) *)

I. GENERAL REMARKS

This Symposium was held in the building of the Consiglio Nazionale delle Ricerche (National Research Council of Italy) in Rome, September 5–9, 1965. It was organized by the World Academy of Art and Science (WAAS) in close cooperation with the Accademia Nazionale di Agricoltura (Italian Academy of Agricultural Sciences) and the Consiglio Nazionale delle Ricerche (National Research Council of Italy), and was also sponsored and partly financed by UNESCO.

The main purpose of the Symposium was to discuss *New Approaches* to the salinity problem and the use of saline water and sea-water for normal agriculture and for productivizing deserts.

The 104 scientists who participated in it came from the following countries:

Argentine, Australia, Canada, Federal Republic of Germany, France, India, Iran, Israel, Italy, Kuwait, Malta, Morocco, Netherlands, Netherland Antilles, New Zealand, Poland, Spain, Sudan, Sweden, Switzerland, Tunis, U.A.R., U.S.A. and U.S.S.R.

Apart from these 24 countries the following Specialized Agencies of UNO and International Non-governmental Scientific Organizations were officially represented:

UNESCO (The United Nations Educational and Scientific Organization);
FAO (The Food and Agricultural Organization of UNO);
WHO (The World Health Organization of UNO); and
WMO (The World Meteorological Organization of UNO);
ISCU (International Council of Scientific Unions);
IUBS (International Council of Biological Unions);

---

* Held simultaneously with a symposium on "Causes of Conflicts" (see World Academy of Art and Science, Vol. III "Conflict Resolution and World Education" ed. by STUART MUDD, 1966, pp. XI + 244).

IBP   (International Biological Program);
and the newly established Council of
WU   (The Transnational World University of WAAS).

II. ORGANIZATION OF THE SYMPOSIUM

Presidents:

Representing the World Academy of Art and Science (WAAS):
its Vice President, Professor HUGO OSVALD (Sweden);
Representing the National Academy of Agriculture of Italy:
its President, H.E. Professor GIUSEPPE MEDICI;
Representing the National Research Council of Italy:
its President, Professor VINCENZO CAGLIOTI;
Representing the National Committee on Nuclear Energy (CNEN):
its Vice President, Professor CARLO SALVETTI.

Honorary Secretaries (Convenors)

Dr. HUGO BOYKO (Israel), Honorary Secretary General of WAAS, (International Convenor)
Dr. GABRIELE MALAGUTI (Italy), Chancellor, Accademia Nazionale di Agricultura;
Professor ALESSANDRO DE PHILIPPIS (Italy), Director, Centro di Sperimentazione Agraria e Forestale   (Convenors for Italy).

Chairmen of the Sessions:

1st Session (General Lectures and Principles):
    Professor L. LINTAS (Italy), President, Consiglio Superiore di Agricoltura;
    Professor HUGO OSVALD (Sweden), Vice President of WAAS.
2nd Session (Experiments with Saline Water and Sea-water):
    Eng. PIERO CASINI (Italy), President, Associazione Nazionale Bonifiche;
    Professor V. J. CHAPMAN (New Zealand), University of Auckland;
    Dr. EUCLIDE GIULIANI (Italy),Director, Associazione Nazionale Bonifiche.
3nd Session (Desalination):
    Professor H. HEIMANN (Isreal), Israel Institute of Technology, Haifa;
    Professor P. C. RAHEJA (India), Director, Central Arid Zone Research Institute, Jodhpur; (now (1968) Managing Director, Lake Nasser Water Resources Research and Development Project c/o FAO, Cairo, UAR);
    Professor C. SALVETTI (Italy), Vice-President, National Committee on Nuclear Energy.

Official Delegates of:

UNESCO: Dr. THA HLA, Director, Dept. for the Advancement of Science;
FAO: Dr. J. DE MÉREDIEU, Chief, Water Resources and Irrigation Branch;

WHO: Dr. J. DE MÉREDIEU, Rome;
WMO: Dr. ARNE FORSMAN, Geneva;
ICSU: Dr. F. W. G. BAKER, London;
IBP: Dr. F. W. G. BAKER, London;
IUBS: Professor CARL-GÖREN HEDÉN, Stockholm;
WU: Professors GAETANO MARTINO, Rome and STUART MUDD, Phila-
              delphia, for the Council of the World University.

Resolution Committee (elected by the Plenum at the beginning of the
Symposium):
    Professor V. J. CHAPMAN (New Zealand)
    Professor A. DE PHILIPPIS (Italy)
    Professor P. C. RAHEJA (India)

Social Events and Excursions: Various social events were organized by the Ministry for Foreign Affairs, the Lord Mayor of Rome and others, and visits were
arranged to some of the magnificent buildings and Art treasures of Rome.
    A most instructive excursion was organized by the National Committee
on Nuclear Energy (CNEN).
    The tour to the Nuclear Center of the "Cassaccia" of C.N.E.N. and to the
plants for desalination and use of nuclear energy in agriculture concluded
the Symposium.

III. RESOLUTIONS

The lectures dealt with the principles as well as the experiments in various
countries, and most of them were accompanied by illustrative slides. Almost all the lectures led to vivid discussions and as a final result, the following resolutions were unanimously accepted by the Plenum:

    1) The participants in the International Symposium on irrigation with
saline water and sea-water with and without desalination express their
deepest thanks to the Executive of the World Academy of Art and Science,
to the Accademia Nazionale di Agricoltura and to the Consiglio Nazionale
delle Ricerche of Italy for having organized and made possible the Symposium, and they assure these august bodies that the deliberations have been
exceedingly fruitful and should lead to further and more efficient utilization
of saline land and water and the wider production of fresh water from saline
sources. Particular thanks are also expressed to UNESCO for their sponsor-
ship and support.

    2) The participants in the International Symposium on irrigation with
saline water and sea-water with and without desalination, organized by the
World Academy of Art and Science, the Accademia Nazionale di Agricoltura

and the Consiglio Nazionale delle Ricerche, and sponsored by UNESCO, firmly believe that *results already achieved by irrigation with highly saline water indicate clearly that those arid areas at least where such a water supply is available, can be rendered capable of crop production, and they therefore strongly recommend to international and national organizations concerned with human welfare that adequate provision should be made for the necessary expansion of research and field trials.*

The following fields are seen as of particular importance:

a) Ecological, physiological, genetic, biochemical and agrotechnical studies on promising economic plants under direct irrigation with highly saline and sea-water under various climatic conditions;

b) Research in the laboratories and in the field on *biological* desalination of soil and water;

c) Microbiological studies connected with these lines of research.

3) (Proposed by Dr. MOSTAFA BENAIE, Iran). The participants in the International Symposium on the problems of irrigation with saline water and desalination of salt water organized by the World Academy of Art and Science, the Accademia Nazionale di Agricoltura and the Consiglio Nazionale delle Ricerche believe that the world's food resources can be materially increased, not only by salt water irrigation of arid areas, but also by the *desalination of extensive deltaic areas* in the humid tropics, and they therefore urge FAO, UNESCO and SWPC (South-West Pacific Commission) and philantropic foundations to make available funds for research and reclamation of such lands, and also to make governments aware of the great importance of these lands.

Informatory Notes about the World Academy of Art and Science and the World University *

* The main points are taken from the draft proposal for a WAAS Information Brochure prepared by JOHN MCHALE, Honorary Secretary of the American Division of WAAS.

A. THE WORLD ACADEMY

The establishment of the World Academy of Art and Science (WAAS) fulfilled a need to which attention was drawn at the International Conference on Science and Human Welfare in October 1956.

On the 24th of December, 1960, the World Academy was declared as founded by forty scientists of wide international renown as Charter Members.

The main purposes of WAAS are essentially:

1. To create a Transnational Forum for the interchange of knowledge and information, and the study, independent of national barriers, of problems whose settlement is vital to the well being of mankind, and

2. To act as a dispassionate advisory body to the leading international organizations, for the benefit of mankind as a whole.

Organization:

The following Officers have been elected at the Third Plenary Meeting in Rome, September 1965:

Honorary President:	Lord J. BOYD-ORR (Scotland)
President:	HUGO N. BOYKO (Israel)
Vice-Presidents:	JULIAN ALEKSANDROWICZ (Poland)
	GEORGE E. GORDON CATLIN (England)
	Sir JOHN C. ECCLES (Australia)
	IVAN MALEK (Czechoslovakia)
	GIUSEPPE MEDICI (Italy)
	STUART MUDD (USA) (Stand-by President)
	HERMANN JOSEPH MULLER (USA)
	HUGO OSVALD (Sweden)
	BORIS PREGEL (USA)
	ALBERT SZENT-GYORGYI (USA)
	MORRIS L. WEST (Australia)

Legal Advisor:	Max Habicht (Switzerland)
Chairman of Committees:	
Organization:	Carl-Göran Hedén (Sweden)
Semantics and	
Communication:	Lloyd L. Morain (USA)
Publications:	Stuart Mudd (USA)
Fellowship:	Hugo Osvald (Sweden)
Membership:	John McHale (USA)
Honours:	Alessandro de Philippis (Italy)

According to the statutes, membership in the World Academy of Art and Science is restricted to 300 Fellows, with voting rights. Fellows over 75 years of age are not included in this maximum number.

The transnational nature of WAAS is characterized not only theoretically but also factually by its composition. The Fellows and Members reside in 42 different countries and represent all major groups, races and creeds of mankind.

At the last Plenary Meeting it was decided in principle to establish an additional Category of Members in order to associate a greater number of suitable personalities for active cooperation and to have a spiritually close group in reserve for the election of Fellows.

The criteria for the election of *Fellows* are:

Personalities of international reputation who are known for their wisdom and humanitarian thinking, and for their knowledge and comprehensive understanding beyond their own working field.

The criteria for the election of *Members* are:

Personalities active in Science or Art who are concerned with improving the human state and who play significant roles in the areas where they are active.

Regional division of WAAS

In accordance with the decisions of the Third Plenary Meeting (Rome, Sept. 1965) with regard to adapting the organizational structure of WAAS to the growth and geographical distribution of our Membership and to the dynamic increase of our activities, Regional Divisions of WAAS are being established.

The American Division has been incorporated as an Educational Institution with an official charter from the Regents of the State of New York. A meeting of the American Division of WAAS was held on May 22, 1966 in New York and the following officers were elected:

Boris Pregel (New York, N.Y.) President and Treasurer
Stuart Mudd (Philadelphia-Haverford, Pa.) Vice-President
John McHale (Carbondale, Ill.) Secretary.

In view of its valuable educational activities, contributions to the American Division are tax deductible.

Internal Communication

Contact between the Centers and the Fellows and Members who reside in all regions of the globe is established through journeys of the Officers, Group meetings and Plenary meetings, apart from the Symposia, publications (in particular the WAAS-Newsletter) and extensive correspondence.

The first Plenary Meeting was held (in the Palace of the Royal Academy of Sciences) in Brussels, July, 1961;

The second took place in Stockholm in July-August 1963 on the premises of the Royal Swedish Academy of Engineering, with the exception of the Opening Session which was held in the House of Parliament.

The third was convened in Rome on the premises of the National Research Council of Italy, in September 1965;

The fourth will be held in New York on the premises of the New York Academy of Sciences, October 1969.

Apart from the first Plenary Meeting, all are organized in connection with scientific Symposia on problems of vital importance for mankind. In particular:

The Plenary Meeting in Stockholm with the Symposium on "Global Impacts of Applied Microbiology" organized by IAMS (International Association of Microbiological Societies) and co-sponsored by WAAS; it was presided over by our Fellow ARNE TISELIUS, Nobel Laureate and President of the Nobel Foundation.

The Plenary Meeting in Rome was connected with two Symposia; namely:

1) "On Causes of Conflicts and World Education", presided over by our Fellows GEORGE E. G. CATLIN (U.K.), one of the founders of Political Science, and GAETANO MARTINO, Rector of the University of Rome;

2) "On Irrigation with Saline Water or Seawater, with and without Desalination", presided over by our Fellow Senator G. MEDICI, who was at the preparatory time of the Symposium the Minister of Commerce and Industry in Italy.

The former Symposium was organized by WAAS, the latter in cooperation with the National Research Council of Italy and the Academy of Agricultural Sciences of Italy, and was sponsored and partly financed by UNESCO.

The provisional title of the forthcoming Symposium in connection with the Fourth Plenary Meeting is "Conference on the Planetary Society" – (Energy, Environment and Society in Transition). The organization of this Symposium, planned on a large scale, is in the hands of the American Division of WAAS, together with the American Geographical Society.

The World Academy publishes besides its "Newsletter" (two to four numbers yearly) and occasional papers, a series of books, the "WAAS-Series", on topics of particular importance.

Up to date the following volumes have been published:

Vol. I., 1961:   "Science and the Future of Mankind", edited by HUGO BOYKO, pp. VIII + 380, 8 figs.

Vol. II., 1964: "The Population Crisis and the Use of World Resources", edited by STUART MUDD.
Associate Editors: HUGO BOYKO, ROBERT C. COOK, LARRY NG, W. TAYLOR THOM Jr.
pp. XIX + 563, 43 figs.

Vol. II., 1965: "The Population Crisis, Implications and Plans for Action", paperback edition, edited by LARRY K. N. NG and STUART MUDD.
Associate Editors: HUGO BOYKO, ROBERT C. COOK, FRANK J. DITTER, JAMES E. McLENNAN and W. TAYLOR THOM Jr.
pp. XI + 364.

Vol. III., 1966: "Conflict Resolution and World Education", edited by STUART MUDD, pp. XI + 294.

Vol. IV., 1968: "Saline Irrigation for Agriculture and Forestry" (Proceedings of the UNESCO-WAAS-ITALY Symposium, Rome, 5–9 September 1965), edited by HUGO BOYKO, pp. XXIV + 325, 46 figs. and maps.

Vol. V., 1968: "Towards World Community", edited by JOHN NEF, pp. XII + 153, 2 photographs.

"The editions are primarily published by Dr. W. Junk N.V., Publishers, 13 van Stolkweg, The Hague, the Netherlands. American reprint editions and the paperback edition of Vol. II are published by the Indiana University Press, Bloomington, Indiana 47401, U.S.A."

B. THE WORLD UNIVERSITY

The IIIrd Plenary Meeting of the World Academy of Art and Science (Rome, Sept. 1965) accepted the plan of its President to create a disseminated transnational "World University".

As a first step to implement the plan, a Council was constituted (see page 302). The year 1967 decleared by UNESCO as "The Year of Cooperation" has been chosen for the creating of this World University as a lasting contribution of the World Academy of Art and Science to UNESCO's appeal.

A concise description of the basic plan may be presented here:

Many detailed projects have been worked out for the establishment of a World University. These usually involve the idea of an actual physical center, a kind of international but locally determined campus, where people from many countries and races could meet and study.

Under the circumstance of the World Academy of Art and Science, a close intellectual cooperation, not bound by geographical limitations of any kind, seems more promising. In this way it should be possible to enlarge the traditional concepts of academic work and to combine its efforts, undertaken in different places of our globe, into worldwide research on problems affecting humanity as a whole.

Through such a disseminated transnational World University also the excellent work of "Scientific Years and Decades", sponsored by the Specialized Agencies of UNO and the Scientific Unions, could achieve a lasting value, complementing their scientific one, and constitute a further step in the realization of the ethical aims of science.

The criteria for selecting a unit (Department of a University, Scientific Institute, etc.) as one of the units of the World University and thus as international centre for this specific research, are as follows:

1) The research subject must have a potential of strong global impact on Human Welfare;

2) It must require an international cooperation or coordination;

3) The respective institution must be particularly suited for just this subject with regard to research personnel and/or equipment and geographical position;

4) The governing body of the selected center must have approved the respective agreement, and the Council of the World University will have to give its approval, in order to avoid duplication.

Any Institution of Higher Learning, Department, etc., can suggest a research subject for which, in their opinion, they could serve as a world center for transnational research team work. The *research subject* can be restricted or may have a wider scope, in any case it *must have a potential of a very strong global impact on Human Welfare and must require by its very nature an international cooperation or coordination*.

The same applies to independent Institutions or research units of Learned Societies of high international scientific reputation.

Any Institution willing to act as a world center for one subject should be prepared to cooperate in one or more suitable projects for which other Institutions will act as centers.

After approval of its research subject, the Institution in question will be asked to indicate those Institutions which are apt to cooperate in this specific research scheme as "Cooperating Centres".

The Members of this Council, elected at the Plenary Meeting are listed below. After its constitution, the First Council, together with the Board of Officers of the World Academy of Art and Science, selected out of numerous proposals a few subjects as the first units of the World University, and agreements with the authorities of the respective Institutions of Higher Learning were concluded.

Gradually, this worldwide network of cooperating universities should become increasingly close-knit, and the work on the various research projects will lead to the publication of annual or bi-annual reports, either by the World University or in close cooperation with International Specialized Agencies, etc.

The World Academy and the World University will closely cooperate with relevant agencies of UNO.

Professor HAROLD D. LASSWELL of Yale University has accepted the Chairmanship of the Executive Committee of the World University.

Agreements with a number of Institutions of Higher Learning were signed in the first year and additional ones are now being negotiated.

The text of these agreements is worded as follows and adapted to each specific case:

"The Executive Committee of the World Academy of Art and Science (WAAS) and the Council of the World University (WU)

A) acknowledge the interdisciplinary nature of research on (to be inserted), its global impact and its need of international or transnational cooperation or coordination, and

B) acknowledge the Department (Institute) Alpha of the University Beta as the International Centre of the proposed Research Project; The Department (Institute) Alpha undertakes to organize research in this specific field on a transnational basis with the cooperation of "Collaborating Centres" in other countries.

If there should be any unforeseeable collision between the autonomic rights of the University Beta and the disseminated World University of WAAS, then the autonomic rights of the University Beta shall have the priority.

The University Beta will by this cooperation acquire the right to send one of its leading scientists as Delegate-Observer to the Council of the World University.

Any agreements on financial matters, e.g. for Specific Projects, Grants, Fellowships etc. will be subject to specific agreement. The same applies to post-graduate Fellowships, to be accorded by the World University to scientists working within the framework of the WU-research projects."

Funds were appropriated for the disposal of the Executive Committee for organizatory purposes and to serve, to a certain extent, for the financing of research work, for Fellowships, etc.

The ultimate aim of such a close, gradually developing network of Institutions of Higher Learning is, apart from the concerned scientific approach to vital problems of mankind, to achieve a complementary and well integrated world education.

The Council:

The Members of the first Council of the World University are:

Lord JOHN BOYD-ORR (Brechin, Scotland)
Dr. HUGO N. BOYKO (Rehovot, Israel)
Prof. CHESTER CARLSON (New York, N.Y.)
Prof. GEORGE E. G. CATLIN (London, England)
Prof. EINAR DU RIETZ (Uppsala, Sweden) (deceased)
Prof. Sir JOHN C. ECCLES (Canberra, Australia)
Prof. CARL-GÖRAN HEDÉN (Stockholm, Sweden)
Dr. ROBERT M. HUTCHINS (Santa Barbara, California)
Prof. HAROLD D. LASSWELL (New Haven, Connecticut) (Chairman)
Prof. CHO-MING LI (Hongkong)
Prof. P. C. MAHALANOBIS (Calcutta, India)
Academician IVAN MALEK (Praha, Czechoslovakia)
Prof. GAETANO MARTINO (Rome, Italy) (deceased)
Arch. JOHN MCHALE (Carbondale, Illinois) (Hon. Secretary)
LLOYD L. MORAIN (San Francisco, California)
Prof. EMILY MUDD (Haverford, Pennsylvania)
Prof. STUART MUDD (Philadelphia, Pennsylvania)
Prof. HERMANN JOSEPH MULLER (Milwaukee, Wisconsin) (deceased)
Prof. Dato Sir ALEXANDER OPPENHEIM (Kuala Lumpur, Malaysia)
Prof. HUGO OSVALD (Knivsta, Sweden)
Prof. LINUS PAULING (Big Sur, California)
Prof. BORIS PREGEL (New York, N.Y.)
Prof. ALI-AKBAR SIASSI (Teheran, Iran)
Prof. ALBERT SZENT-GYORGYI (Boston, Massachusetts)
Prof. HAROLD TAYLOR (New York, N.Y.)
Prof. W. TAYLOR THOM Jr. (Trenton, New Jersey)

INTRODUCTORY NOTE BY THE EXECUTIVE COMMITTEE

This Declaration has been drafted by Professor Harold D. LASSWELL, Yale University, Chairman of the Executive Committee of the World University, and is presented here in its revised form, worked out in close cooperation with Professor Boris PREGEL, the President of the American Division of the World Academy of Art and Science, New York, and approved by the Executive Committee of the World University. The initiative taken by Dr. Hugo BOYKO led to the formation of the World Academy of Art and Science in 1960. From the beginning Dr. BOYKO had the World University in mind as a major undertaking of WAAS. The institution is now in process of formation. The *Declaration* is an outline of the objectives and structures of the World University as conceived by those principally responsible for its development.

Declaration of the World University of the World Academy of Art & Science

I.

The idea of a world university is to foster the growth of knowledge and to cultivate enlightened judgment in all that concerns the needs and aspirations of man.

In long perspective, the conception of a world university has been timely since the epoch when early civilizations began to accelerate the accumulation of knowledge and the rate and scale of interdependence. The growth of knowledge depends on individual motivation and talent, and on situations that encourage creativity. The impersonality of knowledge enables it to become a legacy common to all.

Today the timeliness of the idea of a world university is beyond reasonable reservation. The expansion of science and technology has put at our disposal an unparalled instrument of fulfilment or destruction; if man is to take the future evolution of body, mind and civilization in his own hands it is imperative to find more effective ways of integrating what he knows with what he does.

The world university begins with no unalterable blueprint. The approach is exploratory and self-appraising. The university will endeavor to supplement, without duplicating, the institutions presently devoted to higher education, inquiry and consultation.

Above all, the world university proposes to identify and to serve the common interest of mankind, not by hovering at a height hypothetically distant from humanity, but by offering a means whereby inquiring minds can relate themselves and their intellectual specialities to a conception of human dignity that is open to continual clarification in the light of the changing environment.

No human institution can be sure of choosing the procedures through which its long range aspirations for humanity can be realized and not betrayed by failures of motivation, judgment or luck. The world university is initially committed to a series of informed guesses about the arrangements that will harmonize performance with goal.

2.

The University is a unique establishment: it is responsible to The World Academy of Art and Science, which is a worldwide, non-official institution composed of individuals coming from diverse national background who have been chosen for eminence in the natural and social sciences and in the humanistic studies. The Fellows of the World Academy are a sample of the world's most significant contributors to knowledge. National communities are under-represented whose members do not as yet participate fully in the universalizing civilization of science and technology. Nevertheless, as presently composed, the Academy provides auspices for the World University in which multi-national experience and individual responsibility are combined to mitigate the pressure of more parochial interests. The Fellows of the Academy cut across the diversities of tradition, language and social structure which, unless fused by creative imagination and effort, dissolve the latent commonwealth of man into warring congeries of special interest groups.

Attached to the World Academy of Art and Science, the World University is formally responsible to a community of scholars who share an inclusive concern for man as man. The members of the Academy endeavor to support and to become effective instruments for the realization of a world order in which human dignity is honored in deed as in word.

3.

The World University is committed to a flexible structure whose primary units (centers), though specialized, are contextually oriented. To prescribe that the primary units shall be specialized is to recognize differentiations within the field of modern knowledge. Contextuality implies concern with the potential impact of knowledge, however detailed, on man. Hence the program of a primary unit of the World University includes the consideration of social consequences and of policy implications. The units are interdisciplinary among adjacent fields of specialization; they also include specialized competence in examining the interplay between knowledge and the social and physical environment.

4.

The primary units (centers) of the World University are from the principal sub-divisions of science and scholarship. It is redundant to itemize the current categories by which these sub-divisions are referred to. For the present the broad picture is enough: specialists focus on man and other biological forms: on the physical environment, including the earth, other solar satellites and the galaxies; and on the social environment, with its ever changing institutions.

In order to encourage flexibility the World University will not begin with faculties in compartmentalized traditional divisions. The primary units (centers) may be institutes, departments, schools or programs. The primary units are encouraged to initiate "cooperating centers" within the comprehensive framework of the university. A primary unit (center) is eligible to join as many world university co-operating centers as it desires. For example, a primary unit that specializes on communication among the members of a species of primates may opt for membership in cooperating centers on primates, on communication, and so on. A unit on microbiology may belong to cooperating centers on microbiology, on systematic biology, and so on.

5.

Primary units and cooperating centers will be served by the central organs of the World University. Of fundamental importance in this connection is assistance in realizing interdisciplinary and contextual objectives.

A means of furthering the common purpose will be travelling World University Fellows who will receive hospitality at the appropriate centers. The most distinctive Fellowship program will be designed to provide the primary units with specialized talent to participate in seminars on the social consequences and policy implications of knowledge. For example: a specialist in law and organization who is concerned with the immediate and the eventual significance of microbiology for public policy may spend his Fellowship period at two or three research centers where, among other activities, he participates in the continuing seminar on the social consequences of science. The continuing seminar may develop memoranda and audio-visual materials of sufficient importance to be circulated through the entire university, and ultimately to the official and unofficial public.

6.

If the policy implications of knowledge for the world community are to be given adequate consideration the World University must provide mechanisms by which attention is directed to major problems or problem areas. The World University will initiate or give effect to proposals to establish "commissions" to serve as task forces to explore emerging or neglected issues. Example: the growth of modern technology has led to largescale pollution of soil, water and air, with cumulative danger to life. Perhaps World University commissions can expedite the solution of the scientific and technical problems involved, and formulate programs for applying the new technology in ways that will contribute most to building the world community.

Task force projects will be developed within the framework, and in harmony with, the programs of the World Academy of Art and Science. The conferences of the Academy, for instance, have frequently identified two broad goals of world policy which are bound to affect the specific initiatives of the World University: *Security*, or the elimination of war and of scarcities attributable to food-population imbalances: *freedom*, or the protection of wide areas of responsible choice from invasion by biological or cultural engineering.

7.

As the preceding paragraphs imply, the World University and the World Academy include among their aims the cultivation of more equal participation in scientific and technical civilization throughout the globe. The University will give attention to programs designed to multiply regional centers of strength in advanced education, research and consultation. In this connection it may be feasible to encourage the formation of information storage and retrieval networks to serve the scientific and scholarly needs of users everywhere.

8.

The World University's commitment to a contextual viewpoint and to more balanced cultural development will be furthered by originating and disseminating programs of instruction that contribute to modes of education adapted to the urgent needs and opportunities of the world community. To some extent education is a matter of following a syllabus of recommended reading. In part it is exposure to audio-visual experiences that give vividness and context to detail. Education is more: it implies experience with people and procedures as well as with articulated principles. Hence the World University encourages careers in which life in the laboratory or on field expeditions is accompanied or punctuated by opportunities to take a hand in world task projects at the margin where knowledge and institution-building merge with one another. Alienation among specialists, and between specialists and non-specialists, can be ameliorated by discovering common interests in the course of joint activity. Ego-segregating tendencies can be partially offset by encouraging the self-integrating tendencies that are likewise present in man's basic potential.

9.

The World University is committed to a structure that combines unity of purpose with dispersal of units. The idea of a world university does not allow it to be a religion with a Holy Place, or an instrument of cultural unbalance. Therefore the University does not contemplate the erection of a huge, centralized campus, although such a facility would possess some symbolic advantages. We believe that the cultivation of enlightenment, and the use of enlightened judgment on public problems, are better symbolized by co-centers throughout the globe than by a single super-campus.

The World University's commitment to dispersal is in no way incompatible with obtaining whatever facilities are best adapted to its several tasks. No doubt edifices will be acquired in many parts of the world to provide conference and task force headquarters or to store and exhibit special collections. However, since the World University is a moral expression of the actual or latent unity of the world community of knowledge, the fundamental policy is to strengthen the whole rather than to aggrandize its instrument. The expectation is that world regions will presently provide sites for co-centers adapted to the needs of the locality.

The "unity-with-dispersal" principle is announced at the beginning of the World University's life in order to strengthen the self-correcting tendencies that work against the dispositions to over-centralize. Studies of organization have identified several factors which tend to swing the balance toward centralization. For example, the immediate convenience of central officials and staff typically favors centralization: thus propinquity cuts down the time required to obtain stored information or consultation. There is the seductive appeal of a monumental edifice and a many-bodied staff, which by arousing respect in others guild the inner image of the self. In our era of rapid communication and of greater self-awareness "unity-with-dispersal" is a viable policy.

10.

The President of the World Academy of Art and Science will be the honorary and supervising head of the World University. The Chairman of the Executive Committee of the World University will be the active head. The Chairman will be responsible for the planning and general administration of World University operations.

We emphasize the importance of planning for immediate and long term future contingencies, and of continuing self-appraisal of past and present relationships between objectives and performance. If immediate acts are to harmonize with ultimate aims, long-range objectives must be defined with sufficient clarity to provide practical guidance. The Executive Committee is composed of the President, the Chairman, the Treasurer, the Executive Secretary, and three other members. With the exception of the President, who serves *ex officio* in his capacity as President of the World Academy of Art and Science, the Executive Committee chooses its officers.

In order to ensure continuity during the early stages of formation, the Executive Committee of the World University is authorized to name its own members for a six year period beginning in 1968. Thereafter the members are chosen at a joint meeting of The World University Council and the officers of the World Academy. The Council meets at the request of the President. The Executive Committee of the World University will nominate two of its members every two years to serve six year terms each.

11.

The Executive Committee is authorized to admit primary units as members of the World University when an agreement with a responsible authority provides adequate facilities for cooperation in the programs of the World University. The Executive Committee is authorized to establish World Task Forces (commissions) and to accept or develop the facilities required by University programs.

The primary units, the cooperating centers, and the task forces of the World University are authorized to send a member each to the Council of the World University.

The Executive Committee reports regularly to the Council of the World

University, cooperating with a Council to prepare an audit and a review of the activities of the University for report to the World Academy.

The Executive Committee of the World University will provide for the appointment of World University Fellows, cooperate in the activities of the World Academy, and in general, engage in activities that further the aims of the University.

The Executive Committee will maintain a commission on cooperation with Academic and Professional Associations in order to facilitate joint operations with the World Academy.

The Executive Committee is authorized to obtain funds for the World University and to assist in obtaining support for its primary units. The sources of funds of the World University are to be publicly announced at periodic intervals. One aim is to maintain sufficient diversity among sources of support to sustain both the appearance and the fact of independence.

12.

Recapitulating the *goal*:

The idea of a World University is to foster the growth of knowledge and to cultivate enlightened judgment in all that concerns the needs and aspirations of man.

The distinctive viewpoint is contextual, seeking to evaluate the social consequences and policy implications of specialized knowledge.

One aim is to facilitate more equal participation throughout the globe in the universalizing civilization of science and technology.

The ever-changing structure of the World University will seek to combine unity with dispersal.

Recapitulating the *structure*: The World University is an instrument of the World Academy of Art and Science whose President is *ex-officio* President of the World University and a member of its Executive Committee. The Council of the University is composed of officers of the World Academy, members nominated by the President and ratified by the Council, together with members designated by the primary units (centers) of the University. After the first six year period, the Executive Committee nominates to the World University Council two members of the Executive Committee to serve for six years each.

The Executive Committee of The World University is immediately responsible for operating the institution. The Chairman is the principal planning, appraising and general administrative official. The President of the World Academy of Art and Science, *ex officio* President of the World University and Chairman of the Council of the University, is the honorary and supervisory head of the institution. The Executive Committee selects its officers, admits Primary Units and authorizes the establishment of World

University cooperating centers (to any of which any primary unit may belong) and World Task Forces (commissions).

It is taken for granted that the structure of the World University will be continually adapted to the advancement of knowledge, the examination of social consequences, and the creative consideration of policy implications for the future of mankind.

Index

W.A.A.S. IV